Studies in Advanced Mathematics

Clifford Algebras in Analysis and Related Topics

Studies in Advanced Mathematics

JOHN RYAN
University of Arkansas

Clifford Algebras in Analysis and Related Topics

CRC Press
Taylor & Francis Group
Boca Raton London New York

CRC Press is an imprint of the
Taylor & Francis Group, an **informa** business

CRC Press
Taylor & Francis Group
6000 Broken Sound Parkway NW, Suite 300
Boca Raton, FL 33487-2742

© 1996 by Taylor & Francis Group, LLC
CRC Press is an imprint of Taylor & Francis Group, an Informa business

First issued in paperback 2019

No claim to original U.S. Government works

ISBN-13: 978-0-367-44884-4 (pbk)
ISBN-13: 978-0-8493-8481-3 (hbk)

Visit the Taylor & Francis Web site at
http://www.taylorandfrancis.com

and the CRC Press Web site at
http://www.crcpress.com

Library of Congress Number 95-23667

Library of Congress Cataloging-in-Publication Data

Clifford algebras in analysis and related topics / edited by John Ryan
    p. cm. – (Studies in advanced mathematics)
   "Based on a conference held in Fayetteville, Arkansas during the weekend, April 8-10th, 1993" – Introd.
  Includes bibliographical references.
  ISBN 0-8493-8481-8
  1. Clifford algebras. 2. Mathematical analysis. I. Ryan, John, 1955-
II. Series.
QA199.C56   1996
515′.1 – dc20
                                                95-23667
                                                  CIP

Contents

1

Acknowledgment

This contributed volume is derived from a conference held in Arkansas.

It is my pleasure to acknowledge the help of the Department of Mathematical Sciences, the Graduate School, and the J. William Fulbright College of Arts and Sciences of the University of Arkansas, Fayetteville, for their generous funding of the conference on which this contributed volume is based. During a part of the time this volume was in preparation the editor was supported by a grant from the Arkansas Science and Technology Authority. I am grateful to A.S.T.A. for their support.

Thanks are also due to the other members of the University of Arkansas Spring Lecture Series Committee, for their help and support in a year of preparation for the meeting. Specifically, thanks are due to Itrel Monroe, chairman of the committee, and John Duncan, Mihalis Maliakas, and William Summers. A special thanks is due to Janet Miller for her tremendous secretarial support both before and during the meeting.

I am very grateful to William Pezzaglia, who advertised the conference by placing abstracts of the talks and various announcements in the Clifford Algebra e-mail Bulletin Board, clf_alg@stars.sfsu.edu, and to Jaime Keller for publishing a report on the conference in the journal *Advances in Applied Clifford Algebra*.

I would also like to thank Steven Krantz for his strong encouragement and constant support of this project. I am also indebted to Wayne Yuhasz, and Nora Konopka of CRC Press and Lori Pickert of Archetype for all their help.

All papers appearing in this volume were refereed. For their editorial assistance I am truly grateful to Pascal Auscher, Tao Qian, Chun Li, Alan McIntosh, Marius Mitrea, Palle Jorgensen, Grant Welland, Zhijian Wu, Stephen Semmes, Daniel Luecking, John Akeroyd, John Duncan, Itrel Monroe, and Mihalis Maliakas.

For their patient assistance with my limitations with TEX and e-mail, I am indebted to my colleagues Dan Luecking and Bob Miller.

I am indebted to Brenda Barrett for the typing of all of my own contributions to this project.

Thanks are also due to John Duncan, Gary Tee, A. Gsponer and Andreas Sparschuh for their assistance with some historical details appearing in the introduction.

A special thanks is due to Cherie Moore of Washington University, Saint Louis, for the gargantuan task of placing this volume in camera-ready form.

Also, my heartfelt thanks go to my wife Christine, and my daughters Tammy and Danielle, for their constant support throughout the nearly two years this project has taken to come to fruition.

John Ryan, Arkansas, September, 1994.

Conference Participants

John Adams Department of Physics, San Francisco State University, San Francisco, California 94132.
E-mail: adams@stars.sfsu.edu

Josefina Alvarez Department of Mathematical Sciences, New Mexico State University, Las Cruces, New Mexico 88003-0001.
E-mail: jalvarez@nmsu.edu

Enrique Ramírez de Arellano Departamento de Matemáticas, CINVESTAV del IPN, Mexico, D.F., MEXICO.
E-mail: eramirez@math.cinvestav.mx

Pascal Auscher Département de Mathématiques, Université de Rennes I, Campus de Beaulieu, 35042 Rennes Cedex, FRANCE.
E-mail: auscher@univ-rennes1.fr

Ruediger Belch Rutherford College, The University of Canterbury, Kent CT2 7NX, UNITED KINGDOM.
E-mail: rb5@ukc.ac.uk

Don Burkhardt Department of Physics, University of Georgia, Athens, Georgia 30602.

Guy David Mathématique Bât 425, Université de Paris-Sud, 9140 Orsay Cedex, FRANCE.
E-mail: gdavid@anh.matups.fr

Daniel Dix Department of Mathematics, University of South Carolina at Columbia, Columbia, South Carolina 29208.
E-mail: dix@bigcheese.math.scarolina.edu

Matt Enjalran Department of Physics, University of Massachusetts at Amherst, Amherst, Massachusetts 01003.
E-mail: menjalran@phast.umass.edu

John Gilbert Department of Mathematics, University of Texas at Austin, Austin, Texas 78712.
E-mail: gilbert@math.utexas.edu

Angela Grant Department of Mathematics, Lincoln University, 820 Chestnut, Jefferson City, Missouri 65102.

Bjorn Jawerth Department of Mathematics, University of South Carolina at Columbia, Columbia, South Carolina 29208.
E-mail: bj@loki.math.scarolina.edu

0-8493-8481-8/96/$0.00+$.50
© 1996 by CRC Press

Palle E.T. Jorgensen Department of Mathematics, University of Iowa, Iowa City, Iowa 52242-1466.
E-mail: jorgen@math.uiowa.edu

Carlos Kenig Department of Mathematics, University of Chicago, 5734 University Avenue, Chicago, Illinois 60637.
E-mail: cek@math.uchicago.edu

Frank Kippig Fachbereich Mathematik, Bergakademie Freiberg, Bernhard-von-Cotta-Staße 2, 0-9200 Freiberg, Saxony, GERMANY.
E-mail: kippig @mathe.ba-freiberg.d400.de

Joe Lakey Department of Mathematics, University of Texas at Austin, Austin, Texas 78712.
E-mail: lakey@math.utexas.edu

Chun Li School of Mathematics, Physics, Computing, and Electronics, Macquarie University, North Ryde, New South Wales 2109, AUSTRALIA.
E-mail: chun@macadam.mpce.mq.edu.au

Pertti Lounesto Institute of Mathematics, Helsinki University of Technology, SF-02150 Espoo, Helsinki, FINLAND.
E-mail: lounesto@dopey.hut.fi

Alan McIntosh School of Mathematics, Physics, Computing, and Electronics, Macquarie University, North Ryde, New South Wales 2109, AUSTRALIA.
E-mail: alan@macadam.mpce.mq.edu.au

Gary Miller Department of Mathematics, University of New Brunswick, Fredericton, New Brunswick E3B 5A3, CANADA.
E-mail: gary@math.unb.ca

Marius Mitrea Department of Mathematics, University of Minnesota at Minneapolis, Minnesota 55455.
E-mail: mitrea@math.umn.edu

Robert E. Mullins Department of Mathematics and Computer Science, William Wehr Physics Building, Marquette University, Milwaukee, Wisconsin 53233.
E-mail: bobm@mscs.mu.edu

Tao Qian Department of Mathematics, Statistics, and Computing Science, University of New England, Armidale, New South Wales 2351, AUSTRALIA.
E-mail: tao@neumann.une.edu.au

William M. Pezzaglia, Jr. Department of Physics, California State University at Hayward, Hayward, California 94542.
E-mail: bpezzag@mcs.csuhayward.edu

John Ryan Department of Mathematical Sciences, University of Arkansas at Fayetteville, Fayetteville, Arkansas 72701.
E-mail: jryan@comp.uark.edu

Stephen Semmes Department of Mathematics, Rice University, Houston, Texas 77251.
E-mail: semmes@math.rice.edu

Michael Shapiro Departamento de Matemáticas, ESFM del IPN, Edificio 9, Unidad ALM, 07300, Mexico, D.F., MEXICO.
E-mail: mshapiro@godel.esfm.ipn.mx

Zhongwei Shen Department of Mathematics, Purdue University, West Lafayette, Indiana 47907-1395.
E-mail: shenz@math.purdue.edu

Don Small Department of Mathematics, University of New Brunswick, Fredericton, New Brunswick E3B 5A3, CANADA.
E-mail: don@math.unb.edu

Karel Stroethoff Department of Mathematics, University of Montana, Missoula, Montana 59812.
E-mail: ma_kms@selway.umt.edu

Terrance Tao Department of Mathematics, Princeton University, Princeton, New Jersey 08544.
E-mail: tao@math.princeton.edu

Toma Tonev Department of Mathematics, University of Montana, Missoula, Montana 59812.
E-mail: ma_tt@lewis.umt.edu

Rodolfo Torres Department of Mathematics, University of Michigan at Ann Arbor, Ann Arbor, Michigan 48109.
E-mail: rodolfo.torres@math.lsa.umich.edu

Nikolai Vasilevski Departamento de Matemáticas, CINVESTAV del IPN, Mexico, D.F., MEXICO.
E-mail: nvasilev@math.cinvestav.mx

Greg Verchota Department of Mathematics, Syracuse University, Syracuse, New York 13244.
E-mail: verchota@math.syr.edu

Grant Welland Department of Mathematics, University of Missouri at Saint Louis, 8001 Natural Bridge Road, Saint Louis, Missouri 63121.
E-mail: welland@eads.umsl.edu

Zhijian Wu Department of Mathematics, University of Alabama at Tuscaloosa, Tuscaloosa, Alabama 35487.
E-mail: zwu@mathdept.as.ua.edu

Zhenyuan Xu Department of Mathematics, Ryerson Polytechnic University, 350 Victoria Street, Toronto, Ontario M5B 2K3, CANADA.
E-mail: zxu@acs.ryerson.ca

Keith Yale Department of Mathematics, University of Montana, Missoula, Montana 59812.
E-mail: ma_iky@relway.umt.edu

Chiping Zhou Department of Mathematics, University of Hawaii at Honolulu Community College, 874 Dillingham Boulevard, Honolulu, Hawaii 96817.

E-mail: chiping@pulua.hcc.hawaii.edu

FIGURE 2.1

'CLIFFORD ALGEBRAS IN ANALYSIS'
April 8–10 1993
UNIVERSITY OF ARKANSAS, FAYETTEVILLE

Front Row (Left to Right):
Chun Li (Sydney, Australia), Xhenyun Xu (Toronto, Canada), Terrance Tao (Princeton), Garry Miller (New Brunswick, Canada), Don Small (New Brunswick, Canada), Pertti Lounesto (Helsinki, Finland), Ruediger Belch (Kent, UK), Naoki Kimura (Arkansas), Alan McIntosh (Sydney, Australia).

Second Row:
Debora Gayle (San Francisco), Enrique Ramirez De Arellano (Mexico City, Mexico), Palle Jorgensen (Iowa), Don Burkhardt (Georgia), Toma Tonev (Montana), William M. Pezzaglia, Jr. (San Francisco).

Third Row:
Pascal Auscher (Rennes, France), Chiping Zhou (Hawaii), John Adams (San Francisco), John Gilbert (Texas), Josefina Alvarez (New Mexico), Keith Yale (Montana), Frank Kippig (Freiberg, Germany), Michael Shapiro (Mexico City, Mexico), Tao Qian (Armidale, Australia), Xiangsheng Xu (Arkansas).

Back Row:
Serge Tabachnikov (Arkansas), Zhijian Wu (Alabama), Guy David (Paris, France), Karel Stroethoff (Montana), John Ryan (Arkansas), Nikolai Vasilevski (Mexico City, Mexico), Itrel Monroe (Arkansas), Zhongwei Shen (Indiana), Daniel Dix (South Carolina).

Photographer: Matthew Enjalran (Amherst, MA)
Location: OLD POST OFFICE (Fayetteville, AR)
Plaque of Senator William Fulbright on left side

3

Introduction

John Ryan

Clifford analysis started as an attempt to generalize one-variable complex analysis to higher dimensions using Clifford algebras generated from
Euclidean space. More recently, deep and unexpected links to classical
harmonic analysis, several complex variables, and representation theory
have been discovered. In the early stages the subject was developed exclusively in three and four dimensions using the quaternionic division algebra,
which is an example of a Clifford algebra. Later it was realized that results obtained in the quaternionic setting, particularly the generalization
of Cauchy's integral formula, did not exclusively rely on the division algebra property of the quaternions, but that it is sufficient for an algebra to
contain a vector subspace where all non-zero vectors are invertible in the
algebra. In the Clifford algebra setting, this invertibility corresponds to
the usual Kelvin inversion of vectors in Euclidean space. This fact is not
too surprising, given that Clifford algebras are specifically designed to help
describe the geometric properties of quadratic forms on vectors spaces, see
for instance, [2]. For some time, it has been understood by most people
working with Clifford analysis that most results so far obtained in quaternionic analysis more or less automatically extend to all finite dimensions
using Clifford algebras.

In fact, Clifford algebras are remarkably simple algebras to set up. Loose-
ly speaking, one would like to consider R^n as a subspace of some algebra,
so that under the algebra multiplication we have that $x^2 = -\|x\|^2$ for each
vector x in R^n. If no other constraints are introduced, the minimal algebra
satisfying this requirement is an example of a universal Clifford algebra.
When $n = 1$, we obtain the complex number system. When $n = 2$, we
obtain the quaternions. For $n > 2$, the algebra is no longer a division
algebra. However, each non-zero vector x in R^n is invertible in the algebra,
with multiplicative inverse $x^{-1} = \dfrac{-x}{\|x\|^2}$. This inverse corresponds, up to a
sign, to the Kelvin inverse of the vector x.

Clifford algebras were introduced in the nineteenth century by mathe-

0-8493-8481-8/96/$0.00+$.50
© 1996 by CRC Press

maticians and mathematical physicists in various attempts to provide a
good foundation to geometric calculus in Euclidean space. An historical
account of this development and Clifford's role here is given in the introduction of [10] and some papers therein, e.g., [26]. Clifford was a student
of Maxwell's. He is one of the youngest scientists ever to be elected a
Fellow of the Royal Society, and he was Professor of Applied Mathematics at University College, London. He died of tuberculosis at the age of
thirty-three in 1879. His interest in the algebra that bears his name arose,
in part, from his attempts to place Maxwell's work on electromagnetism
in a more mathematically rigorous setting. His paper [11] describing these
algebras appeared in the *American Journal of Mathematics* one year before
his death.

The subject of Clifford analysis has been discovered and independently
rediscovered about ten times in the last century. The earliest known work
on the subject is by A.C. Dixon [16]. Alfred Cardew Dixon (1865–1936)
was Professor of Mathematics at Queen's University, Belfast, from 1901
to 1930. He was a Fellow of the Royal Society, and President of the London Mathematical Society from 1931 to 1933. Later, C. Lanczos described
the rudiments of quaternionic analysis in his doctoral thesis [31]. In the
1920's, Felix Klein independently rediscovered the area, [30]. In the 1930's
and 1940's, the Swiss mathematician Rudolph Fueter and his students published about fifteen papers on the subject, e.g., [18]. Most of these papers
appeared in the journal *Commentarii Mathematici Helvetici*. An excellent
summary of this work is given in a paper of A. Sudbery [56], and a more
detailed account is given in some lecture notes of E. Bareiss [4]. Surprisingly, many topics covered by Fueter and his collaborators have not been
touched upon in more recent books on the subject, though Gursey and
Tze [23] have used some of these results in their study of Yang Mills field
equations.

At much the same time as Fueter's work appeared, Moisil and Theodorescu [39] worked on closely related results. This appears to have been the
start of a period of research by Romanian mathematicians into aspects of
Clifford analysis and related topics which spanned a period of over thirty
years; see for instance, [27,41] and references therein. It was also during
the 1930's that possible links to mathematical physics were first noted.
In particular, the differential operator arising in the generalized Cauchy-
Riemann equations, and the "conjugate" of this operator combine to give
the Laplacian in Euclidean space. This is in complete analogy to the fact
that in one-variable complex analysis, the operators $\dfrac{d}{dz}$ and $\dfrac{d}{d\bar{z}}$ combine
to give the Laplacian in two-dimensional space. Earlier, Dirac [17] had
used a matrix representation of a Clifford algebra to introduce a factorization of the wave operator, or d'Alambertian, in terms of two first-order
differential operators. For this reason, the differential operator arising in

the generalized Cauchy-Riemann equations of Clifford analysis is often referred to as the Dirac operator. This operator corresponds to the $d + \delta$ operator acting on differential forms over R^n, where d is deRham's exterior derivative and δ its adjoint. However, it should be pointed out that the alternating algebra does not possess the correct algebraic structure to admit a meaningful Cauchy integral formula, akin to the one from Clifford analysis. Essentially, the alternating algebra generated from R^n does not incorporate Kelvin inversion.

It was also at this time that it was first noted that the analysis so far developed in the quaternion setting generalized to the Clifford algebra setting; see for instance, a paper of Haefeli [24]. Some further work in this direction was developed in the 1950's by Sce, [46].

In a four-year period in the late 1960's and early 1970's independent papers by Richard Delanghe [14], T.E. Littlewood and C.D. Gay [35], David Hestenes [25], and Viorel Iftimie [27] were published. Each of these papers illustrated how many aspects of one-variable complex analysis extend to Euclidean space using Clifford algebras. Here, fundamental, but implicit, use is made of Kelvin inversion to set up Cauchy's integral formula, Laurent and Taylor series, etc. In particular, in [27] Iftimie sets up basic results on Cauchy transforms over domains in R^n, and establishes Plemelj formulae for Hölder continuous functions defined over compact Liapunov surfaces in Euclidean space. Using the Plemelj formulae, he is able to show that the square of the singular Cauchy transform over such a surface is, when acting on Hölder continuous functions, the identity map. This is in complete analogy to the case in complex analysis. Consequently, the stage was set for applying Clifford analysis to study boundary value problems. More recently, these results have been extended to L^p-spaces over the boundaries of Lipschitz domains in R^n; see for instance [33,34].

Clifford analysis provides an extremely rich framework for generalizing many results from one-variable complex analysis. A review of the basic results of Clifford analysis is given in [21,Ch.4]. One subtle difference is that the generalized analytic functions, which are often called monogenic functions, are defined on domains in R^n, and usually take values in the Clifford algebra generated from that space or some spinor subspace of the algebra. One apparent limitation to the theory is that the pointwise multiplication of two monogenic functions is, in general, not a monogenic function. This follows from the noncommutativity of the algebra. Though it should be pointed out that in [51], a very natural product is introduced which reduces in the two-dimensional setting to the usual product. Other basic properties of one-variable complex analysis do not hold in the Clifford analysis setting, e.g., the Riemann mapping theorem.

The term "Clifford analysis" was first coined in the late 1970's, when the editor of this volume used it as a title of a manuscript. The manuscript was referenced by Sommen in [50], and most of the main results for this manuscript appeared in [42]. Later, the term was used by Brackx, Delanghe, and Sommen, [8], for the title of the first book in the area.

During the 1970's and 1980's, research into Clifford analysis started to become significantly less sporadic and isolated. Richard Delanghe began to build a research group at Ghent State University, Belgium, which has become the largest group currently working in the area. In particular, Frank Sommen [51] independently rediscovered a result of Littlewood and Gay [35] showing that real analytic functions defined on domains in R^{n-1} have Cauchy-Kowalweski extensions to monogenic functions defined in some neighborhood in R^n. Although this result is extremely simple, as are many basic results in Clifford analysis, it has a basic impact of linking problems in real analysis, in R^{n-1}, to function theory over domains in one higher dimension. In particular, in [52] Sommen uses this idea to link up Clifford analysis in R^n with the Fourier transform over R^{n-1}. It is in this work and in his later work on plane wave decompositions, [53], that it is realized that this analysis requires both the use of complex numbers and their generalization, the real Clifford algebras. In particular, both algebras are used fundamentally to set up projection operators to describe the decomposition of special classes of functions defined on R^{n-1} into classes of monogenic functions defined on upper- and lower-half-space in R^n. For this reason, it becomes necessary to introduce complex Clifford algebras. For mathematicians, this effectively, and efficiently, dispenses with objections raised by some physicists to the use of complex Clifford algebras; see for instance, remarks made in [26].

The projection operators mentioned in the previous paragraph are, in fact, Fourier transforms of the Plemelj formulae/operators for upper- and lower-half-space. Moreover, the singular Cauchy transform over R^{n-1} is the vector sum over R^{n-1} of the Riesz transforms described by Stein and Weiss in [54,55]. So, this singular Cauchy transform can be seen as a generalization of the Hilbert transform over the line. It follows that the work of Stein and Weiss, [54,55], on H^p-spaces in R^n using conjugate harmonic functions fits perfectly into the context of Clifford analysis. This point is well described in [20,Ch.2]. In fact, conjugate harmonic functions are vector-valued harmonic functions whose derivatives are symmetric, and have vanishing trace. Such a system of equations is called a Riesz system, and is a special case of the Cauchy-Riemann equations arising in Clifford analysis.

It was during the mid-1980's that R. Coifman had the idea that many hard problems in classical harmonic analysis could either be simplified or solved using Clifford analysis. This arose in the context of the Coifman-McIntosh-Meyer theorem, [13], which establishes the L^2-boundedness of the double-layer potential operator over Lipschitz graphs in R^n. This was a landmark result in classical harmonic analysis which was cited at the time in a report to the American Mathematical Society listing three recent dramatic examples of progress in theoretical mathematics. The original proof for the case $n = 2$ uses the complex number system, but in higher dimensions the Calderón rotation method is used. Coifman suggested that the two-dimensional proof can be mimicked in higher dimensions using Clifford algebras and Dirac operators, giving rise to a more natural proof. This was carried out for Lipschitz graphs with small Lipschitz constant by Margaret Murray, [40]. The argument was completed for all Lipschitz constants by Alan McIntosh [36]. A key idea here is that the double-layer potential operator over a sufficiently smooth surface is the real, or scalar, part of the singular Cauchy transform over the surface in R^n. This idea and these results had the impact of opening up the field to a much broader spectrum of mathematical interests.

Originally, the L^2-boundedness of the double-layer potential operator was worked out over Liapunov surfaces. So, the surface is C^1 with a Hölder continuous derivative. This added smoothness gives sufficient cancellation for one to deduce that the operator is weakly singular. It follows that the operators $\frac{1}{2}I \pm DLP$ are Fredholm, where I is the identity and DLP is the double-layer potential operator. Some more work reveals that these operators are injective, and so they are invertible. Consequently, it becomes an easy matter to use invertibility to produce solutions to the interior and exterior Dirichlet problems for such domains.

When one replaces Liapunov surfaces by Lipschitz surfaces, one no longer has the cancellation property mentioned in the previous paragraph. So the Fredholm operator theory is no longer available, and one needs to find different techniques. The first step in solving the Dirichlet problem over Lipschitz surfaces is to establish the L^2-boundedness of the double-layer potential over such surfaces. Several proofs of this result have now appeared, and some of them make use of Clifford analysis. One main advantage of the Clifford algebra-based proofs is that they bring to light the functional calculus of Dirac operators over Lipschitz surfaces, and unify much of the existing theory.

Also, in the 1980's, Ahlfors rediscovered results of Vahlen [58] and showed [1] that Möbius transformations in R^n could be described using a group of 2×2 matrices with values in a Clifford algebra. This inspired some authors to use Vahlen matrices to find analogues of Schwarzian derivatives on R^n [9,44].

The analogue of the Vahlen group over Minkowski space is the Lie group $SU(2, 2)$. This group is used, [29], to describe the conformal covariance of the Dirac operator and its iterates over Minkowski space. This too, can be placed in the context of Clifford algebras. These ideas, together with ideas described in a paper of Imaeda's [28] and many other references given in this introduction, have inspired the editor of this volume to study intertwining operators for conformally covariant operators over Euclidean space and $\mathbf{C}^n$, and to study Clifford analysis over very general types of cells of harmonicity in $\mathbf{C}^n$, together with links with several complex variables, see for instance, [43,45], and references therein.

The introduction of Vahlen matrices has inspired some authors in the early 1990's to develop Clifford analysis over hyperbolic space. Also, following ideas presented in [29] and mentioned in the previous paragraph, it would seem desirable to see further work done on the links between Clifford analysis and twistor theory. Work in this direction has been initiated in the last chapter of [15]. It would also seem likely that work previously done on automorphic forms and involving the use of Lie groups such as $SU(n,n)$, $Sp(n,R)$, and $Sp(n,\mathbf{C})$ could also be developed using Vahlen matrices over Minkowski-type spaces. Hopefully, some new and interesting results are awaiting discovery here. In addition, it would be nice to see closer ties developed between Clifford analysis and the study of Dirac operators over general spin manifolds, and to see links with the Atiyah-Singer index theorem as developed in [5,6,20,32].

The area of Clifford analysis which has seen the most rapid growth in recent years has been the one inspired by Coifman on applications to classical harmonic analysis and the theory of singular integrals. Besides the references that we have cited so far, there is also the work of Auscher and Tchamitchian [3], Gaudry, Long, and Qian [19], and Mitrea [38], where Clifford algebra-valued Haar wavelets/martingales are used to deduce the L^2-boundedness of the double-layer potential operator over Lipschitz surfaces in R^n. This extends to Euclidean space a proof due to Coifman, Jones, and Semmes, of the same result in the complex plane using complex-valued Haar wavelets; see [12].

A very good summary of these results, together with the Clifford $T(b)$ theorem is given in the Master's thesis of Terrance Tao [57]. Further very interesting results involving Clifford analysis within singular integral theory have been developed in recent times by Stephen Semmes [47,48,49]. Besides these recent developments on singular integrals and their applications to boundary value problems, Gürlebeck and Sprössig [22] have also considered related problems over Liapunov surfaces. Their approach also considers the use of colocation methods and other numerical techniques.

It should be pointed out that this review of the development of Clifford analysis, though intended to be fairly thorough, is by no means complete. Firstly, it is almost certain that there are still some long-forgotten papers in the area which will eventually be rediscovered. This seems inevitable, as such papers keep turning up with a fair regularity. Also, constraints of space and time prevented us from pointing out some further interesting developments and works in this area, or related areas.

This volume is based on a conference held in Fayetteville, Arkansas during the Easter weekend, April 8–10th, 1993. The conference was entitled "Clifford Algebras in Analysis", and the principal speaker was Alan McIntosh, of Macquarie University, Australia. Though there have been three other conferences on Clifford algebras and their applications in mathematical physics, [7,10,37], including one which took place one month after this one, this is the first conference, together with the proceedings, which deals almost exclusively with the impact of Clifford analysis on harmonic analysis. We were fortunate to be able to gather a highly-distinguished group of researchers in classical harmonic analysis and Clifford analysis for the meeting. It is hoped that the meeting and this volume will help set the pace for future research in this fascinating and growing area of mathematics. To this end, we have included a selection of open problems provided by many researchers with interests in this area. The idea for such a list came from a similar problem book in function theory developed by Walter Hayman and David Brannan. It is hoped that the problem book produced here will be added to with the passage of time, and will be addressed in future publications and conferences.

Bibliography

[1] L. V. Ahlfors, *Möbius transformations in R^n expressed through 2×2 matrices of Clifford numbers*, Complex Variables **5**, (1986), 215–224.

[2] M. F. Atiyah, R. Bott, and A. Shapiro, *Clifford modules*, Topology **3**, (1965), 3–38.

[3] P. Auscher and Ph. Tchamitchian, *Bases d'ondelettes sur les courbes corde-arc, noyou de Cauchy et Espaces de Hardy associeés*, Revista Matematica Iberoamericana **5**, (1989), 139–170.

[4] E. Bareiss, *Functions of a hypercomplex variable*, lecture notes based on talks given by Rudolph Fueter, University of Zürich, 1948–1949.

[5] N. Berline, E. Getzler, and M. Vergne, *Heat Kernels and Dirac Operators*, Springer-Verlag, Heidelberg, 1992.

[6] B. Booss-Bavnbeck and K. Wojciechowski, *Elliptic Boundary Problems for Dirac Operators*, Birkhauser, Basel, 1993.

[7] F. Brackx, R. Delanghe, and H. Serras, eds., *Clifford Algebras and their Applications in Mathematical Physics*, Deinze 1993, Kluwer, Dordrecht, 1993.

[8] F. Brackx, R. Delanghe, and F. Sommen, *Clifford Analysis*, Pitman, London, 1982.

[9] T. K. Carne, *The Schwarzian derivative for conformal maps*, J. Reine Angew. Math. **408**, (1990), 10–33.

[10] J. S. R. Chisholm and A.K. Common, eds., *Clifford Algebras and their Applications in Mathematical Physics*, Riedel, Dordrecht, 1986.

[11] W. K. Clifford, *Applications of Grassmann's extensive algebra*, Amer. J.of Math. **1**, (1978), 350–358.

[12] R. Coifman, P. Jones, and S. Semmes, *Two elementary proofs of the L^2-boundedness of Cauchy integrals on Lipschitz curves*, J. of the A.M.S. **2**, (1989), 553–564.

[13] R. Coifman, A. McIntosh, and Y. Meyer, *L'intégrale de Cauchy définit un opérateur borné sur L^2 pour les courbes lipschitziennes*, Annals of Mathematics **116**, (1982), 361–387.

[14] R. Delanghe, *On regular-analytic functions with values in a Clifford algebra*, Math. Ann. **185**, (1970), 91–111.

[15] R. Delanghe, F. Sommen, and V. Soucek, *Clifford Algebra and Spinor-Valued Functions : A Function Theory for the Dirac Operator*, Kluwer, Dordrecht, 1992.

[16] A. C. Dixon, *On the Newtonian potential*, Quarterly Journal of Mathematics **35**, (1904), 283–296.

[17] P. A. M. Dirac, *The quantum theory of the electron, I*, Proceedings of the Royal Society A **117**, (1928), 610–624.

[18] R. Fueter, *Die Funktionentheorie der Differentialgleichungen $\Delta u = 0$ und $\Delta\Delta u = 0$ mit Vier Reallen Variablen*, Commentarii Mathematici Helvetici **7**, (1934–1935), 307–330.

[19] G. Gaudry, R. Long, and T. Qian, *A martingale proof of L^2-boundedness of Clifford-valued singular integrals*, Annali di Matematica, Pura Appl. **165**, (1993), 369–394.

[20] J. Gilbert and M. Murray, *Clifford Algebras and Dirac Operators in Harmonic Analysis*, Cambridge University Press, Cambridge, 1991.

[21] R. P. Gilbert and J. Buchanan, *First-Order Elliptic Systems : A Function Theoretic Approach*, Academic Press, San Diego, 1983.

[22] K. Gürlebeck and W. Sprössig, *Quaternionic Analysis and Elliptic Boundary Value Problems*, Birkhäuser Verlag, Basel, 1990.

[23] F. Gursey and H. C. Tze, *Complex and quaternionic analyticity in chiral and gauge theories, 1*, Annals of Physics **128**, (1980), 29–130.

[24] H. G. Haefeli, *Hypercomplexe Differentiale*, Commentarii Mathematici Helvetici **20**, (1947), 382–420.

[25] D. Hestenes, *Multivector functions*, J. Math. Anal. Appl. **24**, (1968), 467–473.

[26] D. Hestenes, *A unified language for mathematics and physics*, Clifford Algebras and Their Applications in Mathematical Physics, Reidel, Dordrecht, 1–23, 1986.

[27] V. Iftimie, *Fonctions hypercomplexes*, Bull. Math. de la Soc. Sci. Math. de la R.S. de Roumanie **9**, (1965) 279–332.

[28] K. Imaeda, *A new formulation of classical electrodynamics*, Nuovo Cimento **32B**, (1976), 138–162.

[29] H. P. Jakobsen and M. Vergne, *Wave and Dirac operators, and representations of the conformal group*, J. Funct. Anal. **24**, (1977), 52–106.

[30] F. Klein, *Vorlesungen über nichteuklidische Geometrie*, Springer, Berlin, 1928.

[31] C. Lanczos, *Die Funktionentheoretischen Beziehungen Der Maxwellschen Aethergleichungen*, Doctoral Thesis, Budapest, 1919.

[32] H. B. Lawson, Jr. and M.-L. Michelson, *Spin Geometry*, Princeton University Press, Princeton, NJ, 1989.

[33] C. Li, A. McIntosh, and S. Semmes, *Convolution singular integrals on Lipschitz surfaces*, J. of the A.M.S. **5**, (1992), 455–481.

[34] C. Li, A. McIntosh, and T. Qian, *Clifford algebras, Fourier transforms, and singular convolution operators on Lipschitz surfaces*, Revista Mathematica Iberoamericana, to appear.

[35] T. E. Littlewood and C. D. Gay, *Analytic spinor fields*, Proc. Roy. Soc. **A313**, (1969), 491–507.

[36] A. McIntosh, *Clifford algebras and the higher-dimensional Cauchy integral*, Approximation Theory and Function Spaces, Banach Center Publications **22**, (1989), 253–267.

[37] A. Micali, R. Boudet, and J. Helmstetter, eds., *Clifford Algebras and their Applications in Mathematical Physics*, Kluwer, Dordrecht, 1992.

[38] M. Mitrea, *Singular Integrals, Hardy Spaces, and Clifford Wavelets*, Lecture Notes in Mathematics **No.1575**, (1994), Springer-Verlag, Heidelberg.

[39] Gr. C. Moisil and N. Theodorescu, *Fonctions holomorphes dans l'espace*, Mathematica (Cluj) **5**, (1931), 142–159.

[40] M. Murray, *The Cauchy integral, Calderón commutation, and conjugation of singular integrals in R^m*, Trans. of the A.M.S. **298**, (1985), 497–518.

[41] D. Pascali, *Representations of quaternionic areolar polynomials in tridimensional space*, Stud. Cerc. Mat. **18**, (1966), 239–242

[42] J. Ryan, *Clifford analysis with generalized elliptic and quasi-elliptic functions*, Appl. Anal. **13**, (1982), 151–171.

[43] J. Ryan, *Cells of harmonicity and generalized Cauchy integral formulae*, Proc. of the London Math. Soc. **60**, (1990), 295–318.

[44] J. Ryan, *Generalized Schwarzian derivatives for generalized fractional linear transformations*, Ann. Polon. Math. **LVII**, (1992), 29–44.

[45] J. Ryan, *Some applications of conformal covariance in Clifford analysis*, this volume.

[46] M. Sce, *Sulle Serie de Potenzi Nei Moduli Quadratic*, Lincei Rend. Sci. Fis. Mat. et Nat. **23**, (1957), 220–225.

[47] S. Semmes, *Differentiable function theory on hypersurfaces in R^n (without bounds on their smoothness)*, Indiana Univ. Math. J. **39**, (1990), 985–1004.

[48] S. Semmes, *Analysis vs. geometry on a class of rectifiable hypersurfaces in R^n*, Indiana Univ. Math. J. **39**, (1990), 1005–1035.

[49] S. Semmes, *Chord-arc surfaces with small constant, I*, Adv. in Math. **85**, (1991), 198–223.

[50] F. Sommen, *Spherical monogenics and analytic functionals on the unit sphere*, Tokyo J. Math. **4**, (1981), 427–456.

[51] F. Sommen, *A product and an exponential function in hypercomplex function theory*, Appl. Anal. **12**, (1981), 13–26.

[52] F. Sommen, *Microfunctions with values in a Clifford algebra, II*, Scientific Papers of the College of Arts and Sciences, University of Tokyo **36**, (1986), 15–37.

[53] F. Sommen, *Plane wave decomposition of monogenic functions*, Ann. Polon. Math., **49**, (1988), 101–114.

[54] E. M. Stein and G. Weiss, *On the theory of harmonic functions of several variables, I: The theory of H^p-spaces*, Acta Math, **103**, (1960), 25–62.

[55] E. M. Stein and G. Weiss, *Introduction to Fourier Analysis on Euclidean Spaces*, Princeton University Press, Princeton, NJ, 1971.

[56] A. Sudbery, *Quaternionic analysis*, Math. Proc. of the Cambridge Philosophical Soc., **85**, (1979), 199–225.

[57] T. Tao, *Convolution operators generated by right-monogenic and harmonic kernels*, Master's Thesis, Flinders University, Australia, 1992.

[58] K. Th. Vahlen, *Über Bewegungen und Complexe Zahlen*, Math. Ann. **55**, (1902), 585–593.

4

Problem Book

H. Blaine Lawson (State University of New York at Stonybrook)

Is there a reproducing kernel of Cauchy type for solutions to the Dirac equation over (spin) manifolds with negative curvature?

John Ryan (University of Arkansas, Fayetteville)

Suppose M is a smooth, real, n-dimensional, compact manifold in $\mathbf{C}^n$, satisfying

(i) $M \cap N(\underline{z}) = \{\underline{z}\}$

(ii) $TM_z \cap N(\underline{z}) = \{\underline{z}\}$

for each $\underline{z} \in M$ $(N(\underline{z}) = \{\underline{z}' \in \mathbf{C}^n : (z - z')^2 = 0\}$—the null cone in $\mathbf{C}^n)$.

Suppose Ω is a domain in $\mathbf{C}^n$ with $M \subseteq \Omega$ and $h : \Omega \to \mathbf{C}$ is a complex harmonic function so that $h(z)$ is holomorphic and

$$\sum_{j=1}^{n} \frac{\partial^2 h}{\partial z_j^2} = 0.$$

Is it true that for each such M we have that $h|_M$ satisfies a maximum principle?

Pascal Auscher (Université de Rennes, France)

It is known that the Clifford Haar-type b-wavelets for the $T(b)$ theorem can be defined when $b(x)$ is not only *accretive*, i.e.,

$$b(x) = b_0(x)e_0 - 1\sum_{i=1}^{n} b_i(x)e_i, b_0(x) \geq \delta_0 > 0, \qquad A.0$$

but *pseudo-accretive*, i.e.,

$$\left| \frac{1}{|Q|} \int_\phi b(x)dx \right| \geq \delta_0 > 0 \qquad A.1$$

0-8493-8481-8/96/$0.00+$.50
© 1996 by CRC Press

and even *para-accretive:* $\exists \delta_0 > 0$, $\delta_1 < 1$, for each cube Q, there exists a sub-cube R such that

$$\left| \frac{1}{|R|} \int_R b(x)dx \right| \geq \delta_0 > 0 \quad \text{and} \quad |R| > \delta_1 |Q|. \qquad A.2$$

Smooth Clifford b-wavelets can be obtained under conditon (A.0) with arbitrary high regularity.

Smooth Clifford b-wavelets can also be obtained under condition (A.1), but the regularity seems to be related to the smallness of δ_0: the smaller δ_0, the smaller r (r is then the Hölder regularity of the b-wavelets).

Problem 1: What is the exact relation between regularity of the b-wavelets and δ_0 of condition (A.1)?

Problem 2: Construct smooth Clifford b-wavelets under para-accretivity condition (A.2).

T. Tao (Princeton University)

Let Ω be an open subset of $\mathbf{R}^m$, and let u be a scalar-valued bounded harmonic function on Ω. Does there always exist a bi-monogenic function f on Ω such that $[f]_0 = u$ (i.e., the scalar part of f is u) when

(a) Ω is a sphere, or a rectangular box? (Answer: Yes, explicit condition possible)

(b) Ω is a bounded star-like region?

(c) Any generalizations (e.g., Ω with null m^{th} homotopy group)?

Also, can f always be chosen so that $Range(f) \subseteq span(e_0, e_1, \ldots, e_m)$?

(Editor's comment: When Ω is star-shaped, one can construct a left-monogenic function whose real part is u. See, for instance: A. Sudbery Quaternionic Analysis, Math. Proc. of the Cambridge Philosophical Society 86 (1979) 199–225, or J. Ryan Complexified Clifford analysis, Complex Variables 1 (1982) 151–171. Also, where Ω is a Lipschitz domain the answer is yes. See M. Mitrea 'Clifford algebras and boundary estimates for harmonic functions' Clifford Algebra and their Applications in Mathematical Physics, ed by F. Brackx, R. Delanghe and H. Serras, Kluwer, 1993.)

Palle E.T. Jorgensen (University of Iowa)

Let $\Omega \subset \mathbf{R}^n$, $n > 1$, be open and bounded, and let $D = \sum_1^n E_j \frac{\partial}{\partial x_j}$ be the corresponding Dirac/Clifford operator, acting on vector functions which are C^∞ and compactly supported in Ω. Then D is a symmetric Hilbert space operator with dense domain and corresponding adjoint D^*. Give a geometric description of the domain of D^*. In Jorgensen's talk, a selfadjoint extension, D_A, was described, $D \subset D_A \subset D^*$. Find the spectrum of D_A, and relate it directly to the geometry of Ω.

Let the operator D_A be defined as above, and associated with some $\Omega \subset \mathbf{R}^n$. Suppose $n = 2$, and Ω is one of the following:

Ex. 1.

Ex. 2.

Ex. 3.

Then answer the problems for these special cases. For Ex. 3, $\left\{ \frac{1}{r-1} \frac{\partial}{\partial x_j} \right\}$, $j = 1, 2$ (two symmetric operators in $L^2(\Omega)$), have commuting selfadjoint extension operators in the scalar space $L^2(\Omega)$. Relate the joint spectrum of these to the spectrum of D_A.

Nikolai Vasilevski and Michael Shapiro
(ESFM del I.P.N., Mexico City, Mexico)

Let Ω be a domain in $\mathbf{H}$, the quaternions. Also, let

$$(Df)(x) = \sum_{k=0}^{3} \frac{\partial f}{\partial t} + i \frac{\partial f}{\partial x} + j \frac{\partial f}{\partial y} + k \frac{\partial f}{\partial z}$$

$$(D^\psi f)(x) = \sum_{k=0}^{3} \frac{\partial f}{\partial t} + \frac{\partial f}{\partial x} i + \frac{\partial f}{\partial y} j + \frac{\partial f}{\partial z} k$$

for $f : \Omega \to \mathbf{H}$. Introduce the Bergman kernel function of the domain Ω

$$B(x, \xi) = (D_\xi \bar{D}_x g)(x, \xi) = D_x \cdot D_\xi g(x, \xi),$$

where $g(x, \xi)$ is the classical Green function of Ω, and the Bergman operator

$$(Bf)(x) = \int_\Omega B(x, \xi) f(\xi) d\xi : L_2(\Omega) \to L_2(\Omega).$$

Let $X_0(x)$ be a characteristic function at a subdomain Ω_0 of Ω.

Problem: Calculate $\Delta = sp \left(B - X_0(x) I \right)^2$.

This problem is an essential step for describing various algebras generated by Bergman operators. It is known that $\Delta \subset [0, 1]$, and *most probably*: $\Delta = [0, 1]$.

Each $f : \Omega \to \mathbf{H}$ can be represented in the "complex" form

$$f = f^{(1)} + f^{(2)} j,$$

where

$$f^{(k)} : \Omega \subset \mathbf{R}^4 = \mathbf{C}^2 \to \mathbf{C}.$$

We have

$$D = 2 \left(\frac{\partial}{\partial \bar{z}_2} + k \frac{\partial}{\partial \bar{z}_2} \right),$$

and thus the set of hyperholomorphic functions contains (but does not coincide with) the set of all holomorphic (in the sense of complex analysis) mappings $f = (f^{(1)}, f^{(2)}) : \Omega \subset \mathbf{C}^2 \to \mathbf{C}$.

Problem: Find connections between the Bergman function $B(x, \xi)$ of a domain Ω and the two-dimensional complex analysis Bergman function of a domain Ω.

Zhijian Wu (University of Alabama, Tuscaloosa)

Let

$$A = \left\{ f : f(x, y) \text{ monogenic in } \mathbf{R}^{n+1}_+ : \frac{1}{4} \int \int_{\mathbf{R}^{n+1}_+} |f(x, y)|^2 dx dy < \infty \right\}.$$

A is a subspace of $L^2(\mathbf{R}^{n+1}_+)$. One can find the orthonormal projection $P : L^2(\mathbf{R}^{n+1}_+) \to A$. P can be expressed as an integral operator.

Question: Can we express $I - P$ as an integral operator?

Example: If $n = 1$, the answer for this question is Yes. In fact, if $\varphi \in C_0^\infty(\mathbf{R}_+^2)$,

$$(I - P)(\varphi)(w) = \int_{\mathbf{R}_+^2} \frac{\bar{\partial}\varphi(z)\,Im\,z}{(w - z)(w - \bar{z})}\,dxdy.$$

John Ryan (University of Arkansas, Fayetteville), and Tao Qian (University of New England, Australia)

The surfaces of star-shaped domains have global parametrizations that may allow us to solve some boundary value problems first on these domains, and then transfer the solutions to more general domains. The following question then arises: What are the domains in $\mathbf{R}^n$, $n \geq 3$, that are the ranges under conformal mappings of the star-shaped ones? When $n = 2$, simply connected domains are the ranges under conformal mappings of the star-shaped ones. In the higher-dimensional case, the only conformal mappings are the elements of the Möbius transform group, and they are not monogenic functions.

Tao Qian (University of New England, Australia)

Singular integral theory with holomorphic (monogenic) kernels has been developed on Lipschitz surfaces and Lipschitz perturbations of the n-torus (see the expository papers of A. McIntosh and T. Qian in this collection).
What is the analogue on n-dimensional solid balls? What is the analogue on n-dimensional complex balls?

Pertti Lounesto (Helsinki University of Technology, Finland)

The Maxwell equations can be condensed into one equation by Clifford bivectors (at least in an isotropic and homogeneous media). The Maxwell equations are also conformally covariant (as photons are massless). Furthermore, the solutions of sourceless Maxwell equations are monogenic. Under a Möbius transformation $x \to g(x) = (ax + b)/(cx + d)$, a monogenic function $F(x)$ is transformed as follows

$$F(x) \to G(x) = \frac{c\tilde{x} + d}{|cx + d|^n} F(g(x)),$$

where $G(x)$ is also monogenic. However, for a bivector field $F(x)$, the transformed field $G(x)$ is not, in general, a bivector. Is it possible to transform the electromagnetic bivector field $F(x)$ under Möbius transformations so that the transformed field is also a bivector?

Klaus Gürlebeck (Technical University of Chemnitz, Germany)

It is known in complex function theory that under certain conditions a continuous function r defined on a closed curve Γ can be factorized in the following form,

$$r(t) = r_-(t)t^\kappa r_+(t) \forall t \in \Gamma, \tag{1}$$

where r_- allows a holomorphic extension into the exterior domain and r_+ has a holomorphic extension into the interior domain with boundary Γ. The value κ is an integer. If r is a rational function, one can show in a constructive way that there exists a factorization (1) with rational functions r_- and r_+.

Problem: Is it possible to find a similar factorization (explicitly?), also for functions defined on the boundary Γ of a bounded domain G, in R^n, with values in (real) Clifford algebras? The most interesting case for applications is the case of quaternionic-valued functions.

Wolfgang Sprößig (Bergakademie Freiberg Technische Universität, Freiberg, Germany)

Suppose ω is a closed, bounded rectangle in R^2. Then $\omega \times R \subseteq R^3$ is called a channel-domain. For a channel-domain with density φ we have the electric field

$$\underline{E} = \sum_{i=1}^{3} E_i e_i,$$

the magnetic field

$$\underline{H} = \sum_{i=1}^{3} H_i e_i,$$

and the electric conductivity κ, dielectric constant ε, and permeability μ. We consider in the channel the solution of the following stationary Maxwell equations:

$$div\,\varepsilon\underline{E} \;=\; 0 \qquad div\,\mu\underline{H} \;=\; 0 \qquad rot\,\underline{E} \;=\; 0 \qquad rot\,\underline{H} \;=\; \kappa\underline{E}.$$

Prove the existence, uniqueness, and regularity of the solution, if the normal components of the solution is given on the boundary of the channel; so that $\underline{s} \cdot \underline{H} = g$ for g belonging to a suitable function space.

Pertti Lounesto (Helsinki University of Technology, Finland)

Does the Clifford algebra version of the Bott periodicity theorem dictate results in Clifford analysis which vary from dimension to dimension? For instance, are there results in three dimensions which do not hold in seven dimensions, even though there might be similar results in $11 = 8 + 3$ dimensions?

Marius Mitrea (University of South Carolina)

Recall the generalized Hardy spaces $\mathcal{H}^p(\Omega)$ discussed by Kenig in [Ke] for Ω a (special) Lipschitz domain in the complex plane. For $1 < p < \infty$, they also have a natural (and in many respects, satisfactory) extension to higher dimensions within the Clifford algebra framework (cf., e.g., [Mi]). In this setting, establish a Riesz boundary behavior theory for the end-point case $p = 1$.

References:

[Ke] C.E. Kenig, *Weighted H^p spaces on Lipschitz domains*, Am. J. Math. **102**(1980), 129–163.

[Mi] M. Mitrea, *Clifford Wavelets, Singular Integrals, and Hardy Spaces*, Lecture Notes in Math., 1575, Springer-Verlag (1994).

Josefina Alvarez (New Mexico State University)

Let T be a Calderón-Zygmund operator in the sense of R. Coifman and Y. Meyer. That is to say, assume that the distribution kernel $k(x, y)$ of T satisfies the pointwise condition

$$|k(x, y) - k(x, z)| \leq C \frac{|y - z|^\delta}{|x - z|^{n+\delta}}$$

if $2|y - z| < |x - z|$, for some $0 < \delta \leq 1$. Concerning the continuity of the operator T on Hardy spaces, the following results are known.

(i) T maps continuously H^p into L^p for $\frac{n}{n+\delta} < p \leq 1$, and this result is optimal.
(ii) T maps continuously $H^{\frac{n}{n+\delta}}$ into $L^{\frac{n}{n+\delta},\infty}$.
(iii) T maps continuously $H^{p,\infty}$ into $L^{p,\infty}$, for $\frac{n}{n+\delta} < p \leq 1$.

Problem: Is the result in (iii) optimal? If not, what continuity result can be formulated for $p = \frac{n}{n+\delta}$?

Zhenyuan Xu (Ryerson Polytechnical University, Canada)

It is well known that the index plays a very important rôle in the study of boundary value problems in complex analysis. For instance, consider the

Riemann-Hilbert problem

$$\frac{\partial}{\partial \bar{z}} w = 0, \quad in \ \Omega,$$

$$R_e \left[\lambda \left(\bar{z} \right) w(z) \right] = \gamma(z), \quad on \ \Gamma = \partial \Omega,$$

where Ω is a unit disk with center at the origin. The index is defined by

$$\kappa = \frac{1}{2\pi} \Delta_\Gamma \arg \lambda(z).$$

Then, for $\kappa \geq 0$, the Riemann-Hilbert problem is solvable for any Hölder continuous functions $\lambda(z)$ and $\gamma(z)$. Moreover, the solution linearly depends on $2\kappa + 1$ real arbitrary constants. For $\kappa < 0$, the Riemann-Hilbert problem is not solvable, except when $\lambda(z)$ and $\gamma(z)$ satisfy $-2\kappa - 1$ consistent conditions. Is there an analogue of the index for Clifford-valued functions in R^m?

Daniel B. Dix (University of South Carolina)

I will employ the notation described in my paper.

(1) Consider a homogeneous "scalar" Dirac equation $\mathcal{D}m = T(x, v)m$, where $v(k', t)$ is a $\mathbf{C}_n$-valued function of $k' \in \mathbf{R}^n$ and $t \in \mathbf{R}$. $T(x, v)$ is a linear operator, depending parametrically on x and v, which acts on $m(x, t, k)$, and yields a distribution on $\mathbf{R}^{n+1}$, parameterized by x and t, which is an appropriate right-hand side for the inhomogeneous Dirac equation. Suppose this equation has a solution with asymptotic behavior

$$m(x, t, k) \, e_0 + \sum_{h=0}^{\infty} \left[\sum_{(l) \in n^h / S_h} W_{(l)}(k) q_{h,(l)}(x, t) \right], \quad \|k\| \to \infty, |k_0| > \varepsilon.$$

The coefficients $Q_{h,(l)}(x, t)$ of this asymptotic expansion satisfy a complicated coupled system of nonlinear evolution equations. Can simple examples of the operator $T(x, v)$ and a linear evolution of $v(k', t)$ be found such that this infinite coupled system of evolution equations effectively reduces to a coupled system involving only finitely many of the coefficients and their partial derivatives in the x variables? If so, then this would be a truly multidimensional example of a nonlinear system of partial differential equations solvable by a Clifford inverse scattering method. In such an example, how would one define the forward scattering transform?

(2) In complex analysis, we have the notion of the sheaf of holomorphic functions on a complex manifold. In particular, we can discuss holomorphic functions defined on a neighborhood of ∞ on the Riemann

sphere. Is there some analogue of this in Clifford analysis? In particular, how would one make sense out of the "sheaf of monogenic functions" on $S^{n+1} = \mathbf{R}^{n+1} \cup \{\infty\}$? Is there a monogenic analogue of (some substantial portion of) the theory of Riemann surfaces?

Stephen Semmes (Rice University, Houston, Texas)
Generalizations of Complex Analysis to Higher Codimensions

There are two particularly prominent methods for generalizing classical complex analysis in the plane to $\mathbf{R}^n$ for $n > 2$. The first is to use a Cauchy-Riemann system, like the classical Riesz system (of vector fields which are curl- and divergence-free) or Clifford holomorphicity. These are first-order linear elliptic systems of partial differential equations. The second main approach is to look at quasi-regular mappings, which are (roughly speaking) maps for which the maximal stretching of the differential at any point is bounded by a constant multiplied by the minimal stretching at that point. These two approaches are very different in style—the first is better suited for linear analysis, while the second is more geometric in focus—but they are also at opposite extremes in terms of dimensions. This point is illustrated by the following observation. Let $f(x)$ be a Clifford-holomorphic function on some domain in $\mathbf{R}^n$. For each point x_0 in the domain, the differential of f at x_0 is controlled by its restriction (as a linear mapping) to any hyperplane through the origin in $\mathbf{R}^n$. This follows from the definition of Clifford holomorphicity, which provides a formula for the derivative in any given direction v in terms of the derivatives in the remaining $n - 1$ directions. By contrast, if $f(x)$ is quasi-regular, then the differential of f at x_0 is controlled by its restriction to any line through the origin, by definition of quasi-regularity.

I like to think of Clifford analysis (and other Riesz systems) as being "codimension-1 complex analysis" on $\mathbf{R}^n$. This is also related to the usual integration formulas in Clifford analysis (like Cauchy's theorem) which involve integrals over hypersurfaces. Similarly, quasi-regular mappings define a kind of codimension-$(n - 1)$ complex analysis.

Problem: Find interesting kinds of "codimension-d" complex analysis on $\mathbf{R}^n$ for other choices of d.

It is not clear exactly what this means, but there are some basic principles. By definition, "complex analysis" should deal with a class of functions or mappings which are distinguished by a condition on their first derivatives. In codimension-d complex analysis, the differential of a "holomorphic" object should be controlled (somehow) by its restriction to any codimension-d-plane through the origin. One might hope for nice integral formulae, but for codimension-d submanifolds. There should be some interesting interplay between "holomorphic" objects on the complement of a d-dimensional submanifold and their boundary behavior. These principles should be viewed more as illustrative than definitive, and I am certainly not saying that there is at most one reasonable codimension-d complex analysis.

When $d > 1$, I don't believe that there is a nice codimension-d complex analysis that is based on a first-order linear system and which has roughly the same analytic features as for Riesz systems and Clifford analysis when $d = 1$. I envision two types of theories, one which is nonlinear, not so algebraic, and more brutally geometric, and another which is linear, has interesting integral formulae, and which probably uses differential forms more seriously and is more degenerate analytically when $d > 1$ than when $d = 1$. Of course, we see part of this dichotomy already in the contrast between Clifford analysis and quasi-regular mappings in $\mathbf{R}^n$ when $n > 2$.

One reason for raising the issure of codimension-d complex analysis is that I would like to have a nice higher-codimension version of some of the results in [DS], [S1], [S2], and [S3]. In particular, in [S1] and [S2] there are some integration-by-parts computations which are used to good advantage, and I would like to have suitable extensions of these arguments to higher codimensions. In these extensions, there should be integral formulae with topological content, just as the Cauchy formula in codimension-1 contains the information of when a point lies in a given domain or its exterior. In the higher-codimension case, the analogous topological issue is the linking number of pairs of spheres (and other submanifolds). There are, of course, classical integral formulae for computing linking numbers (see [F], especially p.79ff), but I have never managed to use them to obtain interesting analytic information, as occurs in codimension-1 in [S1] and [S2].

I am not convinced that singular integral operators will have such an important role in higher-codimension (linear) complex analysis as in codimension-1. I am more optimistic about approximations to the identity and square functions estimates. In codimension-1 complex analysis, there are some interesting approximations to the identity on hypersurfaces which are built out of the Cauchy kernel and which contain interesting geometric information about the surface (see [S3], especially (1.0) on p.1010). One can imagine analogous objects in higher codimensions, using differential forms and Clifford algebras. These higher-dimensional analogues could be associated to $(d-1)$-spheres which link a given codimension-d submanifold (on which we are to have the approximation to the identity), just as in the codimension-1 case the approximation to the identity is defined using pairs of points which lie in different components of the complement of the hypersurface (i.e., linking 0-spheres). Unfortunately, I have not succeeded in producing anything that I can work with analytically. See [DS, Sect.8, Ch.3, Part III] for different remarks on the same general topic.

References

[DS] G. David and S. Semmes, "Analysis of and on Uniformly Rectifiable Sets", *Mathematical Surveys and Monographs* **38**(1993), Am. Math. Soc.

[F] H. Flanders, *Differential Forms, with Applications to the Physical Sciences*, Academic Press, 1963.

[S1] S. Semmes, A criterion for the boundedness of singular integrals on hypersurfaces, *Trans. Am. Math. Soc.* **311**(1989), pp.501–513.

[S2] S. Semmes, Differentiable function theory on hypersurfaces in $\mathbf{R}^n$ (without bounds on their smoothness), *Ind. Math. J.* **39**(1990), 985–1004.

[S3] S. Semmes, Analysis vs. geometry on a class of rectifiable hypersurfaces in $\mathbf{R}^n$, *Ind. Math. J.* **39**(1990), pp.1005–1035.

Rodolfo H. Torres (University of Michigan at Ann Arbor) and Grant Welland (University of Missouri at Saint Louis)

Let D be a bounded Lipschitz domain in R^3, and let N be the unit outward normal to the boundary of the domain. Let k be a complex number, and let $\Phi(X) = -\dfrac{e^{ik|X|}}{4\pi|X|}$ be the fundamental solution for the Helmholtz operator $\Delta + k^2 I$.

Problem: Find the spectrum in $L^2_T(\partial D)$ (square integrable vectors fields on ∂D) of the operator M given by

$$MF(P) = p.v. \int_{\partial D} N(P) \times curl\,(\Phi(P - Q)F(Q))\,d\sigma(Q).$$

This operator arises in the study of boundary value problems for Maxwell equations using the method of layer potentials. The knowledge of the spectrum will be important in solving some transmission problems. For transmission problems in Lipschitz domains, see, for example,

(1) L. Escauriaza, E. Fabes, and G. Verchota, "On a regularity theorem for weak solutions to transmission problems with internal Lipschitz boundaries", *Proc. Am. Math. Soc.* **115**(1992), pp.1069–1076.

(2) M. Mitrea, R. Torres, and G. Welland, "Regularity and approximation results for the Maxwell problem in C^1 and Lipschitz domains", in this proceedings.

(3) R. Torres, "A transmission problem in the scattering of electromagnetic waves by a penetrable object", preprint.

(4) R. Torres and G. Welland, "The Helmholtz equation and transmission problems with Lipschitz interfaces", *Ind. Univ. Math. J.* **42**(1993), to appear.

William M. Pezzaglia, Jr. (Physics Department, Santa Clara University)

Metric Signature-Dependent Structure of Integral Solutions

Spacetime can apparently be equally represented by $\mathbf{R}^{1,3}$ or by $\mathbf{R}^{3,1}$. One may even go so far as to propose a physical principle: *that tangible phenomena are invariant under a change of signature from the* "*west coast*" $(---+)$ *to the* "*east coast*" $(+++-)$. However, the Clifford algebras generated are known to be inequivalent: $\mathbf{R}(4) = End\mathbf{R}^{3,1}$ vs. $\mathbf{H}(2) = End\mathbf{R}^{1,3}$. Only in the first case can the Klein-Gordon equation (1a) below be factored into the meta-monogenic Dirac form, eq. (1b),

$$\left(\Box^2 = m^2\right)\phi = 0, \tag{1a}$$

$$\Box\Psi = m\Psi, \tag{1b}$$

$$\Psi = (\Box + m)\phi. \tag{1c}$$

If $G(x)$ is a Green function for the Klein-Gordon eq. (1a), then $\mathcal{G} = (\Box + m)G$ is the corresponding Green function for the Dirac eq. (1b). This leads to a convenient Cauchy integral solution,

$$(\Box - m)\mathcal{G}(x) = \delta(x), \tag{1d}$$

$$\Psi(y) = \oint d\Sigma^\mu \mathcal{G}(y, x)e_\mu \Psi(x). \tag{1e}$$

In the $(---+)$ signature, one could make the alternative factorization,

$$\left(\Box^2 + m^2\right)\phi = 0, \tag{2a}$$

$$\Box\Psi = m\Psi\Gamma, \tag{2b}$$

$$\Psi = \Box\phi + m\phi\Gamma, \tag{2c}$$

where $\Gamma^2 = -1$ is an element of the algebra (note that there is no commuting i in the four-dimensional algebra). The sign difference between eq. (1a) and eq. (2a) is due to the change in metric signature. In the particular case of $\Gamma = e_1 e_2 e_3 e_4$, or $\Gamma = e_1 e_2$, eq. (2b) is sometimes called the *Hestenes-Dirac equation.*

Question 1: What is the analogy of eq. (1e) for the Hestenes-Dirac eq. (2b)? Does it require some condition on the element Γ, or a restriction on the form of the wave function Ψ?

Question 2: If the integral solution exists, and is of different form than eq. (1e), then how does one reconcile this with the physical assumption stated above, namely, that the metric choice should *not* make a difference?

Question 3: To be complete, one should consider the *Greider-Dirac equation*: $\Box\Phi = m\Phi\Gamma$, where, different than eq. (2b), we have $\Gamma^2 = +1$. Does the integral solution differ in content and/or form from eq. (1e)?

Heinz Leutwiler (University of Erlangen-Nürnberg, Germany)

Let C_n denote the (universal) Clifford algebra generated by the elements $e_1, \ldots, e_n$, subject to the conditions $e_k^2 = -1$ and $e_h e_k = -e_k e_h$ ($h \neq k$). Harmonicity (in the sense of Hodge) of the 1-form $\omega = Re(f dx)$, $f = \sum_{k=0}^{n} u_k e_k$, $dx = \sum_{k=0}^{n} dx_k e_k$, if calculated with respect to the hyperbolic metric on $\mathbf{R}_+^{n+1}$, leads to the equation $(H_n) : x_n Df(x) + (n-1)u_n(x) = 0$, where $x = \sum_{k=0}^{n} x_k e_k$ varies over some open subset Ω of $\mathbf{R}^{n+1}$. Here, $D = \sum_{k=0}^{n} \dfrac{e_k \partial}{\partial x_k}$ denotes the Dirac/Fueter operator. In case $n = 2$, where $C_2 = \mathbf{H}$ (the quaternions), it has been verified (see *Complex Variables* **20**(1992), pp.19–51) that every (H_2)-solution f, defined in some neighborhood U of 0, admits a generalized power series expansion in terms of the following homogeneous, polynomial (H_2)-solutions:

$$E_n^k(x) = \sum_{\substack{\mu_0 + \mu_1 + \cdots + \mu_n = k \\ \mu_\nu \in \{0, 1\}}} e_1^{\mu_0} x e_1^{\mu_1} x \ldots e_1^{\mu_{n-1}} x e_1^{\mu_n}$$

$$(x = x_0 + x_1 e_1 + x_2 e_2), \ 0 \leq k \leq n + 1.$$

Problem 1: If f is an (H_2)-solution in $U \backslash \{0\}$, does there exist a "generalized Laurent series expansion", i.e., an expansion in terms of $E_n^k(x)$ ($0 \leq k \leq n+1$), $E_{-1}^0(x) = x^{-1}$, and $E_{-n}^k(x) := x^{-1} E_{n-2}^k \left(x^{-1}\right) x^{-1}$ ($o \leq k \leq n-1$)?

Problem 2: Find a basis for the **R**-vector space of homogeneous, polynomial (H_n)-solutions in case $n > 2$. What about the corresponding expansions?

Problem 3: Is there a Cauchy-type formula for the solutions of (H_n)?

(Editor's comment: For problem 3, see also the problem of H. Blaine Lawson.)

William M. Pezzaglia, Jr. (Physics Department, Santa Clara University)

Integral Solutions Involving "Dextrad" (right-side applied) Kernels

An integral solution of the meta-monogenic Dirac equation $(\Box - m)\Psi = 0$ can sometimes be written,

$$\Psi\left(\vec{X}',t'\right) = \int d^3\vec{X}\, F\left(\vec{X}',t';\vec{X},t\right)\Psi\left(\vec{X},t\right), \tag{1a}$$

independent of t, where $(\Box - m)F = 0$. When the solution Ψ is restricted, it was shown in my paper that there exists an alternative right-side applied *dextrad* kernel, such that

$$\Psi\left(\vec{X}',t'\right) = \int d^3\vec{X}\, \Psi\left(\vec{X},t\right) H\left(\vec{X}'.t';\vec{X},t\right). \tag{1b}$$

This was useful in constructing a *Path Integral Formulation*.

Question 1: What is the general relationship between kernel F of eq. (1a) and the dextrad kernel H of eq. (1b)?

Question 2: The more general Cauchy solution would be of the form

$$\Psi(y) = \oint d\sum^{\mu} \mathcal{G}(y,x)e_\mu \Psi(x), \tag{2}$$

where the Green function obeys $(\Box - m)\mathcal{G} = \delta$. Does there exist in a "dextrad" equivalent of perhaps one of the following forms?

$$\Psi(y) = \oint d\Sigma^{\mu} e_\mu \Psi(x)\mathcal{K}(y,x), \tag{3a}$$

$$\Psi(y) = \oint d\Sigma^{\mu} \Psi(x)e_\mu \mathcal{K}(y,x), \tag{3b}$$

$$\Psi(y) = \oint d\Sigma^{\mu} \Psi(x)\mathcal{K}(y,x)e_\mu. \tag{3c}$$

Question 3: Can a $\mathcal{K}$ always be found given $\mathcal{G}$? Must Ψ be restricted for $\mathcal{K}$ to exist (probably yes)?

R.P. Gilbert (University of Delaware)

Consider the differential equation

$$\bar{\partial}w - \sum_A C_A(x)H_A w(x) = 0, \tag{1}$$

where the $C_A(x)$ are hypercomplex functions and H_A is a mapping defined as follows:

$$H_A := H_{\alpha_i} \ldots H_{\alpha_p}, \qquad A := \{\alpha_1, \ldots, \alpha_p\}$$

$$H_i e_i \to -e_i, \qquad H_i e_j = e_j \ for \ i \neq j, \ i, j \geq 2$$

$$\bar{\partial} := \partial_{x_1} + e_2 \partial_{x_2} + \cdots + e_n \partial_{x_n}.$$

What are necessary and sufficient conditions for the differential equation (1) to have the "Liouville property", i.e., when each bounded, continuous solution may be expressed as a linear combination of $m \leq n$ solutions $\{w_1, \ldots, w_m\}$? See R.P. Gilbert and J. Buchanan *First-order elliptic systems: A function theoretic approach* (Ch.4), Academic Press, 1983, for further discussion.

Chiping Zhou (University of Hawaii at Honolulu Community College)

Riemann-Hilbert problems and Neumann problems have been formulated and solved in both the classical sense and the L^p sense for the Dirac equation $Dw = F$ in the upper half-space of $\mathbf{R}^m$ (see the papers of Z. Xu and C. Zhou in this collection). We know that the general linear elliptic equation with principal part D can be written as

$$\mathcal{L}w(x) := Dw(x) - \sum_A C_A(x)\mathcal{J}_A w(x) = F(x), \tag{1}$$

where C_A and F are Clifford-valued functions, $\mathcal{J}_i(e_k) = e_k - 2\delta_{ik}e_i$, and $\mathcal{J}_A = \mathcal{J}_{\alpha_1}\mathcal{J}_{\alpha_2}\ldots\mathcal{J}_{\alpha_h}$ for $A = \{\alpha_1, \ldots, \alpha_h\} \subset \{1, \ldots, m\}$ (see B. Goldschmidt, *Beiträge zur Algebra und Geometrie* **13**(1982), 21–24).

Question: Can we properly formulate and then solve the Riemann-Hilbert and/or Neumann problems for solutions of (1)? In particular, can we do this when $\mathcal{L} = D_\lambda := D - \lambda$ $\lambda \in \mathbf{R}$ (this is an interesting case because $D_\lambda D_{-\lambda} = D_{-\lambda}D_\lambda = -\Delta - \lambda^2 I$ is the Helmholtz operator)?

Bernhelm Booss-Bavnbeck (Roskilde, Denmark) and Krzysztoff P. Wojciechowski (Indiana University-Purdue University)

What is the natural "Grassmannian" of elliptic self-adjoint boundary value problems for the Dirac operator on the 4-ball and other even-dimensional manifolds with boundary?

T. Wolff (University of California at Berkeley)

Let D be the open unit disc in R^n, $\overline{D}$ its closure, ∂D its boundary.

Consider the generalized Cauchy-Riemann system

$$dw = 0$$

$$d^*\omega = 0$$

where $\omega = \sum_i \omega_i dx_i$ is a 1-form on D, and $d\omega = \sum_{ij} \frac{\partial f_i}{\partial x_j} dx_j \wedge dx_i$, $d^*\omega = \sum_i \frac{\partial f_i}{\partial x_i} dx_1 \wedge \ldots \wedge dx_n$.

Is it possible to have a solution (other than the zero solution) which *extends smoothly to* $\overline{D}$ and vanishes on a subset of ∂D with *positive $n-1$-dimensional Lebesgue measure?*

This is of course equivalent to asking whether there is a nonconstant harmonic function $u : D \to R$ which extends smoothly to $\overline{D}$ and so that ∇u vanishes on a subset of ∂D with positive $n-1$-dimensional measure. Some background:

1) If the solution ω vanishes on an open subset of ∂D instead of a subset with positive measure then it must vanish identically, even if *a priori* it is only assumed continuous up to the boundary. This follows from the Schwartz reflection principle and the uniqueness statement in the Cauchy-Kowalewsky theorem.

2) If $n = 2$ then the answer to the question is again no, even if ω is only assumed continuous up to the boundary. This follows from the second F and M Riesz theorem applied to the analytic function $\omega_1 - i\omega_2$.

3) If $n \geq 3$ and if ω is only assumed continuous (or Holder continuous) up to the boundary then the answer is yes (T. Wolff, to appear in the conference proceedings in honor of E. M. Stein, Princeton University Press). Further related work has been done by A. B. Aleksandrov with P. Kargaev and by J. Bourgain with T. Wolff.

An approach to the question via Carleman type inequalities was proposed some time ago by D. Jerison. This is discussed on p. 639 ff. of my survey article in the *Journal of Geometric Analysis,* vol. 3, number 6.

5

Clifford Algebras, Fourier Theory, Singular Integrals, and Harmonic Functions on Lipschitz Domains

Alan McIntosh[*][†]

5.1 Introduction

This is a slightly expanded version of my lectures at the conference on Clifford Algebras in Analysis held in Fayetteville, Arkansas. I present an exposition of basic material on Clifford algebras, monogenic functions, and singular integrals with monogenic kernels. Then follow results from the paper "Clifford algebras, Fourier transforms, and singular convolution operators on Lipschitz surfaces" by Chun Li, Tao Qian and myself [LMcQ]. Here is an edited excerpt from its introduction.

In the Fourier theory of functions of one variable, it is common to extend a function and its Fourier transform holomorphically to domains in the complex plane $\mathbb{C}$, and to use the power of complex function theory. This depends on first extending the exponential function $e^{ix\xi}$ of the real variables x and ξ to a function $e^{iz\zeta}$ which depends holomorphically on both the complex variables z and ζ. Our thesis is this. The natural analog in higher dimensions is to extend a function of m real variables monogenically to a function of $(m+1)$ real variables (with values in a complex Clifford algebra), and to extend its Fourier transform holomorphically to a function of m complex variables. This depends on first extending the exponential function $e^{i<\mathbf{x},\xi>}$ of the real variables $\mathbf{x} \in \mathbb{R}^m$ and $\xi \in \mathbb{R}^m$ to a function $e(x,\zeta)$ which depends monogenically on $x = x_0 e_0 + \mathbf{x} \in \mathbb{R}^{m+1}$ and holomorphically on $\zeta = \xi + i\eta \in \mathbb{C}^m$.

We explore this thesis for functions ϕ whose monogenic extensions are

---

[*]The author was supported by the Australian Government through the Australian Research Council.

[†]Permission has been obtained from Revista Matemática Iberoamericana to quote material from [LMcQ].

1991 *Mathematics Subject Classification.* 42B20, 47A60, 42B10, 42B15

bounded by a constant multiple of $|x|^{-m}$ on a cone $C_{\mu+}^0 = \{x = x_0 e_0 + \mathbf{x} \in \mathbb{R}^{m+1} : x_0 > -|\mathbf{x}|\tan\mu\}$. The Fourier transforms b of these functions satisfy $b(\xi)(|\xi|e_0 - i\xi) = 0$ and extend holomorphically to bounded functions on certain cones $S_\nu^0(\mathbb{C}^m)$ in $\mathbb{C}^m$ (for all $\nu < \mu$). Conversely, every bounded holomorphic function b on $S_\mu^0(\mathbb{C}^m)$ which satisfies $b(\xi)(|\xi|e_0 - i\xi) = 0$ for all $\xi \in \mathbb{R}^m$, is the Fourier transform of a function ϕ whose monogenic extension is bounded by $c|x|^{-m}$ on $C_{\nu+}^0$ (for all $\nu < \mu$).

If ϕ is a right–monogenic function which is bounded by $c|x|^{-m}$ on $C_{\mu+}^0$, then the singular convolution operator T_ϕ defined by

$$(T_\phi u)(x) = \lim_{\delta \to 0} \int_\Sigma \phi(x + \delta e_0 - y)\, n(y)\, u(y)\, dS_y$$

is a bounded linear operator on $L_p(\Sigma)$ for $1 < p < \infty$ [LMcS]. Here Σ is the Lipschitz surface consisting of all the points $x = g(\mathbf{x})e_0 + \mathbf{x} \in \mathbb{R}^{m+1}$, where $\mathbf{x} \in \mathbb{R}^m$, and g is a real–valued Lipschitz function which satisfies $\|\mathbf{D}g\|_\infty \leq \tan\omega$ for some $\omega < \mu$. We have embedded $\mathbb{R}^m$ in a complex Clifford algebra with at least m generators in the usual way, and identified the extra basis element e_0 of $\mathbb{R}^{m+1}$ with the identity e_0 in the Clifford algebra. Clifford multiplication is used in the above integrand, in which $n(y)$ denotes the unit normal (which is defined at almost all $y \in \Sigma$).

So the Fourier transform b of ϕ can be thought of as the Fourier multiplier corresponding to T_ϕ. But we can also think of the mapping from b to $T_\phi \in \mathcal{L}(L_p(\Sigma))$ as giving us a bounded H_∞ functional calculus of a differential operator $\frac{1}{i}\mathbf{D}_\Sigma$, and write $T_\phi = b(\frac{1}{i}\mathbf{D}_{vee\Sigma})$.

For example, if $b(\xi) = \frac{1}{2}(e_0 + i\xi/|\xi|)$, then $\phi(x) = \overline{x}/\sigma_m|x|^{m+1}$ and T_ϕ is the boundary value of the Cauchy integral for monogenic functions.

The scalar part of this operator is the singular double–layer potential operator. We consider how these results relate to its use in the study of boundary value problems for harmonic functions on the open set Ω_+ above Σ.

These lectures are devoted to some applications of Clifford algebra to analysis that I understand and have found useful. There are many more aspects which are explored in the books by Brackx, Delanghe and Sommen [BDS], Gilbert and Murray [GM], Gürlebeck and Sprössig [GS], Lawson and Michelsohn [LM], Delanghe, Sommen and Souček [DSS] and others, as well as the forthcoming book by Ryan.

5.2 Lecture 1

I have pleasure in thanking the University of Arkansas, and all those who helped to organise this conference, especially John Ryan, for their kind invitation and support. I also wish to acknowledge the contribution of Tao Qian and Chun Li with whom I have collaborated on this topic, as well as Raphy Coifman, Stephen Semmes, John Ryan, and the many other people who have contributed to my understanding.

Throughout these lectures, for m a positive integer, the real vector space $\mathbb{R} \oplus \mathbb{R}^m$ is embedded in a real or complex finite dimensional algebra $\mathcal{A}$ which has an identity element e_0. We identify $(\lambda, \xi) \in \mathbb{R} \oplus \mathbb{R}^m$ with $\lambda e_0 + \xi \in \mathcal{A}$.

To say that $\mathcal{A}$ is a real or complex algebra means that it is a real or complex vector space on which an associative product uv of elements $u, v \in \mathcal{A}$ is defined, and the distributive laws $(cu+v)w = c(uw)+vw$, $w(cu+v) = c(wu) + wv$ hold for all $u, v \in \mathcal{A}$ and $c \in \mathbb{R}$ or $\mathbb{C}$. The identity e_0 is the (unique) element in $\mathcal{A}$ which satisfies $e_0 u = u e_0 = u$ for all $u \in \mathcal{A}$.

At all times we make the following assumption about the algebra $\mathcal{A}$. The identity

$$\xi^2 = -|\xi|^2 e_0 = -(\xi_1{}^2 + \xi_2{}^2 + \ldots + \xi_m{}^2) e_0 \tag{I}$$

is satisfied by all $\xi \in \mathbb{R}^m$.

An easy consequence of (I) is the identity

$$\xi\eta + \eta\xi = -2 \langle \xi, \eta \rangle e_0 = -2 (\xi_1\eta_1 + \xi_2\eta_2 + \ldots + \xi_m\eta_m) e_0$$

(because $2(\xi\eta + \eta\xi) = (\xi + \eta)^2 - (\xi - \eta)^2 = (-|\xi + \eta|^2 + |\xi - \eta|^2) e_0 = -4\langle\xi, \eta\rangle e_0$).

In the case when $\mathcal{A}$ is a complex algebra, the embedding of $\mathbb{R} \oplus \mathbb{R}^m$ in $\mathcal{A}$ induces the embedding of $\mathbb{C} \oplus \mathbb{C}^m$ in $\mathcal{A}$, where $(\lambda, \xi + i\eta) \in \mathbb{C} \oplus \mathbb{C}^m$ is identified with $\lambda e_0 + \xi + i\eta \in \mathcal{A}$. Then all $\zeta = \xi + i\eta \in \mathbb{C}^m$ satisfy

$$\zeta^2 = -|\zeta|_*^2 e_0$$

where $|\zeta|_*^2$ is defined by

$$|\zeta|_*^2 = \sum_{j=1}^{m} \zeta_j{}^2 = |\xi|^2 - |\eta|^2 + 2i \langle \xi, \eta \rangle.$$

(I would prefer to denote this by $|\zeta|^2$, but refrain from doing so in order not to cause confusion with $\| \zeta \|^2 = |\xi|^2 + |\eta|^2$.)

If, in addition to the identity (I), $\mathcal{A}$ is generated by the elements $\xi \in \mathbb{R}^m$, then $\mathcal{A}$ is called a *real or complex Clifford algebra* with m generators, and denoted by $\mathbb{R}_{(m)}$ or $\mathbb{C}_{(m)}$. These m generators are usually taken to be the standard basis elements of $\mathbb{R}^m$ and denoted by $e_1, e_2, \ldots, e_m$. Clearly they satisfy

$$e_j{}^2 = -e_0 \quad \text{and}$$

$$e_j e_k = -e_k e_j$$

when $1 \leq j \leq m$, $1 \leq k \leq m$, $j \neq k$. These algebras were introduced and studied by W.K. Clifford in 1878 [C].

The Clifford algebras $\mathbb{R}_{(1)}$ and $\mathbb{R}_{(2)}$ can be identified with the complex numbers (on identifying e_1 with i), and the quaternions (on identifying e_1, e_2, $e_1 e_2$ with i, j, k). Notice, however, that we are mostly considering complex algebras, and are deliberately not identifying any of the basis elements with the complex number i.

For the remainder of this lecture, $\mathcal{A}$ is a complex algebra in which (I) is satisfied.

5.2.1   (A) Spectral theory of $\xi \in \mathbb{R}^m$

Our aim in this first section is to develop the spectral theory of vectors $\xi \in \mathbb{R}^m$ when $\mathbb{R}^m$ has been embedded in the complex algebra $\mathcal{A}$.

Let $\xi \in \mathbb{R}^m$. The *complex spectrum* $\sigma(i\xi)$ of $i\xi$ in the algebra $\mathcal{A}$ is

$$\sigma(i\xi) = \{ \lambda \in \mathbb{C} : (\lambda e_0 - i\xi) \text{ does not have an inverse in } \mathcal{A} \}$$
$$= \{ \pm|\xi| \}.$$

This is easily seen after checking that

$$(\lambda e_0 - i\xi)(\lambda e_0 + i\xi) = \lambda^2 e_0 - i^2 \xi^2 = (\lambda^2 - |\xi|^2) e_0$$

so that

$$(\lambda e_0 - i\xi)^{-1} = \frac{\lambda e_0 + i\xi}{\lambda^2 - |\xi|^2}$$

whenever the denominator is not zero.

The spectrum is simple in the following sense. Define, for $\xi \in \mathbb{R}^m$, $\xi \neq \mathbf{0}$,

$$\chi_{\pm}(\xi) = \frac{1}{2} \left(e_0 \pm \frac{i\xi}{|\xi|} \right).$$

Using $(i\xi)^2 = |\xi|^2$ we obtain

$$\chi_+(\xi)^2 = \chi_+(\xi) \,, \; \chi_-(\xi)^2 = \chi_-(\xi) \,, \; \chi_+(\xi)\chi_-(\xi) = 0 = \chi_-(\xi)\chi_+(\xi).$$

Further,

$$i\xi \chi_{\pm}(\xi) = \chi_{\pm}(\xi)\, i\xi,$$
$$e_0 = \chi_+(\xi) + \chi_-(\xi) \quad \text{and}$$
$$i\xi = |\xi|\, \chi_+(\xi) + (-|\xi|)\, \chi_-(\xi),$$

this being the spectral representation of $i\xi$ as a linear combination of its spectral idempotents, $\chi_{\pm}(\xi)$.

For any polynomial $P(\lambda) = \sum a_k \lambda^k$ of one variable with complex coefficients, we have

$$P\{i\xi\} \;=\; \sum a_k (i\xi)^k \;=\; P(|\xi|)\chi_+(\xi) \;+\; P(-|\xi|)\chi_-(\xi).$$

Thus it is natural to associate with the polynomial P of one variable, the polynomial p of m variables, defined by

$$\begin{cases} p(\xi) & = P\{i\xi\} \;=\; P(|\xi|)\chi_+(\xi) \;+\; P(-|\xi|)\chi_-(\xi), \quad \xi \neq 0 \\ p(0) & = P(0). \end{cases}$$

Indeed, to every complex–valued function B of one real variable, we associate the function b of m real variables, defined at $\xi \in \mathbb{R}^m$ by

$$b(\xi) \;=\; B\{i\xi\} \;=\; B(|\xi|)\chi_+(\xi) \;+\; B(-|\xi|)\chi_-(\xi)$$

if $|\xi|$ and $-|\xi|$ are both in the domain of B, and by $b(0) = B(0)$ if 0 is in the domain of B. Thus we have defined a *functional calculus* of $i\xi$, in that we have defined an algebra homomorphism from complex–valued functions of one real variable to the algebra $\mathcal{A}$, which agrees with the natural definition for polynomials of $i\xi$. In this sense we have a spectral theory of $i\xi$ when $\xi \in \mathbb{R}^m$.

Note that this homomorphism maps the characteristic functions $\chi_{\mathbb{R}^+}$ and $\chi_{\mathbb{R}^-}$ of the half lines $\mathbb{R}^+ = \{\lambda \in \mathbb{R} : \lambda > 0\}$ and $\mathbb{R}^- = \{\lambda \in \mathbb{R} : \lambda < 0\}$ to the spectral idempotents $\chi_\pm(\xi)$, corresponding to the eigenvalues $\pm|\xi|$. That is, $\chi_{\mathbb{R}^+}\{i\xi\} = \chi_+(\xi)$ and $\chi_{\mathbb{R}^-}\{i\xi\} = \chi_-(\xi)$ when $\xi \neq 0$.

5.2.2 (B) Spectral theory of $\zeta \in \mathbb{C}^m$

Let us extend this procedure so as to deal with holomorphic functions of complex variables.

For ζ in $\mathbb{C}^m$, the complex spectrum of $i\zeta$ in the algebra $\mathcal{A}$ is

$$\begin{aligned} \sigma(i\zeta) &= \{\,\lambda \in \mathbb{C} : (\lambda e_0 - i\zeta) \text{ does not have an inverse in } \mathcal{A}\,\} \\ &= \{\,\pm|\zeta|_*\,\} \end{aligned}$$

where $\pm|\zeta|_* = \pm\sqrt{|\zeta|_*^2}$.

The corresponding spectral idempotents are

$$\chi_\pm(\zeta) \;=\; \frac{1}{2}\left(e_0 + \frac{i\zeta}{\pm|\zeta|_*}\right)$$

when $|\zeta|_*^2 \neq 0$, in which case the spectrum is again simple in the same sense as before. Thus we again have

$$\chi_+(\zeta)^2 \;=\; \chi_+(\zeta)\,, \quad \chi_-(\zeta)^2 \;=\; \chi_-(\zeta)\,, \quad \chi_+(\zeta)\chi_-(\zeta) \;=\; 0 \;=\; \chi_-(\zeta)\chi_+(\zeta)\,,$$

$$\begin{cases} i\zeta \chi_{\pm}(\zeta) & = \chi_{\pm}(\zeta)\,i\zeta \\ e_0 & = \chi_+(\zeta) + \chi_-(\zeta) \\ i\zeta & = |\zeta|_*\,\chi_+(\zeta) + (-|\zeta|_*)\,\chi_-(\zeta), \end{cases}$$

and indeed, for any polynomial $P(\lambda) = \sum a_k \lambda^k$ of one variable with complex coefficients, we have

$$P\{i\zeta\} \;=\; \sum a_k (i\zeta)^k \;=\; P(|\zeta|_*)\,\chi_+(\zeta) \;+\; P(-|\zeta|_*)\,\chi_-(\zeta).$$

When $\zeta \neq 0$ and $|\zeta|_*^2 = 0$ however, the complex spectrum $\sigma(i\zeta) = \{0\}$ is not simple, for then

$$P\{i\zeta\} \;=\; P(0) + P'(0)\,i\zeta \;=\; a_0 + a_1 i\zeta.$$

Suppose now that B is a complex–valued holomorphic function of one complex variable. Define the associated holomorphic function b of m complex variables by

$$b(\zeta) \;=\; B\{i\zeta\} \;=\; \begin{cases} B(|\zeta|_*)\,\chi_+(\zeta) + B(-|\zeta|_*)\,\chi_-(\zeta) \;,\; |\zeta|_* \neq 0 \\ B(0) + B'(0)\,i\zeta, \; |\zeta|_* = 0 \end{cases}$$

if $\pm|\zeta|_*$ are both in the domain of B.

This time we have defined a *holomorphic functional calculus* of $i\zeta$, in that we have defined an algebra homomorphism from holomorphic complex–valued functions B of one complex variable to $b(\zeta) = B\{i\zeta\}$ in the algebra $\mathcal{A}$, which agrees with the natural definition for polynomials. In this sense we have a spectral theory of $i\zeta$ when $\zeta \in \mathbb{C}^m$.

The fact that the mapping $B \mapsto b$ is an algebra homomorphism is easily verified. What this means is that, if $B \mapsto b$ and $F \mapsto f$, then $cB + F \mapsto cb + f$ (when $c \in \mathbb{C}$), and $BF \mapsto bf$.

The most important examples for our purposes are given by the following exponential functions. For $t \in \mathbb{R}$, let $E_t(\lambda) = e^{-t\lambda}$, $\lambda \in \mathbb{C}$. The associated functions of one variable are $e_t(\zeta) = e(t, \zeta)$, where

$$\begin{aligned} e(t, \zeta) \;=\; e^{-t\{i\zeta\}} \;&=\; e^{-t|\zeta|_*}\,\chi_+(\zeta) + e^{t|\zeta|_*}\,\chi_-(\zeta) \\ &=\; \cosh(t|\zeta|_*)e_0 \;-\; \sinh(t|\zeta|_*)\frac{i\zeta}{|\zeta|_*} \end{aligned}$$

when $|\zeta|_*^2 \neq 0$, and $e_t(\zeta) = e_0 - ti\zeta$ when $|\zeta|_*^2 = 0$.

On using the fact that the mapping $B \mapsto b$ is an algebra homomorphism, we see that

$$e(t, \zeta)\,e(s, \zeta) \;=\; e(t + s, \zeta) \quad \text{and}$$

$$\frac{d}{dt}\,e(t, \zeta) \;=\; -i\zeta\,e(t, \zeta) \;=\; -e(t, \zeta)\,i\zeta.$$

Other important examples, defined for each complex number α, are the functions $R_\alpha(\lambda) = (\lambda - \alpha)^{-1}$, $\lambda \neq \alpha$. Then $R_\alpha\{i\zeta\} = (i\zeta - \alpha)^{-1} = (i\zeta + \alpha)\,/(|\zeta|_*^2 - \alpha^2)$, $|\zeta|_*^2 \neq \alpha^2$.

Let us now restrict our attention to holomorphic functions defined on certain cones in $\mathbb{C}$ and $\mathbb{C}^m$. In $\mathbb{C}$, we consider the open sectors

$$S_{\mu+}^0(\mathbb{C}) = \{ \zeta \in \mathbb{C} : |\arg \zeta| < \mu \}, \ S_{\mu-}^0(\mathbb{C}) = -S_{\mu+}^0(\mathbb{C})$$

and the open double sector, $S_\mu^0(\mathbb{C}) = S_{\mu+}^0(\mathbb{C}) \cup S_{\mu-}^0(\mathbb{C})$, where $0 < \mu < \pi/2$.

Our higher–dimensional analogue of $S_\mu^0(\mathbb{C})$ may seem somewhat strange at first. It is the open cone

$$S_\mu^0(\mathbb{C}^m) = \{ \zeta = \xi + i\eta \in \mathbb{C}^m : |\zeta|_*^2 \notin (-\infty, 0] \text{ and } |\eta| < \tan \mu \, Re|\zeta|_* \}.$$

Up until now, we did not need to specify which of the square roots of $|\zeta|_*^2$ was denoted by $\pm$, but here, and whenever we know that $|\zeta|_*^2 \notin (-\infty, 0]$, we take $Re|\zeta|_* > 0$.

Note that $S_\mu^0(\mathbb{C}) = S_\mu^0(\mathbb{C}^1)$ when $m = 1$.

The following result is an exercise, the solution of which appears as theorem 2.1 of [LMcQ].

THEOREM 5.2.1
For $\zeta = \xi + i\eta \in S_\mu^0(\mathbb{C}^m)$,

(i) $Re|\zeta|_*$, $||\zeta|_*|$, $\|\zeta\|$ *and* $|\xi|$ *are all equivalent, with the constants appearing in the equivalences depending on μ. For example,*

(ii)

$$||\zeta|_*| \leq \sec \mu \, Re|\zeta|_* \ ;$$

$$|\zeta|_* \in S_{\mu+}^0(\mathbb{C}) \ ;$$

(iii) $\|\chi_\pm(\zeta)\| \leq \frac{1}{\sqrt{2}\cos \mu}$.

Let $H_\infty(S_\mu^0(\mathbb{C}))$ denote the Banach space of all bounded complex–valued holomorphic functions B defined on $S_\mu^0(\mathbb{C})$ with the norm $\| B \|_\infty = \sup \{ |B(\lambda)| : \lambda \in S_\mu^0(\mathbb{C}) \}$. Also let $H_\infty(S_\mu^0(\mathbb{C}^m)) = H_\infty(S_\mu^0(\mathbb{C}^m), \mathcal{A})$ denote the Banach space of all bounded holomorphic functions b defined on $S_\mu^0(\mathbb{C}^m)$ with values in $\mathcal{A}$ under the norm $\| b \|_\infty = \sup\{ \| b(\zeta) \| : \zeta \in S_\mu^0(\mathbb{C}) \}$. Here $\mathcal{A}$ is given any norm for which

$$\| \lambda e_0 + \zeta \|^2 = |\lambda|^2 + \| \zeta \|^2 = |\lambda|^2 + \sum_{k=1}^{m} |\zeta_k|^2$$

when $\zeta = \lambda e_0 + \zeta \in \mathbb{C} \oplus \mathbb{C}^m$. (Note that the functions in $H_\infty(S_\mu^0(\mathbb{C}))$ are complex–valued, while functions in $H_\infty(S_\mu^0(\mathbb{C}^1))$ take their values in $\mathcal{A}$, even when $m = 1$.)

The following result is an immediate consequence of the preceding one.

THEOREM 5.2.2

The mapping $B \mapsto b$ from $H_\infty(S^0_\mu(\mathbb{C}))$ to $H_\infty(S^0_\mu(\mathbb{C}^m))$ defined by

$$b(\zeta) = B\{i\zeta\} = B(|\zeta|_*)\chi_+(\zeta) + B(-|\zeta|_*)\chi_-(\zeta)$$

is a bounded algebra homomorphism, and

$$\|b\|_\infty \leq \frac{\sqrt{2}}{\cos\mu}\|B\|_\infty$$

for all $B \in H_\infty(S^0_\mu(\mathbb{C}))$.

Another exercise is to show that this mapping is one–one by constructing its left–inverse.

5.3 Lecture 2

In this lecture $\mathcal{A}$ is a complex algebra in which (I) is satisfied.

5.3.1 (C) Spectral theory of commuting matrices

As a prelude to the study of the spectral theory and functional calculi of Dirac operators, let us first consider real diagonal matrices. This material is not used explicitly in the later lectures, but it provides a simple framework in which to explore the distinctions between the complex spectrum and the joint spectrum of an operator, and its corresponding functional calculi.

Let $\mathbf{T} = (T_1, T_2, \ldots, T_m) = \sum_{j=1}^m T_j e_j$ where each T_j is the real $N \times N$ diagonal matrix

$$T_j = \begin{bmatrix} \xi_j^1 & 0 & \cdots & 0 \\ 0 & \xi_j^2 & & \vdots \\ \vdots & & \ddots & 0 \\ 0 & \cdots & 0 & \xi_j^N \end{bmatrix}$$

where $\xi_j^k \in \mathbb{R}$ for $1 \leq k \leq N$, $1 \leq j \leq m$. Write

$$\mathbf{T} = \sum_{j=1}^m T_j e_j = \begin{bmatrix} \xi^1 & 0 & \cdots & 0 \\ 0 & \xi^2 & & \vdots \\ \vdots & & \ddots & 0 \\ 0 & \cdots & 0 & \xi^N \end{bmatrix} = \sum_{j=1}^m \begin{bmatrix} \xi_j^1 & 0 & \cdots & 0 \\ 0 & \xi_j^2 & & \vdots \\ \vdots & & \ddots & 0 \\ 0 & \cdots & 0 & \xi_j^N \end{bmatrix} e_j.$$

There are two types of spectra of $\mathbf{T}$, namely the joint spectrum of the m–tuple $(T_1, T_2, \ldots, T_m)$ and the complex spectrum of the single operator

$\mathbf{T}$ in the algebra $\mathcal{B}$ of matrices with entries in $\mathcal{A}$. This is a complex algebra with identity

$$I = \begin{bmatrix} e_0 & 0 & \cdots & 0 \\ 0 & e_0 & & \vdots \\ \vdots & & \ddots & 0 \\ 0 & \cdots & 0 & e_0 \end{bmatrix}.$$

The joint spectrum is the set of joint eigenvalues,

$$\sigma_m(\mathbf{T}) = \{ \xi^1, \xi^2, \ldots, \xi^N \} \subset \mathbb{R}^m.$$

For any function $b : \sigma_m(\mathbf{T}) \to \mathcal{A}$, there is a natural definition of $b(\mathbf{T})$, namely

$$b(\mathbf{T}) = \begin{bmatrix} b(\xi^1) & 0 & \cdots & 0 \\ 0 & b(\xi^2) & & \vdots \\ \vdots & & \ddots & 0 \\ 0 & \cdots & 0 & b(\xi^N) \end{bmatrix}.$$

This defines a functional calculus of $\mathbf{T}$, because the mapping from functions b to matrices $b(\mathbf{T})$ is an algebra homomorphism, and $b(\mathbf{T})$ agrees with the usual definition when b is a polynomial of m variables.

If $\xi^k \neq 0$ for all k, let

$$P_\pm = \chi_\pm(\mathbf{T}) = \begin{bmatrix} \chi_\pm(\xi^1) & 0 & \cdots & 0 \\ 0 & \chi_\pm(\xi^2) & & \vdots \\ \vdots & & \ddots & 0 \\ 0 & \cdots & 0 & \chi_\pm(\xi^N) \end{bmatrix}.$$

It is a consequence of the properties of $\chi_\pm$ that these matrices satisfy

$$P_+^2 = P_+, \quad P_-^2 = P_-, \quad P_+ P_- = 0 = P_- P_+$$

$$I = P_+ + P_-$$

$$i\mathbf{T} = \begin{bmatrix} |\xi^1| & 0 & \cdots & 0 \\ 0 & |\xi^2| & & \vdots \\ \vdots & & \ddots & 0 \\ 0 & \cdots & 0 & |\xi^N| \end{bmatrix} P_+ + \begin{bmatrix} -|\xi^1| & 0 & \cdots & 0 \\ 0 & -|\xi^2| & & \vdots \\ \vdots & & \ddots & 0 \\ 0 & \cdots & 0 & -|\xi^N| \end{bmatrix} P_-$$

$$= |\mathbf{T}| P_+ + (-|\mathbf{T}|) P_-$$

where $|\mathbf{T}|$ is the diagonal matrix with entries ($|\xi^1|, |\xi^2|, \ldots, |\xi^N|$). It follows that

$$(i\mathbf{T})^k = |\mathbf{T}|^k P_+ + (-|\mathbf{T}|)^k P_-$$

for all non-negative integers k.

The second type of spectrum of $\mathbf{T}$ is its usual complex spectrum, where this time $\mathbf{T}$ is regarded as a single operator. As before, we consider $i\mathbf{T}$ rather than $\mathbf{T}$. Then

$$\sigma(i\mathbf{T}) = \{\lambda \in \mathbb{C} : (\lambda I - i\mathbf{T}) \text{ does not have an inverse in } \mathcal{B}\}$$

$$= \{|\xi^1|, |\xi^2|, \ldots, |\xi^N|\} \cup \{-|\xi^1|, -|\xi^2|, \ldots, -|\xi^N|\} \subset \mathbb{R}.$$

Suppose now that $\xi^k \neq 0$ for all k. Given any function $B : \sigma(i\mathbf{T}) \to \mathbb{C}$, define $B\{i\mathbf{T}\}$ by

$$B\{i\mathbf{T}\} = \begin{bmatrix} B\{i\xi^1\} & 0 & \cdots & 0 \\ 0 & B\{i\xi^2\} & & \vdots \\ \vdots & & \ddots & 0 \\ 0 & \cdots & 0 & B\{i\xi^N\} \end{bmatrix}$$

$$= \begin{bmatrix} B(|\xi^1|) & 0 & \cdots & 0 \\ 0 & B(|\xi^2|) & & \vdots \\ \vdots & & \ddots & 0 \\ 0 & \cdots & 0 & B(|\xi^N|) \end{bmatrix} P_+$$

$$+ \begin{bmatrix} B(-|\xi^1|) & 0 & \cdots & 0 \\ 0 & B(-|\xi^2|) & & \vdots \\ \vdots & & \ddots & 0 \\ 0 & \cdots & 0 & B(-|\xi^N|) \end{bmatrix} P_-$$

$$= B(|\mathbf{T}|) P_+ + B(-|\mathbf{T}|) P_- .$$

This defines a functional calculus of $i\mathbf{T}$, because the mapping from functions B to matrices $B\{i\mathbf{T}\}$ is an algebra homomorphism, and $B\{i\mathbf{T}\}$ agrees with the usual definition when B is a polynomial of one variable.

The two functional calculi are related by $B\{i\mathbf{T}\} = b(\mathbf{T})$ in the case when b is the function of m variables associated with B by $b(\xi) = B\{i\xi\}$.

In particular, $\chi_{\mathbb{R}+}\{i\mathbf{T}\} = P_+$ and $\chi_{\mathbb{R}-}\{i\mathbf{T}\} = P_-$. That is, P_+ and P_- are the spectral projections associated with the positive and negative parts of the spectrum of $i\mathbf{T}$, namely $\sigma_+(i\mathbf{T}) = \sigma(i\mathbf{T}) \cap \mathbb{R}^+$ and $\sigma_-(i\mathbf{T}) = \sigma(i\mathbf{T}) \cap \mathbb{R}^-$.

Let us also consider the exponential functions of $i\mathbf{T}$. Recall that, for $t \in \mathbb{R}$, we set $E_t(\lambda) = e^{-t\lambda}$. The associated functions of m variables are

$$e_t(\xi) = e(t, \xi) = e^{-t\{i\xi\}} = e^{-t|\xi|} \chi_+(\xi) + e^{t|\xi|} \chi_-(\xi).$$

$$\text{Therefore, } e(t, \mathbf{T}) = e^{-t\{i\mathbf{T}\}} = e^{-t|\mathbf{T}|} P_+ + e^{t|\mathbf{T}|} P_-$$

which is the group generated by $-i\mathbf{T}$.

Given $u \in \mathbb{C}^m$, define $u_+(t)$ for $t > 0$ by

$$u_+(t) = e^{-t\{i\mathbf{T}\}} P_+ u = e^{-t|\mathbf{T}|} P_+ u.$$

Then $u_+(t)$ has the following properties:

$$\begin{cases} \frac{du_+}{dt}(t) + i\mathbf{T}u_+(t) = 0, & t > 0 \\ \lim_{t \to 0} u_+(t) = P_+ u \\ \lim_{t \to \infty} u_+(t) = 0. \end{cases}$$

In a similar way we see that the function $u_-(t)$ defined for $t < 0$ by

$$u_-(t) = e^{-t\{i\mathbf{T}\}} P_- u = e^{t|\mathbf{T}|} P_- u$$

satisfies

$$\begin{cases} \frac{du_-}{dt}(t) + i\mathbf{T}u_-(t) = 0, & t > 0 \\ \lim_{t \to 0} u_-(t) = P_- u \\ \lim_{t \to \infty} u_-(t) = 0. \end{cases}$$

We shall see equations like this in other contexts later.

5.3.2 (D) Spectral theory of the Dirac operator D

The preceding material can be adapted to deal with much more general situations involving m–tuples of commuting operators with real spectra in a Banach space. See for example my papers with Alan Pryde and Werner Ricker [McP, McPR].

Let us just consider the L_2 theory of

$$\mathbf{D} = (D_1, D_2, \ldots, D_m) = \sum_{j=1}^{m} D_j e_j = \sum_{j=1}^{m} \frac{\partial}{\partial x_j} e_j$$

considered both as an m–tuple, namely the gradient operator, and as a single operator, in which case $\mathbf{D}$ is called the Dirac operator.

Let $\mathcal{X}$ be a finite dimensional left $\mathcal{A}$ module. That is, $\mathcal{X}$ is a finite dimensional complex vector space together with a representation of elements of $\mathcal{A}$ as linear operators on $\mathcal{X}$. If $u \in \mathcal{A}$ and $v \in \mathcal{X}$, the action of u on v is denoted by uv.

We consider $\mathcal{X}$ together with a norm $\|\cdot\|$ and remark that there exists a constant C such that $\|uv\| \leq C \|u\| \|v\|$ for all $u \in \mathcal{A}$ and all $v \in \mathcal{X}$.

We often take $\mathcal{X}$ to be $\mathcal{A}$ itself, with the action as left–multiplication, but it is also common to consider representations of $\mathcal{A}$ on matrix algebras. An account of this is presented, for example, in the book by Gilbert and Murray [GM].

For $1 \leq p < \infty$, let $L_p(\mathbb{R}^m) = L_p(\mathbb{R}^m, \mathcal{X})$ denote the Banach space of (equivalence classes of) measurable functions u from $\mathbb{R}^m$ to $\mathcal{X}$ for which the norm

$$\| u \|_p = \left\{ \int_{\mathbb{R}^m} \| u(\mathbf{x}) \|^p \, d\mathbf{x} \right\}^{\frac{1}{p}} < \infty,$$

and let $L_\infty(\mathbb{R}^m) = L_\infty(\mathbb{R}^m, \mathcal{X})$ be the Banach space of (equivalence classes of) measurable functions u from $\mathbb{R}^m$ to $\mathcal{X}$ for which the norm

$$\| u \|_\infty = \operatorname*{ess\,sup}_{\mathbb{R}^m} \| u(\mathbf{x}) \| < \infty.$$

(Functions which are equal almost everywhere are identified in the usual way.)

For $1 < j < m$, let D_j denote the operator $D_j = \frac{\partial}{\partial x_j}$, considered as an unbounded linear operator in $L_p(\mathbb{R}^m)$ with domain

$$\mathcal{D}(D_j) = \left\{ u \in L_p(\mathbb{R}^m) : \frac{\partial u}{\partial x_j} \in L_p(\mathbb{R}^m) \right\}$$

where the derivative is taken in the sense of distributions. This is a closed operator, meaning that its graph $\{ (u, D_j u) : u \in \mathcal{D}(D_j) \}$ is a closed subset of $L_p(\mathbb{R}^m) \times L_p(\mathbb{R}^m)$. (The theory of closed unbounded operators in Banach spaces can be found in many places. One good reference is the book by Tosio Kato [K].)

For the time being, let us restrict our attention to $p = 2$. In this case it is possible to verify that D_j is closed, and to construct functional calculi of $\mathbf{D}$, by using Fourier theory. Here is a brief survey of the results from Fourier theory that are needed.

The Fourier transform $\hat{u} = \mathcal{F}(u)$ of a function $u \in L_1(\mathbb{R}^m)$ is defined by

$$\hat{u}(\xi) = \mathcal{F}(u)(\xi) = \int_{\mathbb{R}^m} e^{-i \langle \mathbf{x}, \xi \rangle} u(\,d\mathbf{x} \tag{5.3.1}$$

for all $\xi \in \mathbb{R}^m$. The function $\hat{u}$ is a bounded continuous function from $\mathbb{R}^m$ to $\mathcal{X}$ which satisfies $\| \hat{u} \|_\infty \le \| u \|_1$ for all $u \in L_1(\mathbb{R}^m)$.

If $u \in L_2(\mathbb{R}^m) \cap L_1(\mathbb{R}^m)$, then $\hat{u} \in L_2(\mathbb{R}^m)$, and

$$\| \hat{u} \|_2 \approx \| u \|_2$$

for all such u, and so the Fourier transform extends to an isomorphism, also called $\mathcal{F}$, from $L_2(\mathbb{R}^m)$ to $L_2(\mathbb{R}^m)$. (To prove this equivalence, write $u(\mathbf{x}) = \sum_S u_S(\mathbf{x}) e_S$ where $\{ e_S \}$ is a basis in $\mathcal{X}$, and apply Parseval's identity to each component.)

If $u \in L_2(\mathbb{R}^m)$, then $D_j u \in L_2(\mathbb{R}^m)$ if and only if $\xi_j \hat{u}(\xi) \in L_2(\mathbb{R}^m, d\xi)$, and

$$\left(\frac{1}{i} D_j u \right)^{\wedge}(\xi) = \mathcal{F}\left(\frac{1}{i} \frac{\partial u}{\partial x_j} \right)(\xi) = \xi_j \int_{\mathbb{R}^m} e^{-i \langle \mathbf{x}, \xi \rangle} u(\,d\mathbf{x} = \xi_j \hat{u}(\xi).$$

This can be used to show that D_j is a closed linear operator in $L_2(\mathbb{R}^m)$ as claimed above. More generally, suppose that p is a polynomial in m variables with values in $\mathcal{A}$. Then $p(\frac{1}{i}\mathbf{D})u \in L_2(\mathbb{R}^m)$ if and only if $p(\xi)\hat{u}(\xi) \in L_2(\mathbb{R}^m, d\xi)$, and

$$(p(\frac{1}{i}\mathbf{D})u)^\wedge(\xi) \;=\; p(\xi)\hat{u}(\xi) \, ,$$

which can be used to show that the operator $p(\frac{1}{i}\mathbf{D})$ with domain

$$\mathcal{D}(p(\frac{1}{i}\mathbf{D})) = \{u \in L_2(\mathbb{R}^m) \,:\, p(\frac{1}{i}\mathbf{D})u \in L_2(\mathbb{R}^m)\}$$

is also a closed operator in $L_2(\mathbb{R}^m)$.

Once again there are two types of spectra to be considered, namely the joint spectrum of the m–tuple $\frac{1}{i}\mathbf{D} = (\frac{1}{i}D_1, \frac{1}{i}D_2, \dots, \frac{1}{i}D_m)$, and the complex spectrum of the single operator $\mathbf{D} = \sum_{j=1}^{m} D_j\, e_j$.

The joint spectrum of $\frac{1}{i}\mathbf{D} = (\frac{1}{i}D_1, \frac{1}{i}D_2, \dots, \frac{1}{i}D_m)$ is $\sigma_m(\frac{1}{i}\mathbf{D}) = \mathbb{R}^m$. There are various ways to define this which we need not go into here, but essentially what it means is that $\mathbb{R}^m$ is the support of the functional calculus of $\frac{1}{i}\mathbf{D}$.

For any function $b \in L_\infty(\mathbb{R}^m, \mathcal{A})$, there is a natural definition of $b(\frac{1}{i}\mathbf{D})$ defined via the Fourier transform, namely

$$(b(\frac{1}{i}\mathbf{D})\, u\,)^\wedge(\xi) \;=\; b(\xi)\, \hat{u}(\xi).$$

Then $b(\frac{1}{i}\mathbf{D})$ is a bounded linear operator on $L_2(\mathbb{R}^m)$ with

$$\left\| b(\frac{1}{i}\mathbf{D})u \right\|_2 \;\approx\; \| b\,\hat{u} \|_2 \;\le\; c \, \| b \|_\infty \, \| u \|_2$$

for all $u \in L_2(\mathbb{R}^m)$. This, together with the facts (i) that the mapping from functions b to operators $b(\frac{1}{i}\mathbf{D})$ is an algebra homomorphism, and (ii) that there is agreement with the natural definition for polynomials of several variables, means that we have a bounded $L_\infty(\mathbb{R}^m, \mathcal{A})$ functional calculus of $\frac{1}{i}\mathbf{D}$.

The agreement with the natural definition of $p(\frac{1}{i}\mathbf{D})$ for polynomials p of $\frac{1}{i}\mathbf{D}$ is in the sense that, if b and $pb \in L_\infty(\mathbb{R}^m)$, then

$$p(\frac{1}{i}\mathbf{D})\, b(\frac{1}{i}\mathbf{D})\, u \;=\; (pb)(\frac{1}{i}\mathbf{D})\, u \quad \text{for all} \ \ u \in L_2(\mathbb{R}^m)$$

while, if b and $bp \in L_\infty(\mathbb{R}^m)$, then

$$b(\frac{1}{i}\mathbf{D})\, p(\frac{1}{i}\mathbf{D})\, u \;=\; (bp)(\frac{1}{i}\mathbf{D})\, u \quad \text{whenever} \ \ u \in \mathcal{D}(p(\frac{1}{i}\mathbf{D})).$$

It is well known that there is a close connection between the functional calculus of $\frac{1}{i}\mathbf{D}$ and convolution operators. For example, if $\phi \in L_1(\mathbb{R}^m, \mathcal{A})$

and $b = \hat{\phi}$, then, for almost all $\mathbf{x} \in \mathbb{R}^m$,

$$b(\mathbf{D})\, u(\mathbf{x}) = \int_{\mathbb{R}^m} \phi(\mathbf{x} - \mathbf{y})\, u(\mathbf{y})\, d\mathbf{y}.$$

This is L_2 Fourier multiplier theory on $\mathbb{R}^m$.

What are the operators $P_\pm = \chi_\pm(\frac{1}{i}\mathbf{D})$ this time? They satisfy

$$(P_\pm u)^\wedge(\xi) = (\chi_\pm(\tfrac{1}{i}\mathbf{D})\, u)^\wedge(\xi) = \frac{1}{2} \left\{ e_0 \pm \sum_{j=1}^m \frac{i\xi_j}{|\xi|}\, e_j \right\} \hat{u}(\xi)$$

for all $u \in L_2(\mathbb{R}^m)$ and almost all $\xi \in \mathbb{R}^m$. This means that

$$P_\pm u = \frac{1}{2} \left(e_0 \pm \sum_{j=1}^m R_j\, e_j \right) u$$

where R_j are the Riesz transforms, defined for all $u \in L_2(\mathbb{R}^m)$ and almost all $\mathbf{x} \in \mathbb{R}^m$ by

$$R_j u(\mathbf{x}) = \lim_{\varepsilon \to 0} \frac{2}{\sigma_m} \int_{|\mathbf{x}-\mathbf{y}|>\varepsilon} \frac{-x_j + y_j}{|\mathbf{x} - \mathbf{y}|^{m+1}}\, u(\mathbf{y})\, d\mathbf{y}.$$

(See, for example, the book [St] by Eli Stein.) Therefore

$$
\begin{aligned}
P_\pm u(\mathbf{x}) &= \pm\lim_{\varepsilon \to 0} \tfrac{1}{\sigma_m} \int_{|\mathbf{x}-\mathbf{y}|>\varepsilon} \frac{-\mathbf{x}+\mathbf{y}}{|\mathbf{x}-\mathbf{y}|^{m+1}}\, u(\mathbf{y})\, d\mathbf{y} + \tfrac{1}{2}\, u(\mathbf{x}) \\
&= \pm\lim_{\delta \to 0\pm} \tfrac{1}{\sigma_m} \int_{\mathbb{R}^m} \frac{-\mathbf{x}+\delta e_0+\mathbf{y}}{|\mathbf{x}+\delta e_0-\mathbf{y}|^{m+1}}\, u(\mathbf{y})\, d\mathbf{y}
\end{aligned}
$$

for all $u \in L_2(\mathbb{R}^m)$ and almost all $\mathbf{x} \in \mathbb{R}^m$.

(There is a clear connection with the Riesz systems of Stein and Weiss [SW].)

There is also the complex spectrum of the Dirac operator $\mathbf{D}$ to be considered, where $\mathbf{D}$ is now considered as a single closed unbounded linear operator in $L_2(\mathbb{R}^m) = L_2(\mathbb{R}^m, \mathcal{X})$ with domain

$$\mathcal{D}(\mathbf{D}) = \{\, u \in L_2(\mathbb{R}^m) \;:\; \mathbf{D}u \in L_2(\mathbb{R}^m) \,\}.$$

This domain is actually the Sobolev space

$$
\begin{aligned}
H^1(\mathbb{R}^m) &= H^1(\mathbb{R}^m, \mathcal{X}) \\
&= \left\{ u \in L_2(\mathbb{R}^m) \;:\; \frac{\partial u}{\partial x_j} \in L_2(\mathbb{R}^m) \text{ for } 1 \le j \le m \right\}.
\end{aligned}
$$

Let us prove this. Clearly $H^1(\mathbb{R}^m) \subset \mathcal{D}(\mathbf{D})$. On the other hand, suppose that $u \in \mathcal{D}(\mathbf{D})$. Then

$$\xi\, \hat{u}(\xi) \in L_2(\mathbb{R}^m, d\xi) \quad \text{so that}$$

$$\xi_j \, \hat{u}(\xi) \;=\; -\frac{\xi_j}{|\xi|} \frac{\xi}{|\xi|} \, \xi \, \hat{u}(\xi) \;\in\; L_2(\mathbb{R}^m, d\xi) \quad \text{and hence}$$

$$\frac{\partial u}{\partial x_j} \;\in\; L_2(\mathbb{R}^m) \quad \text{for } 1 \le j \le m$$

as required.

We see in this manner that the operator $|\mathbf{D}|$ defined by

$$(|\mathbf{D}| \, u)^{\wedge}(\xi) \;=\; |\xi| \, \hat{u}(\xi)$$

is also a closed unbounded linear operator in $L_2(\mathbb{R}^m)$ with domain

$$\mathcal{D}(|\mathbf{D}|) \;=\; \{\, u \in L_2(\mathbb{R}^m) \;:\; |\mathbf{D}| \, u \in L_2(\mathbb{R}^m) \,\} \;=\; \mathcal{D}(\mathbf{D}) \;=\; H^1(\mathbb{R}^m).$$

Further

$$|\mathbf{D}| \;=\; \mathrm{sgn}\{\mathbf{D}\} \, \mathbf{D} \quad \text{where}$$

$$\mathrm{sgn}\{\mathbf{D}\} \;=\; P_+ - P_- \;=\; \sum_{j=1}^{m} R_j \, e_j$$

is a bounded linear operator on $L_2(\mathbb{R}^m)$ satisfying $(\mathrm{sgn}\{\mathbf{D}\})^2 = I$.

It is easy to show that $\mathbf{D}$ is a self–adjoint operator in $L_2(\mathbb{R}^m)$, and to determine its complex spectrum by using the Fourier transform. We find that

$$\sigma(\mathbf{D}) \;=\; \{\, \lambda \in \mathbb{C} \;:\; (\lambda I - \mathbf{D}) \text{ does not have an bounded inverse } \} \;=\; \mathbb{R}.$$

When λ is not real,

$$((\lambda I - \mathbf{D})^{-1} u)^{\wedge}(\xi) \;=\; \frac{1}{\lambda e_0 - i\xi} \, \hat{u}(\xi) \;=\; \frac{\lambda e_0 + i\xi}{\lambda^2 - |\xi|^2} \, \hat{u}(\xi)$$

so that

$$(\lambda I - \mathbf{D})^{-1} \;=\; (\lambda I + \mathbf{D}) \, (\lambda^2 I + \mathbf{\Delta})^{-1}$$

where

$$\mathbf{\Delta} \;=\; -\mathbf{D}^2 \;=\; \sum_{j=1}^{m} \frac{\partial^2}{\partial x_j{}^2}$$

is the Laplacian on $\mathbb{R}^m$ with domain

$$\mathcal{D}(\mathbf{\Delta}) = H^2(\mathbb{R}^m)$$

$$= \left\{\, u \in H^1(\mathbb{R}^m) : \frac{\partial^2 u}{\partial x_j \partial x_k} \in L_2(\mathbb{R}^m) \text{ for } 1 \le j \le m, 1 \le k \le m \right\}.$$

We remark that $|\mathbf{D}|$ and $-\mathbf{\Delta}$ are non–negative self–adjoint operators in $L_2(\mathbb{R}^m)$, and that $|\mathbf{D}|^2 = -\mathbf{\Delta}$, or equivalently, that $|\mathbf{D}| = \sqrt{-\mathbf{\Delta}}$. For the self–adjoint operator $\mathbf{D}$, there is a natural definition of the operators $B\{\mathbf{D}\}$, whenever B is a bounded complex–valued measurable function defined on $\mathbb{R}$. We have

$$
\begin{aligned}
B\{\mathbf{D}\} &= B(|\mathbf{D}|)P_+ + B(-|\mathbf{D}|)P_- \\
&= B(\sqrt{-\boldsymbol{\Delta}})P_+ + B(-\sqrt{-\boldsymbol{\Delta}})P_-
\end{aligned}
$$

or

$$
(B\{\mathbf{D}\}u)^\wedge(\xi) = B(|\xi|)\,\chi_+(\xi)\,\hat{u}(\xi) + B(-|\xi|)\,\chi_-(\xi)\,\hat{u}(\xi).
$$

Then $B\{\mathbf{D}\}$ is a bounded linear operator on $L_2(\mathbb{R}^m)$ with

$$
\|B\{\mathbf{D}\}u\|_2 \leq c\,\|B\|_\infty\,\|u\|_2
$$

for all $u \in L_2(\mathbb{R}^m)$. This, together with the facts (i) that the mapping from functions B to operators $B\{\mathbf{D}\}$ is an algebra homomorphism, and (ii) that there is agreement with the natural definition for complex–valued polynomials of one variable, means that we have a bounded $L_\infty(\mathbb{R}^m, \mathbb{C})$ functional calculus of $\mathbf{D}$.

Of course, $B\{\mathbf{D}\} = b(\frac{1}{i}\mathbf{D})$ in the case when b is the function of m variables associated with B by $b(\xi) = B\{i\xi\}$.

In particular, the projections associated with the positive and negative parts of the spectrum of $\mathbf{D}$ are

$$
\chi_{\mathbb{R}+}\{\mathbf{D}\} = \chi_+(\tfrac{1}{i}\mathbf{D}) = P_+ = \frac{1}{2}\left(I + \sum R_j\,e_j\right) \quad \text{and}
$$

$$
\chi_{\mathbb{R}-}\{\mathbf{D}\} = \chi_-(\tfrac{1}{i}\mathbf{D}) = P_- = \frac{1}{2}\left(I - \sum R_j\,e_j\right).
$$

Given $u \in L_2(\mathbb{R}^m)$ and $t > 0$, define $u_+(t) \in L_2(\mathbb{R}^m)$ by

$$
u_+(t) = e^{-t\{\mathbf{D}\}}P_+u = e^{-t\sqrt{-\boldsymbol{\Delta}}}P_+u.
$$

Then $u_+(t)$ has the properties

$$
\begin{cases}
\frac{\partial u_+}{\partial t}(t) + \mathbf{D}u_+(t) &= 0, \qquad t > 0 \\
\lim_{t \to 0} u_+(t) &= P_+u \\
\lim_{t \to \infty} u_+(t) &= 0.
\end{cases}
$$

Also, for $t < 0$, the functions $u_-(t) \in L_2(\mathbb{R}^m)$ defined by

$$
u_-(t) = e^{t\{\mathbf{D}\}}P_-u = e^{-t\sqrt{-\boldsymbol{\Delta}}}P_-u
$$

satisfy the properties

$$
\begin{cases}
\frac{\partial u_-}{\partial t}(t) + \mathbf{D}u_-(t) &= 0, \qquad t < 0 \\
\lim_{t \to 0} u_-(t) &= P_-u \\
\lim_{t \to \infty} u_-(t) &= 0.
\end{cases}
$$

Let us now define functions U_+ on $\mathbb{R}^{m+1}_+$ and U_- on $\mathbb{R}^{m+1}_-$ by

$$U_+(te_0 + \mathbf{x}) = u_+(t)(\mathbf{x}), \qquad t > 0, \ \mathbf{x} \in \mathbb{R}^m$$
$$U_-(te_0 + \mathbf{x}) = u_-(t)(\mathbf{x}), \qquad t < 0, \ \mathbf{x} \in \mathbb{R}^m.$$

It is an exercise for the reader to show that U_+ and U_- are C^1 functions on the open half spaces $\mathbb{R}^{m+1}_+$ and $\mathbb{R}^{m+1}_-$ which can be represented by

$$U_\pm(te_0 + \mathbf{x}) = \pm \frac{1}{\sigma_m} \int_{\mathbb{R}^m} \frac{-\mathbf{x} + te_0 + \mathbf{y}}{|\mathbf{x} + te_0 - \mathbf{y}|^{m+1}} u(\mathbf{y}) \, d\mathbf{y}$$

for all $u \in L_2(\mathbb{R}^m)$. So

$$\lim_{t \to 0\pm} U_\pm(te_0 + \mathbf{x}) = P_\pm U(\mathbf{x})$$

for almost all $\mathbf{x} \in \mathbb{R}^m$, and

$$\lim_{t \to \infty\pm} U_\pm(te_0 + \mathbf{x}) = 0$$

for all $\mathbf{x} \in \mathbb{R}^m$.

We have just seen that these functions satisfy the equations

$$\frac{\partial U_\pm}{\partial t}(te_0 + \mathbf{x}) + \sum_{j=1}^{m} e_j \frac{\partial U_\pm}{\partial x_j}(te_0 + \mathbf{x}) = 0, \qquad te_0 + \mathbf{x} \in \mathbb{R}^{m+1}_\pm$$

or in other words, that the functions $U_\pm$ are *left–monogenic* on their respective half planes. Let $\mathcal{R}(P_\pm)$ denote the range of the projection $P_\pm$. Thus the orthogonal spectral decomposition

$$\left\{ \begin{array}{rcccc} L_2(\mathbb{R}^m) &=& \mathcal{R}(P_+) & \oplus & \mathcal{R}(P_-) \\ &=& L_2^+(\mathbb{R}^m) & \oplus & L_2^-(\mathbb{R}^m) \\ u &=& P_+ u & + & P_- u \end{array} \right.$$

is actually a decomposition of $L_2(\mathbb{R}^m)$ into *the Hardy spaces* $L_2^\pm(\mathbb{R}^m) = \mathcal{R}(P_\pm)$ consisting of those L_2 functions with left–monogenic extensions to $\mathbb{R}^{m+1}_\pm$ which decay at ∞.

Since the operator $\text{sgn}\{\mathbf{D}\}$ maps $u = P_+ u + P_- u$ to $\text{sgn}\{\mathbf{D}\} u = P_+ u - P_- u$, it can be thought of as a higher-dimensional Hilbert transform, as well as a sum $\text{sgn}\{\mathbf{D}\} = \sum_{j=1}^{m} R_j e_j$ of Riesz transforms. These formulae constitute a type of higher–dimensional Plemelj formulae.

Results and formulae along the lines developed in this section have been investigated by a number of people. Related results are contained, for example, in the papers of Sommen [S1, S2], Ryan [R1], Peetre and Sjölin [PS] and others.

In later sections we shall generalise this material, and consider the Hardy decomposition of $L_2(\Sigma)$, where Σ is the graph of a Lipschitz function.

5.4   Lecture 3

In this lecture the algebra $\mathcal{A}$ in which $\mathbb{R}^{m+1} = \mathbb{R} \oplus \mathbb{R}^m$ is embedded can be either a real or a complex finite dimensional algebra with identity e_0. We write $x = x_0 e_0 + \mathbf{x} = x_0 e_0 + \sum_{k=1}^m x_k e_k$ for elements of $\mathbb{R}^{m+1}$ (where $\{e_k\}$ is the standard basis of $\mathbb{R}^m$) and for the corresponding elements of $\mathcal{A}$. We still require that (I) $\mathbf{x}^2 = -|\mathbf{x}|^2 e_0$ is satisfied by all $\mathbf{x} \in \mathbb{R}^m$.

The conjugate $\bar{x}$ of $x = x_0 e_0 + \mathbf{x} \in \mathbb{R}^{m+1}$ is $\bar{x} = x_0 e_0 - \mathbf{x} \in \mathbb{R}^{m+1}$.

If $x, y \in \mathbb{R}^{m+1}$, then the following well-known identities are easily verified:

$$\overline{xy} = \bar{y}\,\bar{x}$$

$$x\bar{x} = \bar{x}x = |x|^2$$

$$\langle x, y \rangle = (x\bar{y})_0 = (\bar{x}y)_0$$

and, if $x \neq 0$, then x has the inverse

$$x^{-1} = \frac{\bar{x}}{|x|^2} \in \mathbb{R}^{m+1} \subset \mathcal{A}.$$

Actually, the ability to invert a non–zero vector $x \in \mathbb{R}^{m+1}$ is one reason why it is useful to embed $\mathbb{R}^{m+1}$ in the algebra $\mathcal{A}$.

As before, $\mathcal{X}$ denotes a finite dimensional left $\mathcal{A}$ module.

5.4.1   (E) Monogenic functions

Recall that $\mathbf{D}$ denotes the Dirac operator $\mathbf{D} = \sum_{k=1}^m \frac{\partial}{\partial x_k} e_k$.

Let Ω denote an open subset of $\mathbb{R} \oplus \mathbb{R}^m$, and let ϕ denote a C^1 function defined on Ω with values in $\mathcal{A}$. (A C^k function is a continuous function with continuous partial derivatives of all orders up to k.)

The differential operator $D = \frac{\partial}{\partial x_0} e_0 + \mathbf{D}$ acts on ϕ from the left to give

$$D\phi = \frac{\partial \phi}{\partial x_0} + \sum_{k=1}^m e_k \frac{\partial \phi}{\partial x_k} = \sum_{k=0}^m e_k \frac{\partial \phi}{\partial x_k}$$

and also from the right to give

$$\phi D = \frac{\partial \phi}{\partial x_0} + \sum_{k=1}^m \frac{\partial \phi}{\partial x_k} e_k = \sum_{k=0}^m \frac{\partial \phi}{\partial x_k} e_k.$$

If $D\phi = 0$ then ϕ is called *left–monogenic*, while if $\phi D = 0$ then ϕ is *right–monogenic*.

The equations $D\phi = 0$ and $\phi D = 0$ are both first-order elliptic systems of partial differential equations in m variables (because, non-zero elements $\lambda e_0 + \xi \in \mathbb{R}^{m+1}$ have inverses $(\lambda e_0 - \xi)/|\lambda e_0 + \xi|^2 \in \mathbb{R}^{m+1} \subset \mathcal{A}$.).

REMARK If $m = 1$, then

$$\phi D = D\phi = \left(\frac{\partial}{\partial x_0} e_0 + \frac{\partial}{\partial x_1} e_1 \right) (\phi_0 e_0 + \phi_1 e_1)$$

$$= \left(\frac{\partial \phi_0}{\partial x_0} - \frac{\partial \phi_1}{\partial x_1} \right) e_0 + \left(\frac{\partial \phi_0}{\partial x_1} + \frac{\partial \phi_1}{\partial x_0} \right) e_1$$

so that both the equations $D\phi = 0$ and $\phi D = 0$ reduce to the Cauchy–Riemann equations.

Similarly, if U is a C^1 function defined on Ω with values in $\mathcal{X}$, then $D = \frac{\partial}{\partial x_0} e_0 + \mathbf{D}$ acts on U to give

$$DU = \frac{\partial U}{\partial x_0} + \sum_{k=1}^{m} e_k \frac{\partial U}{\partial x_k} = \sum_{k=0}^{m} e_k \frac{\partial U}{\partial x_k}.$$

If $DU = 0$ then U is called *left–monogenic*.

We shall see in a moment that monogenic functions satisfy a form of Cauchy's formula with the kernel $k(y - x)$ replacing the usual $1/(2\pi(z - \zeta))$ of complex function theory. One consequence of this is the fact that monogenic functions have partial derivatives of all orders. Actually they are harmonic.

THEOREM 5.4.1
Every left–monogenic function is harmonic, as is every right–monogenic function.

PROOF Suppose that $D\phi = 0$. Then

$$\Delta\phi = \sum_{k=0}^{m} \frac{\partial^2 \phi}{\partial x_k{}^2} = \overline{D}D\phi = 0$$

where $\overline{D} = \frac{\partial}{\partial x_0} e_0 - \mathbf{D}$. Similarly, if $\phi D = 0$, then $\Delta\phi = \phi D \overline{D} = 0$. ∎

The concept of a monogenic function has been discovered several times during this century, and various aspects investigated. Possibly the earliest version of Cauchy's theorem for the case $m = 2$ was published by A. C. Dixon in 1904 [Dix]. The first comprehensive development of the theory was made by Brackx, Delanghe and Sommen, and presented in their monograph [BDS] from which the material in this section is adapted.

Of course [BDS] contains a wealth of material which will not even be alluded to in these lectures. It is amazing how much of complex function theory generalises to this situation, due no doubt to the fact that Cauchy's theorem has such a straightforward analogue.

This generalisation is quite distinct from the theory of several complex variables. It will be our aim in the next lecture to indicate how Fourier transforms link the theory of monogenic functions with that of several complex variables.

Let us now present a higher-dimensional version of integration by parts on a bounded *strongly Lipschitz* open subset Ω of $\mathbb{R}^{m+1}$. By this it is meant that its boundary $b\Omega$ is contained in finitely many open sets W such that each set $\Omega \cap W$ is represented by the inequality $X_{m+1} > g(X_1, X_2, \ldots, X_m)$ in some Cartesian coordinate system, where g is a Lipschitz function. (If $b\Omega$ is unbounded, we make the additional assumption that all except one of the sets W is bounded.)

On such a domain, the exterior unit normal $n(y)$ is defined for almost all $y \in b\Omega$, and Gauss's theorem is valid. This is proved, for example, in the book [N] of Nečas, where one can also find results about Sobolev spaces on such domains. A consequence is the following version of integration by parts.

THEOREM 5.4.2
Let Ω be a bounded strongly Lipschitz open subset of $\mathbb{R}^{m+1}$ with boundary $b\Omega$. Suppose that ϕ and U are two C^1 functions defined on an open neighbourhood of $\Omega^- = \Omega \cup b\Omega$, with ϕ taking values in $\mathcal{A}$ and U taking values in $\mathcal{X}$. Then

$$\int_{\Omega} \{(\phi D)U + \phi(DU)\}\, dx \;=\; \int_{b\Omega} \phi(y)n(y)U(y)\, dS_y.$$

PROOF

$$\int_{\Omega} \{(\phi D)U + \phi(DU)\}\, dx \;=\; \int_{\Omega} \sum_{k=0}^{m} \left\{ \frac{\partial \phi}{\partial x_k} e_k U + \phi e_k \frac{\partial U}{\partial x_k} \right\} dx$$

$$=\; \int_{b\Omega} \phi(y)n(y)U(y)\, dS_y$$

by Gauss's Theorem. ∎

The most important functions which are both left– and right–monogenic functions of $x \in \mathbb{R}^{m+1}$ ($x \neq y$) are the Cauchy kernels

$$k_y(x) \;=\; k(x - y) \;=\; \frac{\overline{x - y}}{\sigma_m |x - y|^{m+1}}$$

where σ_m is the m dimensional volume of the unit m-sphere $\{x \in \mathbb{R}^{m+1} : |x| = 1\}$. It is not difficult to verify directly that k_y is left– and right–

monogenic. Or else, when $m \geq 2$, we can write

$$Dk_y(x) = D\overline{D}\left\{\frac{-1}{(m-1)\sigma_m |x-y|^{m-1}}\right\}$$

$$= \Delta\left\{\frac{-1}{(m-1)\sigma_m |x-y|^{m-1}}\right\} = 0$$

to show that k_y is left–monogenic, and something similar to show that k_y is right–monogenic. When $m = 1$, then $k_y(x) = 1/(2\pi(x-y))$.

The following version of Cauchy's theorem and the Cauchy integral formula follow.

THEOREM 5.4.3
Let Ω be a bounded strongly Lipschitz open subset of $\mathbb{R}^{m+1}$ with boundary $b\Omega$ and exterior unit normal $n(y)$ defined for almost all $y \in b\Omega$. Suppose that ϕ is a right–monogenic function from a neighbourhood of Ω^- to $\mathcal{A}$, and U is a left–monogenic function from a neighbourhood of Ω^- to $\mathcal{X}$. Then

$$\int_{b\Omega} \phi(y)n(y)U(y)\,dS_y = 0 \tag{i}$$

$$\int_{b\Omega} \phi(y)n(y)k_x(y)\,dS_y = \frac{1}{\sigma_m}\int_{b\Omega} \phi(y)n(y)\frac{\overline{y-x}}{|y-x|^{m+1}}\,dS_y$$

$$= \begin{cases} \phi(x), & x \in \Omega \\ 0, & x \notin \Omega^- \end{cases} \tag{ii}$$

$$\int_{b\Omega} k_x(y)n(y)U(y)\,dS_y = \frac{1}{\sigma_m}\int_{b\Omega} \frac{\overline{y-x}}{|y-x|^{m+1}}n(y)U(y)\,dS_y$$

$$= \begin{cases} U(x), & x \in \Omega \\ 0, & x \notin \Omega^-. \end{cases} \tag{iii}$$

PROOF [**Proof of (ii)**] If $x \notin \Omega^-$ then, by (i),

$$L.H.S. = \int_{b\Omega} \phi(y)n(y)k_x(y)\,dS_y = 0.$$

Suppose now that $x \in \Omega$. For r small, let S_r denote the sphere with centre x and radius r. The volume of S_r is $\sigma_m r^m$. On applying (i) on the boundary of the region between S_r and $b\Omega$, and then the fact that

$$n(y)\,k_x(y) = \frac{y-x}{|y-x|}\frac{\overline{y-x}}{\sigma_m|y-x|^{m+1}} = \frac{1}{\sigma_m r^m}$$

for all $y \in S_r$, we see that

$$
\begin{aligned}
L.H.S. &= \int_{b\Omega} \phi(y)n(y)k_x(y)\, dS_y = \int_{S_r} \phi(y)n(y)k_x(y)\, dS_y \\
&= \frac{1}{\sigma_m r^m} \int_{S_r} \phi(y)\, dS_y \;\longrightarrow\; \phi(x)
\end{aligned}
$$

as $r \to 0$. The result follows.   ∎

Later on we shall see what happens when $x \in b\Omega$.

It is a good exercise to check that this theorem does indeed reduce to the theorem proved by Cauchy in the case when $m = 1$ and $\mathcal{A} = \mathbb{R}_{(1)}$ is identified with $\mathbb{C}$.

Many of the consequences of Cauchy's original theorem have analogues in this higher dimensional setting. A good reference is [BDS]. In particular, it is easy to see that every right–monogenic function ϕ has partial derivatives of all orders. For example, the first order derivatives are given by

$$
\frac{\partial \phi}{\partial x_k}(x) = \frac{1}{\sigma_m} \int_{b\Omega} \phi(y)n(y)\frac{\partial k_x}{\partial x_k}(y)\, dS_y
$$

when $x \in \Omega$. This formula can be used to derive the higher-dimensional version of Liouville's theorem in the usual way. This states that the only right–monogenic functions which are defined and bounded on all of $\mathbb{R}^{m+1}$ are the constant functions.

Similar statements are satisfied by left–monogenic functions.

We conclude this section by introducing some spaces of right–monogenic functions on certain cones in $\mathbb{R}^{m+1}$.

Let N be a compact set of unit vectors in $\mathbb{R}^{m+1}_+$ which contains e_0. Therefore

$$
\omega = \max_{n \in N} \angle(n,\, e_0) < \frac{\pi}{2}
$$

where $\angle(n,\, e_0)$ denotes the angle between n and e_0. Given $\mu > 0$ such that $\mu + \omega < \frac{\pi}{2}$, let N_μ denote the open neighbourhood of N in the unit sphere defined by $N_\mu = \{\, y \in \mathbb{R}^{m+1}_+ \; : \; |y| = 1 \text{ and there exists } n \in N \text{ such that } \angle(y, n) < \mu \,\}$.

For each unit vector $n \in \mathbb{R}^{m+1}_+$, let C_n^+ denote the open half space

$$
C_n^+ = \{\, x \in \mathbb{R}^{m+1} \; : \; \langle x,\, n \rangle > 0 \,\}\,,
$$

and let $C_{N_\mu}^+$ denote the open cone

$$
C_{N_\mu}^+ = \bigcup_{n \in N_\mu} C_n^+ = \{\, x \in \mathbb{R}^{m+1} \; : \; \langle x,\, n \rangle > 0 \text{ for some } n \in N_\mu \,\}.
$$

Further, let $K(C_{N_\mu}^+)$ denote the function space which consists of all right–monogenic functions on $C_{N_\mu}^+$ that satisfy $\| \phi(x) \| \leq c\,|x|^{-m}$ for some $c \geq 0$. This is a Banach space under the norm

$$\| \phi \|_{K(C_{N_\mu}^+)} \;=\; \frac{\sigma_m}{2} \sup \left\{ |x|^m \, \| \phi(x) \| \; : \; x \in C_{N_\mu}^+ \right\}.$$

Note that, for all $\nu < \mu$, such functions ϕ satisfy

$$\left\| \frac{\partial \phi}{\partial x_k}(x) \right\| \;\leq\; \frac{c_\nu}{|x|^{m+1}} \, \| \phi \|_{K(C_{N_\mu}^+)}$$

for $0 \leq k \leq m$, for all $x \in C_{N_\nu}^+$ and some constant c_ν. This is proved by using the integral formula for $\frac{\partial \phi}{\partial x_k}(x)$ given above, where Ω is the ball with centre x and radius $\frac{1}{2}\sin(\mu - \nu)\,|x|$.

Also let K_N^+ be the linear space of functions ϕ on $\mathbb{R}^m \setminus \{\mathbf{0}\}$ which have right–monogenic extensions to $\phi \in K(C_{N_\mu}^+)$ for some $\mu > 0$, and let

$$M_N^+ \;=\; \{\; \phi \in K_N^+ \; : \; \phi \text{ is left–monogenic (as well as right–monogenic) } \}.$$

Such an extension is uniquely determined by the values of ϕ on $\mathbb{R}^m$. (This is a straightforward consequence of the fact that ϕ is a solution of the elliptic system $\phi D = 0$. This implies that ϕ is analytic in the sense that each component of ϕ has a power series expansion about every point in $C_{N_\mu}^+$.)

Let us now consider convolutions. Given $\phi \in K(C_{N_\mu}^+)$ and $\psi \in M(C_{N_\mu}^+)$, define $(\phi * \psi)(x)$ for $x \in C_n^+ \subset C_{N_\mu}^+$ by

$$(\phi * \psi)(x) \;=\; \int_{\langle y,\, n \rangle = \delta} \phi(x - y)\, n(y)\, \psi(y)\, dS_y$$

where $0 < \delta < \langle x,\, n \rangle$. It follows from Cauchy's theorem and the hypotheses of ϕ being right–monogenic and ψ being left–monogenic, that the integral is independent of the precise surface chosen. On the other hand, it is a consequence of ψ being right–monogenic, that $\phi * \psi$ is right–monogenic, and indeed that

$$\| \phi * \psi \|_{K(C_{N_\nu}^+)} \;\leq\; c_{\nu,\mu} \, \| \phi \|_{K(C_{N_\mu}^+)} \, \| \psi \|_{K(C_{N_\mu}^+)}$$

for all $\nu < \mu$, as is shown in [LMcS].

Thus, if $\phi \in K_N^+$ and $\psi \in M_N^+$, then $\phi * \psi \in K_N^+$. Moreover, if ϕ is also in M_N^+, then so is $\phi * \psi$, which means that M_N^+ is a convolution algebra.

For k defined as usual by $k(x) = \frac{\bar{x}}{\sigma_m |x|^{m+1}}$ we have, by Cauchy's formula, that $\phi * k = \phi$, $k * \psi = \psi$ and, in particular, $k * k = k$.

The functions in K_N^+ and M_N^+ are natural ones to take as convolution kernels for singular integral operators on Lipschitz surfaces with normal vectors in N. We shall characterise them by their Fourier transforms in the next lecture.

5.4.2   (F) Singular convolution integrals on Lipschitz surfaces

The material in this section is adapted from the paper by Chun Li, Stephen Semmes and myself [LMcS].

Henceforth Σ denotes the Lipschitz surface consisting of all points

$$x = g(\mathbf{x})\,e_0 + \mathbf{x} \in \mathbb{R}^{m+1}$$

where $\mathbf{x} \in \mathbb{R}^m$, and g is a real–valued Lipschitz function which satisfies

$$\|\mathbf{D}g\|_\infty = \operatorname*{ess\ sup}_{\mathbf{x}\in\mathbb{R}^m} \left\{ \sum_{j=1}^m \left| \frac{\partial g}{\partial x_j}(\mathbf{x}) \right|^2 \right\}^{1/2} \leq \tan\omega < \infty$$

where $0 \leq \omega < \pi/2$.

The unit normal $n(x)$ to the surface Σ is defined at almost all points $x = x_0 e_0 + \mathbf{x} \in \Sigma$. It is

$$n(x) = \frac{e_0 - \mathbf{D}g(\mathbf{x})}{\sqrt{1 + |\mathbf{D}g(\mathbf{x})|^2}}.$$

Let us take N to be a compact set of unit vectors in $\mathbb{R}^{m+1}_+$ which includes almost all these unit normals together with e_0, and which satisfies

$$\max_{n \in N} \angle(n, e_0) = \omega.$$

Given $\mu > 0$ such that $\mu + \omega < \frac{\pi}{2}$, define N_μ, $C^+_{N_\mu}$, $K(C^+_{N_\mu})$, K^+_N and M^+_N as specified at the end of the previous section.

If $1 \leq p < \infty$ then $L_p(\Sigma) = L_p(\Sigma, \mathcal{X})$ denotes the Banach space of (equivalence classes of) functions u from Σ to $\mathcal{X}$ which are measurable with respect to $dS_x = \sqrt{1 + |\mathbf{D}g(x)|^2}\,dx$, and for which

$$\|u\|_p = \left\{ \int_\Sigma \|u(x)\|^p \, dS_x \right\}^{1/p} < \infty, \quad 1 \leq p < \infty.$$

We are now in a position to define singular convolution operators on Σ. Given $\phi \in K^+_N$ and $u \in L_p(\Sigma)$, define

$$(T_\phi u)(x) = \lim_{\delta \to 0+} \int_\Sigma \phi(x + \delta e_0 - y)\, n(y)\, u(y)\, dS_y$$

for all $x \in \mathbb{R}^m$ where the limit exists. (For each fixed $\delta > 0$, the integral is absolutely convergent.) The function ϕ is called the *convolution kernel* of the operator T_ϕ.

The main result of this section is the following one, which was first for-mulated and proved in [LMcS].

THEOREM 5.4.4
Let $1 < p < \infty$. Given $\phi \in K_N^+$, there is a bounded linear operator T_ϕ on $L_p(\Sigma)$, defined for all $u \in L_p(\Sigma)$ and almost all $x \in \Sigma$ by

$$(T_\phi u)(x) = \lim_{\delta \to 0+} \int_\Sigma \phi(x + \delta e_0 - y)\, n(y)\, u(y)\, dS_y.$$

Moreover, if $\phi \in K(C_{N_\mu}^+)$ for $0 < \mu < \frac{1}{2}\pi - \omega$, then

$$\| T_\phi u \|_p \le C_{\omega,\mu,p}\, \| \phi \|_{K(C_{N_\mu}^+)}\, \| u \|_p$$

for some constant $C_{\omega,\mu,p}$ which depends only on ω, μ and p (and the di-mension m).
    Also, for all $u \in L_p(\Sigma)$,

$$\lim_{\delta \to 0+} \| T_{\phi_\delta} u - T_\phi u \|_p \to 0$$

where $\phi_\delta(x) = \phi(x + \delta e_0)$.
    If, in addition, $\psi \in M_N^+$, then

$$T_{\phi * \psi}\, u = T_\phi\, T_\psi\, u$$

for all $u \in L_p(\Sigma)$.

It suffices to prove the theorem for functions $\phi \in M_N^+$, i.e., for functions which are left– as well as right–monogenic. This is because every function $\phi \in K_N^+$ can be expressed as a finite sum $\phi = \sum e_S\, \phi_S$ where $\phi_S \in M_N^+$ and $e_S \in \mathcal{A}$, as is shown at the end of section (H). (At the time when we wrote the paper [LMcS], we used a more involved method to handle functions ϕ which are right– but not left–monogenic.)
    For k defined as usual by $k(x) = \frac{\bar{x}}{\sigma_m |x|^{m+1}}$, set $P_+ = T_k$. Then $T_\phi P_+ = T_\phi$ for all $\phi \in K_N^+$ and $P_+ T_\psi = T_\psi$ for all $\psi \in M_N^+$. In particular, $P_+{}^2 = P_+$, so that P_+ is a bounded projection in $L_p(\Sigma)$. The range $\mathcal{R}(P_+)$ of this projection is the *Hardy space* $L_p^+(\Sigma) = P_+(L_p(\Sigma))$.
    These definitions are consistent with those given in section (D) for the case when $\Sigma = \mathbb{R}^m$ and $p = 2$ (where we take $g(x) = 0$ and $n(x) = e_0$ for all $x \in \mathbb{R}^m$).
    The most important singular convolution operator is of course the sin-gular Cauchy integral operator C_Σ on Σ, defined for almost all $x \in \Sigma$ by

$$(C_\Sigma u)(x) = 2 \lim_{\varepsilon \to 0} \int_{\{y \in \Sigma\, :\, |x-y|>\varepsilon\}} k(x - y)\, n(y)\, u(y)\, dS_y$$

$$= 2 \lim_{\delta \to 0+} \int_\Sigma k(x + \delta e_0 - y)\, n(y)\, u(y)\, dS_y \; - \; u(x)$$

$$= 2\,(P_+\, u)(x) \; - \; u(x)$$

where $P_+ = T_k$. The equality of the two integrals is shown in [LMcS]. It follows from Calderón–Zygmund theory.

We can also write

$$C_\Sigma\, u \; = \; -2 P_-\, u \; + \; u$$

where P_- is the bounded projection in $L_p(\Sigma)$ defined for almost all $x \in \Sigma$ by

$$(P_-\, u)(x) \; = \; -\lim_{\delta \to 0-} \int_\Sigma k(x + \delta e_0 - y)\, n(y)\, u(y)\, dS_y.$$

The range $\mathcal{R}(P_-)$ of P_- is the Hardy space $L_p^-(\Sigma) = P_-(L_p(\Sigma))$. Since $P_+ + P_- = I$, there is a decomposition of $L_p(\Sigma)$ into Hardy spaces:

$$\left\{ \begin{aligned}
L_p(\Sigma) \;&=\; \mathcal{R}(P_+) \;&\oplus&\; \mathcal{R}(P_-) \\
&=\; L_p^+(\Sigma) \;&\oplus&\; L_p^-(\Sigma) \\
u \;&=\; P_+ u \;&+&\; P_- u \\
C_\Sigma u \;&=\; P_+ u \;&-&\; P_- u
\end{aligned} \right.$$

These are the Plemelj formulae in this context. They were derived and studied by Iftimie about 30 years ago for Hölder continuous functions on Liapunov surfaces [I]. We already saw them in section (D) in the case when $\Sigma = \mathbb{R}^m$ and $p = 2$. In that section we also saw that the Hardy decomposition is related to the functional calculus of the Dirac operator on $\mathbb{R}^m$. We shall do the same for the Dirac operator on Σ in section (J). There we show that P_+ can be expressed as $P_+ = \chi_+(\frac{1}{i}\mathbf{D}_\Sigma)$ in the functional calculus of a Dirac operator $\mathbf{D}_\Sigma$ on the surface Σ.

The space $L_p^+(\Sigma)$ consists of those functions $u \in L_p(\Sigma)$ which have left–monogenic extensions U^+ to the open set $\Omega_+ = \{ X \in \mathbb{R}^{m+1} : X_0 > g(\mathbf{X}) \}$ that decay at ∞, while the space $L_p^-(\Sigma)$ consists of those functions $u \in L_p(\Sigma)$ which have left–monogenic extensions U^- to the open set $\Omega_- = \{ X \in \mathbb{R}^{m+1} : X_0 < g(\mathbf{X}) \}$ that decay at ∞. These extensions are given by the Cauchy integral

$$U^\pm(X) \; = \; (C_\Sigma^\pm u)(X) \; = \; \pm \int_\Sigma k(X - y)\, n(y)\, u(y)\, dS_y$$

for all $X \in \Omega_\pm$.

It is a consequence of the preceding theorem that C_Σ and P_+ are bounded linear operators on $L_p(\Sigma)$ when $1 < p < \infty$. Now

$$k(x - y)\, n(y) = \frac{\overline{(x - y)}\, n(y)}{\sigma_m\, |x - y|^{m+1}},$$

so its scalar part is

$$(k(x-y)\,n(y))_0 = \frac{\langle x-y\,,\,n(y)\rangle}{\sigma_m\,|x-y|^{m+1}},$$

which is the kernel of the *singular double–layer potential operator* $C_{\Sigma 0}$ on Σ.

It follows therefore that this operator is also bounded on $L_p(\Sigma)$. Since this was one of the main reasons for developing this theory, let us state this result as a separate theorem.

THEOREM 5.4.5
Let $1 < p < \infty$. The singular double–layer potential operator $C_{\Sigma 0}$ and $P_{+0} = \frac{1}{2}C_{\Sigma 0} + \frac{1}{2}I$, defined for all $u \in L_p(\Sigma)$ and almost all $x \in \Sigma$ by

$$(C_{\Sigma 0}\,u)(x) = \frac{2}{\sigma_m}\,\lim_{\varepsilon \to 0}\,\int\limits_{\{y \in \Sigma\,:\,|x-y|>\varepsilon\}} \frac{\langle x-y\,,\,n(y)\rangle}{|x-y|^{m+1}}\,u(y)\,dS_y \quad and$$

$$(P_{+0}\,u)(x) = \frac{1}{\sigma_m}\,\lim_{\delta \to 0+}\,\int\limits_{\Sigma} \frac{\langle x+\delta e_0-y\,,\,n(y)\rangle}{|x-y|^{m+1}}\,u(y)\,dS_y$$

are bounded linear operators on $L_p(\Sigma)$ with

$$\|\,C_{\Sigma 0}\,u\,\|_p \le c_{\omega,p}\,\|\,u\,\|_p \quad and$$

$$\|\,P_{+0}\,u\,\|_p \le c_{\omega,p}\,\|\,u\,\|_p$$

for all $u \in L_p(\Sigma)$ and some constant $c_{\omega,p}$ depending only on ω and p.

The proof of these theorems builds upon the work of Zygmund, Calderón, Carleson, Stein, Fefferman, Meyer, Coifman and many other people. In the case when $m = 1$, then C_{Σ} is essentially the Cauchy integral on the graph of a Lipschitz function, considered as a curve in the complex plane. Calderón first proved, in 1977, that this operator is bounded on $L_p(\Sigma)$ when ω is small [Ca]. Subsequently, Coifman, McIntosh and Meyer proved the boundedness for all such curves [CMcM].

The use of the Calderón rotation method leads to the boundedness of the operators T_ϕ in all dimensions in the case when ϕ is an odd function. In particular, this gives the boundedness of the singular Cauchy operator C_{Σ} and the singular double–layer potential operator $C_{\Sigma 0}$. Further, it implies that the singular double–layer potential operator is bounded on $L_p(b\Omega)$, when $b\Omega$ is the boundary of a strongly Lipschitz domain in $\mathbb{R}^{m+1}$, as will be shown in section (L). This fact was used soon after by Verchota to solve the Dirichlet and Neumann problems for harmonic functions with L_2 boundary data on such domains by using layer potentials [V]. (These problems had been solved previously by Dahlberg [D1] and by Jerison and Kenig [JK] using other methods.)

The proof in [CMcM] involved complicated multilinear estimates. Subsequently, other methods were developed which simplified and generalised these results. In particular, there was the $T(b)$ theorem of McIntosh and Meyer [McM] and of David, Journé and Semmes [DJS], as well as the related method of Semmes [S]. There were wavelets [M]. And there was the paper by Coifman, Jones and Semmes [CJS] which presented two elementary proofs of the boundedness of C_Σ in one dimension, the first reducing it to quadratic estimates of Kenig in Hardy spaces, and the second using martingales. Actually, no proofs are really elementary, because they all rely on the power of Calderón–Zygmund theory and Carleson measures in some form.

During the same period, the functional calculus aspect of these results was developed, first by Coifman and Meyer [CM1], and more fully by Tao Qian and myself [McQ]. This is the one-dimensional version of the results to be presented in sections (J) and (M).

Let us return to the consideration of surfaces with dimension $m \geq 2$. Raphy Coifman and Margaret Murray first asked whether the methods of complex analysis could be generalised to higher dimensions using Clifford algebra to prove the boundedness of C_Σ, and hence of $C_{\Sigma 0}$, in $L_p(\Sigma)$. Murray achieved this in her PhD thesis [Mu] for the case of small ω. Subsequently, I showed that this could be done for all ω [Mc]. These proofs involved multilinear estimates.

In order to incorporate this material into a functional calculus, Tao Qian, Chun Li and I first identified the space K_N of convolution kernels ϕ for which the operators T_ϕ are bounded on $L_p(\Sigma)$, and then found the space H_N of their Fourier transforms. See section (M) for the definition of K_N and section (H) for the definition of H_N and the material on Fourier transforms. Essentially, $K_N = K_N^+ \oplus K_N^-$.

Together with Stephen Semmes, we adapted one of the methods in [CJS] to prove the L_p boundedness of T_ϕ when $\phi \in K_N^+$ (the result stated above), and hence for kernels in K_N [LMcS]. At the same time we developed the functional calculus which will be presented in section (J) [LMcQ].

Other methods, such as the Clifford $T(b)$ theorem [DJS], Clifford martingales and Clifford wavelets, have been used by other authors. See for example the papers of Gaudry, Long and Qian [GLQ], Auscher and Tchamitchian [AT], Mitrea [M1], Tao [T] and others.

Let me comment on the proof. It is a well-known fact of harmonic analysis that results of this type can be derived from quadratic estimates in $L_2(\Sigma)$. In [CJS], Coifman, Jones and Semmes showed that it suffices to use the quadratic estimates of Kenig for functions in the Hardy spaces $L_2^+(\Sigma)$ and $L_2^-(\Sigma)$, in order to prove the L_p boundedness of the Cauchy integral on curves. It is surprising that quadratic estimates within these spaces are enough to prove bounds on the whole space, and hence give the decomposition $L_2(\Sigma) = L_2^+(\Sigma) \oplus L_2^-(\Sigma)$, where $\oplus$ is the topological direct sum. See [McQ] for a general discussion of this idea and its relevance for functional calculi. Semmes then gave a straightforward proof of the required square–function estimate in higher dimensions, namely that

$$\| U|_\Sigma \|_2 \;\leq\; c \left\{ \int_\Sigma \int_0^\infty \sum_{j=0}^m \left\| \frac{\partial U}{\partial x_j}(x + te_0) \right\|^2 t\, dt\, dS_x \right\}^{\frac{1}{2}} \qquad \text{(S)}$$

holds for all right–monogenic functions on the open region $\Omega_+ = \{\Sigma + te_0 : t > 0\}$ above Σ which are continuous to the boundary Σ, provided U satisfies the *a priori* estimates $\| U(x) \| \leq C(1 + |x|)^{-m}$ and $\left\| \frac{\partial U}{\partial x_j}(x) \right\| \leq C(1 + |x|)^{-(m+1)}$ for all $x \in \Omega_+ \cup \Sigma$ [LMcS §4]. (The constant c depends on ω but not on C.)

This, together with a similar estimate for functions on $\Omega_- = \{\Sigma + te_0 : t < 0\}$, allows the method of [CJS] to be employed to prove that T_ϕ is bounded on $L_2(\Sigma)$ for all $\phi \in K_N^+$. Calderón–Zygmund theory then implies that T_ϕ is bounded on $L_p(\Sigma)$ for these functions ϕ [LMcS §5]. Indeed, it implies a lot more than this about the boundedness of nontangential maximal functions and the existence of limits almost everywhere. See, for example, the book by Coifman and Meyer [CM] for an account of Calderón–Zygmund theory.

We conclude this section with some remarks about the square–function estimates.

(i)   Our first remark is that the following two expressions are equivalent:

$$\int_\Sigma \int_0^\infty \sum_{j=0}^m \left\| \frac{\partial U}{\partial x_j}(x + te_0) \right\|^2 t\, dt\, dS_x \;\approx\; \int_\Sigma \int_0^\infty \left\| \frac{\partial U}{\partial x_0}(x + te_0) \right\|^2 t\, dt\, dS_x$$

for all right–monogenic functions U on the open region Ω_+. This is a consequence of applying the Rellich estimate for each value of t. It states that

$$\int_\Sigma \sum_{j=0}^m \left| \frac{\partial F}{\partial x_j}(x) \right|^2 dS_x \;\approx\; \int_\Sigma \left| \frac{\partial F}{\partial x_0}(x) \right|^2 dS_x$$

for all harmonic functions on Ω_+ which are continuous to the boundary Σ and satisfy the same *a priori* estimates stated for (S) [JK]. (The Rellich estimate is a consequence of the Rellich identities which can be proved by an appropriate integration by parts on Ω_+. See [N] for a treatment of Rellich identities. An alternative proof using monogenic functions is presented by Mitrea in [M2].)

(ii)   The second remark is that, once (S) has been proved, it is not hard to deduce that the two sides of (S) are actually equivalent:

$$\| U|_\Sigma \|_2 \approx \left\{ \int_\Sigma \int_0^\infty \sum_{j=0}^m \left\| \frac{\partial U}{\partial x_j} (x + te_0) \right\|^2 t\, dt\, dS_x \right\}^{\frac{1}{2}} \tag{S'}$$

for all right–monogenic functions on Ω_+ which decay at ∞.

(iii)   The final remark is to compare (S) with Dahlberg's square–function estimate

$$\| F|_\Sigma \|_2 \leq c \left\{ \int_\Sigma \int_0^\infty |DF (x + te_0)|^2 t\, dt\, dS_x \right\}^{\frac{1}{2}} \tag{D}$$

which holds for all scalar–valued harmonic functions on Ω_+ which decay at ∞ [D2]. It is clear that Semmes' estimate (S) is an immediate consequence of Dahlberg's estimate (D), because every right–monogenic function is harmonic. When I lectured on this material at Yale in 1991, somebody quipped that (S) stands for "simple" and (D) for "difficult", and asked whether (D) can be derived from (S). Later that year Carlos Kenig showed me how to do this by using the Rellich estimate. Marius Mitrea independently proved this in the comprehensive treatment of this material which he developed for his PhD. See for example [M1] and [M2]. (The reason why it is simpler to prove (S) than (D) is that it is easier to treat first-order systems than second-order equations. Clifford algebra allows us to reduce the second-order problem to the first-order one in an efficient way.)

Here is an outline of the proof. Suppose that F is a scalar–valued function which satisfies $\Delta F = 0$ on Ω_+ and decays appropriately at ∞. Define U on Ω_+ by

$$U(x_0 e_0 + \mathbf{x}) = - \int_{x_0}^\infty \overline{D}F(te_0 + \mathbf{x})\, dt.$$

Then U is a right–monogenic function on Ω_+ (because $(\overline{D}F)D = F\overline{D}D = \Delta F = 0$) which has the form $U = Fe_0 + \sum_{j=1}^m U_j e_j$ for some harmonic functions U_j which satisfy $\frac{\partial U_j}{\partial x_0} = -\frac{\partial F}{\partial x_j}$ when $1 \leq j \leq m$. Therefore,

$$\left\| \frac{\partial U}{\partial x_0}(x + te_0) \right\|^2 = \left| \frac{\partial F}{\partial x_0}(x + te_0) \right|^2 + \sum_{j=1}^m \left| \frac{\partial U_j}{\partial x_0}(x + te_0) \right|^2$$

$$= \left| \frac{\partial F}{\partial x_0}(x + te_0) \right|^2 + \sum_{j=1}^m \left| \frac{\partial F}{\partial x_j}(x + te_0) \right|^2$$

$$= \| DF(x + te_0) \|^2$$

so that

$$\| F|_\Sigma \|_2 \leq \| U|_\Sigma \|_2$$

$$\leq c \left\{ \int_{\Sigma} \int_0^{\infty} \sum_{j=0}^{m} \left\| \frac{\partial U}{\partial x_j} (x + te_0) \right\|^2 t \, dt \, dS_x \right\}^{\frac{1}{2}} \quad \text{by (S)}$$

$$\approx \left\{ \int_{\Sigma} \int_0^{\infty} \left\| \frac{\partial U}{\partial x_0} (x + te_0) \right\|^2 t \, dt \, dS_x \right\}^{\frac{1}{2}} \quad \text{by (i)}$$

$$= \left\{ \int_{\Sigma} \int_0^{\infty} \| DF (x + te_0) \|^2 t \, dt \, dS_x \right\}^{\frac{1}{2}}$$

as required. (Note the connections with the seminal work of Stein and Weiss on conjugate harmonic functions [SW].)

5.5   Lecture 4

In this lecture, $\mathcal{A}$ is a finite dimensional complex algebra in which (I) is satisfied, and $\mathcal{X}$ is a finite dimensional left $\mathcal{A}$ module.

5.5.1   (G) Fourier transforms

In section (A), we introduced the functions

$$\chi_{\pm}(\xi) = \frac{1}{2} \left(e_0 \pm \frac{i\xi}{|\xi|} \right)$$

which satisfy

$$\chi_+(\xi)^2 = \chi_+(\xi) , \ \chi_-(\xi)^2 = \chi_-(\xi) , \ \chi_+(\xi)\chi_-(\xi) = 0 = \chi_-(\xi)\chi_+(\xi)$$

and

$$\begin{cases} i\xi\chi_{\pm}(\xi) &= \chi_{\pm}(\xi) \, i\xi \, , \\ e_0 &= \chi_+(\xi) + \chi_-(\xi) \\ i\xi &= |\xi| \, \chi_+(\xi) + (-|\xi|)\chi_-(\xi) \end{cases}$$

for all $\xi \in \mathbb{R}^m$, $\xi \neq 0$.

Using them, we decompose $L_{\infty}(\mathbb{R}^m) = L_{\infty}(\mathbb{R}^m, \mathcal{A})$ as a direct sum

$$\begin{cases} L_{\infty}(\mathbb{R}^m) &= L_{\infty}^+(\mathbb{R}^m) \quad \oplus \quad L_{\infty}^-(\mathbb{R}^m) \\ b(\xi) &= b(\xi)\chi_+(\xi) \quad + \quad b(\xi)\chi_-(\xi) \end{cases}$$

of the two subspaces

$$L_{\infty}^+(\mathbb{R}^m) = \{ \, b \in L_{\infty}(\mathbb{R}^m) \, : \, b\chi_- = 0 \, \} \quad \text{and}$$

$$L_{\infty}^-(\mathbb{R}^m) = \{ \, b \in L_{\infty}(\mathbb{R}^m) \, : \, b\chi_+ = 0 \, \}.$$

We remark that $b\chi_- = 0$ if and only if $b = b\chi_+$, which holds if and only if $b(\xi)\,|\xi| = b(\xi)(i\xi)$. Similarly, $b\chi_+ = 0$ if and only if $b(\xi)\,|\xi| = -b(\xi)(i\xi)$.

The functions in $L_\infty^+(\mathbb{R}^m)$ and $L_\infty^-(\mathbb{R}^m)$ have the following significance in Fourier theory. Their inverse Fourier transforms have right–monogenic extensions to the half spaces $\mathbb{R}_+^{m+1}$ and $\mathbb{R}_-^{m+1}$ which decay at ∞. Let us verify this.

Given $b \in L_\infty^+(\mathbb{R}^m)$, define ϕ_+ on $\mathbb{R}_+^{m+1}$ by

$$\phi_+(x) = \phi_+(x_0 e_0 + \mathbf{x}) = \frac{1}{(2\pi)^m} \int_{\mathbb{R}^m} b(\xi)\, e^{i\langle \mathbf{x}, \xi \rangle}\, e^{-x_0|\xi|}\, d\xi$$

when $x_0 > 0$ and $\mathbf{x} \in \mathbb{R}^m$. This integral is absolutely convergent, and

$$\|\phi_+(x)\| \le \frac{c}{|x_0|^m}\, \|b\|_\infty .$$

Moreover,

$$\frac{\partial \phi_+}{\partial x_0}(x) = \frac{-1}{(2\pi)^m} \int_{\mathbb{R}^m} b(\xi)\, |\xi|\, e^{-x_0|\xi|}\, e^{i\langle \mathbf{x}, \xi \rangle}\, d\xi \quad \text{and}$$

$$\sum_{j=1}^m \frac{\partial \phi_+}{\partial x_j}(x)\, e_j = \frac{1}{(2\pi)^m} \int_{\mathbb{R}^m} b(\xi)\, i\xi\, e^{-x_0|\xi|}\, e^{i\langle \mathbf{x}, \xi \rangle}\, d\xi$$

so that, using the fact that that $b(\xi)\,|\xi| = b(\xi)(i\xi)$, we have

$$\frac{\partial \phi_+}{\partial x_0} + \phi_+ \mathbf{D} = \frac{\partial \phi_+}{\partial x_0} + \sum_{j=1}^m \frac{\partial \phi_+}{\partial x_j}\, e_j = 0.$$

That is, ϕ_+ is the right–monogenic extension to $\mathbb{R}_+^{m+1}$ of the inverse Fourier transform of b.

In a similar way, the inverse Fourier transform of each function $b \in L_\infty^-(\mathbb{R}^m)$ has a right–monogenic extension ϕ_- to $\mathbb{R}_-^{m+1}$ defined by

$$\phi_-(x) = \phi_-(x_0 e_0 + \mathbf{x}) = \frac{1}{(2\pi)^m} \int_{\mathbb{R}^m} b(\xi)\, e^{x_0|\xi|}\, e^{i\langle \mathbf{x}, \xi \rangle}\, d\xi$$

which satisfies

$$\|\phi_-(x)\| \le \frac{c}{|x_0|^m}\, \|b\|_\infty$$

when $x_0 < 0$ and $\mathbf{x} \in \mathbb{R}^m$.

These inverse Fourier transforms can be expressed more concisely in terms of the following exponential functions, which are both left– and right–

monogenic functions of x:

$$
\begin{cases}
e_+(x,\xi) &= e^{i\langle \mathbf{x},\xi\rangle}\, e^{-x_0|\xi|}\, \chi_+(\xi) \\
e_-(x,\xi) &= e^{i\langle \mathbf{x},\xi\rangle}\, e^{x_0|\xi|}\, \chi_-(\xi) \quad \text{and} \\
e(x,\xi) &= e_+(x,\xi) + e_-(x,\xi) = e^{i\langle \mathbf{x},\xi\rangle}\, e^{-x_0\{i\xi\}} \\
&= e^{i\langle \mathbf{x},\xi\rangle}\left\{\cosh(x_0|\xi|) - \sinh(x_0|\xi|)\,\tfrac{i\xi}{|\xi|}\right\}.
\end{cases}
$$

The extension ϕ_+ of the inverse Fourier transform of $b \in L_\infty^+(\mathbb{R}^m)$ is

$$
\phi_+(x) = \frac{1}{(2\pi)^m}\int_{\mathbb{R}^m} b(\xi)\, e_+(x,\xi)\, d\xi = \frac{1}{(2\pi)^m}\int_{\mathbb{R}^m} b(\xi)\, e(x,\xi)\, d\xi
$$

when $x \in \mathbb{R}^{m+1}_+$ (because $b(\xi)\, e_-(x,\xi) = 0$), and the extension ϕ_- of the inverse Fourier transform of $b \in L_\infty^-(\mathbb{R}^m)$ is

$$
\phi_-(x) = \frac{1}{(2\pi)^m}\int_{\mathbb{R}^m} b(\xi)\, e_-(x,\xi)\, d\xi = \frac{1}{(2\pi)^m}\int_{\mathbb{R}^m} b(\xi)\, e(x,\xi)\, d\xi
$$

when $x \in \mathbb{R}^{m+1}_-$.

Let us list the following properties of the inverse Fourier transform:

1. If b is the Fourier transform of $\phi \in L_1(\mathbb{R}^m)$, and $b = b\chi_+$, then $b \in L_\infty^+(\mathbb{R}^m)$, and ϕ_+ is the right–monogenic extension of ϕ to the upper half space $\mathbb{R}^{m+1}_+$.

2. If $b = \chi_+$, then ϕ_+ is the Cauchy kernel

$$
\phi_+(x) = k(x) = \frac{\bar{x}}{\sigma_m\, |x|^{m+1}} = \frac{x_0 e_0 - \mathbf{x}}{\sigma_m\, |x_0 e_0 + \mathbf{x}|^{m+1}}.
$$

3. If $b(\xi)\,\xi = \xi\, b(\xi)$, then $b(\xi)\, e(x,\xi) = e(x,\xi)\, b(\xi)$, so ϕ_+ is left–monogenic as well as right–monogenic.

In practice it is not easy to find the monogenic extension of a given function defined on $\mathbb{R}^m$. The Fourier transform is a useful aid in some cases. Given an integrable function ϕ on $\mathbb{R}^m$, compute its Fourier transform b, decompose b as $b = b_+ + b_-$ where $b_\pm = b\chi_\pm$, and then compute the extension ϕ_+ of the inverse Fourier transform of b_+ to $\mathbb{R}^{m+1}_+$, and the extension ϕ_- of the inverse Fourier transform of b_- to $\mathbb{R}^{m+1}_-$. Having a formula for ϕ_+ on $\mathbb{R}^{m+1}_+$ may allow us to extend it to a neighbourhood of $\mathbb{R}^m$, and similarly for ϕ_-. If so, then $\phi_+ + \phi_-$ is the right–monogenic extension of ϕ.

In particular, if $\| b(\xi) \| \le ce^{-\alpha|\xi|}$, then the inverse Fourier transform of b has a right–monogenic extension to the slab $\{x = x_0 e_0 + \mathbf{x} : |x_0| < \alpha\}$ defined by

$$
\phi(x) = \frac{1}{(2\pi)^m}\int_{\mathbb{R}^m} b(\xi)\, e(x,\xi)\, d\xi, \quad |x_0| < \alpha.
$$

Sommen has investigated the exponential function $e(x, \xi)$ in this type of context [S1, S2]. He has also investigated the function $\mathcal{E}(x, \xi_0 e_0 + \xi)$ obtained by taking the left–monogenic extension of the second variable.

Actually Tao Qian and I spent some time trying to use $\mathcal{E}$ to characterise the Fourier transforms of the functions $\phi \in K_N^+$ by bounded functions which are monogenic on some cone in $\mathbb{R}^{m+1}$. However we eventually realised that the thing to do is to characterise them instead by bounded functions which are holomorphic functions of several complex variables (in the usual sense) on some cone in $\mathbb{C}^m$. We subsequently developed this theory with Chun Li [LMcQ], key ideas of which will be presented in the next section.

First recall some definitions. Let $0 < \mu < \pi/2$. In section (B), we defined holomorphic extensions of the functions $\chi_{\pm}$ to the open cones

$$S_\mu^0(\mathbb{C}^m) = \{ \zeta = \xi + i\eta \in \mathbb{C}^m : |\zeta|_*^2 \notin (-\infty, 0] \text{ and } |\eta| < \tan \mu \, Re|\zeta|_* \}$$

by

$$\chi_{\pm}(\zeta) = \frac{1}{2}\left(e_0 + \frac{i\zeta}{\pm|\zeta|_*} \right)$$

where $|\zeta|_*$ is the square root of

$$|\zeta|_*^2 = \sum_{j=1}^m \zeta_j^{\,2} = |\xi|^2 - |\eta|^2 + 2i \langle \xi, \eta \rangle$$

which satisfies $Re|\zeta|_* > 0$. Recall that

$$\chi_+(\zeta)^2 = \chi_+(\zeta), \ \chi_-(\zeta)^2 = \chi_-(\zeta), \ \chi_+(\zeta)\chi_-(\zeta) = 0 = \chi_-(\zeta)\chi_+(\zeta),$$

$$\begin{cases} i\zeta\chi_{\pm}(\zeta) &= \chi_{\pm}(\zeta)\, i\zeta \\ e_0 &= \chi_+(\zeta) + \chi_-(\zeta) \\ i\zeta &= |\zeta|_* \, \chi_+(\zeta) + (-|\zeta|_*)\, \chi_-(\zeta) \end{cases}$$

and $\chi_{\pm} \in H_\infty(S_\mu^0(\mathbb{C}^m))$ with $\| \chi_{\pm} \|_\infty \leq \frac{1}{\sqrt{2}\cos\mu}$.

Using them, we define the functions

$$\begin{cases} e_+(x, \zeta) &= e^{i\langle x, \zeta \rangle}\, e^{-x_0|\zeta|_*}\, \chi_+(\zeta) \\ e_-(x, \zeta) &= e^{i\langle x, \zeta \rangle}\, e^{x_0|\zeta|_*}\, \chi_-(\zeta) \quad \text{and} \\ e(x, \zeta) &= e_+(x, \zeta) + e_-(x, \zeta) = e^{i\langle x, \zeta \rangle}\, e^{-x_0\{i\zeta\}} \end{cases}$$

which are right– (and left–) monogenic functions of $x \in \mathbb{R}^{m+1}$, and bounded holomorphic functions of $\zeta \in S_\mu^0(\mathbb{C}^m)$.

5.5.2 (H) Correspondence between monogenic functions and holomorphic functions of several complex variables under Fourier transforms

Our aim now is to characterise the Fourier transforms of the functions $\phi \in K_N^+$ which were introduced in section (E).

To do this, we introduce the following cones in $\mathbb{C}^m$. For each unit vector $n = n_0 e_0 + \mathbf{n} \in \mathbb{R}_+^{m+1}$, let

$$
\begin{aligned}
n(\mathbb{C}^m) &= \{\zeta = \xi + i\eta \in \mathbb{C}^m & : |\zeta|_* \notin (-\infty, 0] \text{ and} \\
&& Re\,|\zeta|_*\, e_0 + \eta = \kappa n \text{ for some } \kappa > 0\} \\
&= \{\,\zeta = \xi + i\eta \in \mathbb{C}^m & : |\zeta|_* \notin (-\infty, 0] \text{ and} \\
&& n_0\eta = Re\,|\zeta|_*\, \mathbf{n}\}.
\end{aligned}
$$

It is an exercise to show that this cone can also be represented as

$$
\begin{aligned}
n\ (\mathbb{C}^m) &= \\
&\left\{ \zeta = \xi + i\eta \in \mathbb{C}^m : \xi \neq \mathbf{0} \text{ and } n_0\eta = \left(n_0{}^2\,|\xi|^2 + \langle \xi, \mathbf{n} \rangle^2 \right)^{\frac{1}{2}} \mathbf{n} \right\}
\end{aligned}
$$

in which case this parametrization $\xi \mapsto \zeta = \xi + i\eta$ is smooth with

$$
\left| \det \left(\frac{\partial \zeta_j}{\partial \xi_k} \right) \right| \leq \frac{1}{n_0}
$$

when $\xi \neq \mathbf{0}$. Thus $n(\mathbb{C}^m)$ is a manifold in $\mathbb{C}^m$ with real dimension m.

As at the end of section (E), let N be a compact set of unit vectors in $\mathbb{R}_+^{m+1}$ which contains e_0, and let $\omega = \max_{n \in N} \angle(n, e_0) < \pi/2$. Given $\mu > 0$ such that $\mu + \omega < \frac{\pi}{2}$, let N_μ denote the open neighbourhood of N in the unit sphere defined by

$$
\begin{aligned}
N_\mu &= \\
&\{y \in \mathbb{R}_+^{m+1} : |y| = 1 \text{ and there exists } n \in N \text{ such that } \angle(y, n) < \mu\}.
\end{aligned}
$$

This time we make the additional assumption that N is starlike about e_0 (in the sense that, whenever $n \in N$ and $0 \leq \tau \leq 1$, then $(\tau n + (1 - \tau)e_0)/|(\tau n + (1 - \tau)e_0)| \in N$). Then N_μ is also starlike about e_0.

Define the open cone

$$
N_\mu(\mathbb{C}^m) = \bigcup_{n \in N_\mu} n(\mathbb{C}^m)
$$

and note that

$$
N_\mu(\mathbb{C}^m) \subset S_{\omega+\mu}^0(\mathbb{C}^m).
$$

where $S_{\omega+\mu}^0(\mathbb{C}^m)$ is the cone defined in section (B). (In the special case when $N = \{n \in \mathbb{R}_+^{m+1} : |n| = 1 \text{ and } n_0 \geq \cos\omega\}$, then $N_\mu(\mathbb{C}^m) = S_{\omega+\mu}^0(\mathbb{C}^m)$.)

The cones $N_\mu(\mathbb{C}^m)$ in $\mathbb{C}$ may seem rather strange at first, but the point is that they allow us to completely characterise the Fourier transforms of the spaces K_N^+ and M_N^+. To do this, we need the following function spaces.

Let $H_\infty(N_\mu(\mathbb{C}^m)) = H_\infty(N_\mu(\mathbb{C}^m), \mathcal{A})$ denote the Banach space of all bounded holomorphic functions from $N_\mu(\mathbb{C}^m)$ to $\mathcal{A}$ with the norm

$$\| b \|_\infty = \sup\{\| b(\zeta) \| : \zeta \in N_\mu(\mathbb{C}^m)\}.$$

REMARK In these lectures the word "holomorphic" is used in the traditional sense for functions of several complex variables. At no time is it used in the sense of "Clifford holomorphic" or "monogenic" in the way some authors do.

Decompose $H_\infty(N_\mu(\mathbb{C}^m))$ using the functions χ_+ and χ_- as

$$\begin{cases} H_\infty(N_\mu(\mathbb{C}^m)) &= H_\infty^+(N_\mu(\mathbb{C}^m)) & \oplus & H_\infty^-(N_\mu(\mathbb{C}^m)) \\ b(\zeta) &= b(\zeta)\chi_+(\zeta) & + & b(\zeta)\chi_-(\zeta) \end{cases}$$

where

$$H_\infty^+(N_\mu(\mathbb{C}^m)) = \{ b \in H_\infty(N_\mu(\mathbb{C}^m)) : b\chi_- = 0 \} \quad \text{and}$$
$$H_\infty^-(N_\mu(\mathbb{C}^m)) = \{ b \in H_\infty(N_\mu(\mathbb{C}^m)) : b\chi_+ = 0 \}.$$

Also, let H_N^+ be the algebra of all those functions b on $\mathbb{R}^m \setminus \{0\}$ which have holomorphic extensions to $b \in H_\infty^+(N_\mu(\mathbb{C}^m))$ for some $\mu > 0$, and let

$$A_N^+ = \{ b \in H_N^+ : b(\xi)\xi = \xi b(\xi) \}.$$

One of the reasons why we assume that N is starlike about e_0, is to ensure that the holomorphic extension is uniquely determined by the values of b on $\mathbb{R}^m$.

We are now in a position to state one of our main theorems, namely that the Fourier transform defines an isomorphism between K_N^+ and H_N^+. This result was proved about 1989-90 by Chun Li, Tao Qian and myself [LMcQ].

THEOREM 5.5.1
Let N be a compact set of unit vectors in $\mathbb{R}^{m+1}$ which is starlike about e_0. For every $\phi \in K_N^+$, there is a unique function $b \in H_N^+$ which satisfies Parseval's identity

$$\frac{1}{(2\pi)^m} \int_{\mathbb{R}^m} b(\xi)\,\hat{u}(-\xi)\,d\xi = \lim_{\delta \to 0+} \int_{\mathbb{R}^m} \phi(\delta e_0 + \mathbf{x})\,u(\mathbf{x})\,d\mathbf{x}$$

for all u in the Schwartz space $\mathcal{S}(\mathbb{R}^m)$. So b is the distribution Fourier transform of ϕ, and we write $b = \mathcal{F}_+(\phi)$. We also call ϕ the inverse Fourier transform of b, and write $\phi = \mathcal{G}_+(b)$.

The Fourier transform $\mathcal{F}_+$ is a linear transformation with the following properties:

1. $\mathcal{F}_+$ *is a one-one map of K_N^+ onto H_N^+. That is, for every $b \in H_N^+$ there exists a unique function $\phi \in K_N^+$ such that $b = \mathcal{F}_+(\phi)$.*

2. *If $0 < \nu < \mu < \frac{\pi}{2} - \omega$ and $\phi \in K(C_{N_\mu}^+)$, then $b \in H_\infty^+(N_\nu(\mathbb{C}^m))$ and*

$$\|b\|_\infty \ \leq \ c_\nu \, \|\phi\|_{K(C_{N_\mu}^+)}$$

for some constant c_ν which depends on ν (as well as ω and μ).

3. *If $0 < \nu < \mu < \frac{\pi}{2} - \omega$ and $b \in H_\infty^+(N_\mu(\mathbb{C}^m))$, then $\phi \in K(C_{N_\nu}^+)$ and*

$$\|\phi\|_{K(C_{N_\nu}^+)} \ \leq \ c_\nu \, \|b\|_\infty$$

for some constant c_ν which depends on ν (as well as ω and μ).

4. *If $\phi_t(x) = \phi(x + te_0)$ and $b_t(\xi) = b(\xi)e^{-t|\xi|} = b(\xi)e^{-t\{i\xi\}}$ for $t > 0$, then $\mathcal{F}_+(\phi_t) = b_t$.*

5. $\phi \in M_N^+$ *if and only if $b \in \mathcal{A}_N^+$.*

6. *If $\phi \in K_N^+$, $\psi \in M_N^+$, $b = \mathcal{F}_+(\phi)$ and $f = \mathcal{F}_+(\psi)$, then $bf = \mathcal{F}_+(\phi * \psi)$.*

7. *The mapping $\phi \mapsto b$ is an algebra homomorphism from the convolution algebra M_N^+ onto the function algebra $\mathcal{A}_N^+$.*

PROOF   For a full proof of this result, see theorem 4.1 of [LMcQ]. Here we just give a representation of the inverse Fourier transform $\mathcal{G}_+$ and indicate why it maps H_N^+ to K_N^+. There are some more details in my paper with Chun Li in this volume [LMc].

Suppose that $b \in H_\infty^+(N_\mu(\mathbb{C}^m))$ for some $\mu > 0$. Then, as we saw in the last section, b has an inverse Fourier transform $\phi = \mathcal{G}_+(b)$ with a right–monogenic extension ϕ_+ to $\mathbb{R}_+^{m+1}$ defined by

$$\phi_+(x) = \frac{1}{(2\pi)^m} \int_{\mathbb{R}^m} b(\xi)e(x,\xi)d\xi = \frac{1}{(2\pi)^m} \int_{\mathbb{R}^m} b(\xi)e_+(x,\xi)d\xi \quad x_0 > 0.$$

Our claim is that ϕ has a right–monogenic extension to the larger set $C_{N_\mu}^+ = \cup_{n \in N_\mu} C_n^+$ which satisfies the stated estimate.

Given $n = n_0 e_0 + \mathbf{n} \in N_\mu$, define the right–monogenic function ϕ_n of $x = x_0 e_0 + \mathbf{x} \in C_n^+ = \{x \in \mathbb{R}^{m+1} : \langle x, n \rangle > 0\}$ by

$$\phi_n(x) = \frac{1}{(2\pi)^m} \int_{n(\mathbb{C}^m)} b(\zeta)\, e(x, \zeta)\, d\zeta_1 \wedge d\zeta_2 \wedge \ldots \wedge d\zeta_m$$

$$= \frac{1}{(2\pi)^m} \int_{n(\mathbb{C}^m)} b(\zeta)\, e_+(x, \zeta)\, d\zeta_1 \wedge d\zeta_2 \wedge \ldots \wedge d\zeta_m.$$

The integral is absolutely convergent because, on the surface $n(\mathbb{C}^m)$, the integrand decreases exponentially at ∞. Indeed

$$
\begin{aligned}
\| e_+(x, \zeta) \| &= \left\| e^{i\langle \mathbf{x}, \zeta \rangle} e^{-x_0 |\zeta|_*} \chi_+(\zeta) \right\| \\
&\le e^{-\langle \mathbf{x}, \eta \rangle - x_0 Re|\zeta|_*} \frac{1}{\sqrt{2}\cos(\omega + \mu)} \\
&= e^{-\{\langle \mathbf{x}, \mathbf{n} \rangle + x_0 n_0\} Re|\zeta|_*/n_0} \frac{1}{\sqrt{2}\cos(\omega + \mu)} \\
&= e^{-\langle \mathbf{x}, \mathbf{n} \rangle Re|\zeta|_*/n_0} \frac{1}{\sqrt{2}\cos(\omega + \mu)}
\end{aligned}
$$

when $\zeta = \xi + i\eta \in n(\mathbb{C}^m)$. Therefore

$$
\| \phi_n(x) \| \le \frac{c\, \| b \|_\infty}{\langle x, n \rangle^m}
$$

for all $x \in C_n^+$, because $Re\, |\zeta|_* \approx |\xi|$ and

$$
\| d\zeta_1 \wedge d\zeta_2 \wedge \ldots \wedge d\zeta_m \| = \left| \det \left(\frac{\partial \zeta_j}{\partial \xi_k} \right) \right| d\xi \le \frac{1}{n_0}\, d\xi.
$$

When $n = e_0$, then $\phi_n = \phi_+$. Each of the other extensions ϕ_n also equals ϕ_+ on their common domain, namely $C_n^+ \cap \mathbb{R}_+^{m+1}$. To see this, note that the integrand depends holomorphically on the single complex variable $z = \langle \zeta, \mathbf{n} \rangle$ (on writing $\zeta = z\mathbf{n} + \zeta'$ where $\langle \zeta', \mathbf{n} \rangle = 0$ and holding ζ' constant). Now use the starlike nature of N_μ, and Cauchy's theorem in the z–plane.

Hence there is a unique right–monogenic function ϕ on $N_\mu(\mathbb{C}^m)$ which coincides with each of the functions ϕ_n on C_n^+. This is the inverse Fourier transform $\phi = \mathcal{G}_+(b)$ that we are looking for. For $\nu < \mu$, the bound

$$
\| \phi \|_{K(C_{N_\nu}^+)} \le c_\nu \, \| b \|_\infty
$$

is a consequence of the above estimate for ϕ_n.   ∎

In the special case when $N = \{ e_0 \}$, then $N_\mu(\mathbb{C}^m) = S_\mu^0(\mathbb{C}^m)$ and $K(C_{N_\mu}^+) = C_{\mu+}^0 = \{ x = x_0 e_0 + \mathbf{x} \in \mathbb{R}^{m+1} : x_0 > -|\mathbf{x}| \tan\mu \}$. It is this case which was originally treated in [LMcS].

By knowing the precise characterisation of the Fourier transform, we can deduce properties of functions in K_N^+, and determine whether specific functions on $\mathbb{R}^m$ can be extended to right–monogenic functions on $C_{N_\mu}^+$. As an example, we present the following theorem.

THEOREM 5.5.2
Let $\phi \in K_N^+$. Then ϕ can be expressed as $\phi = \sum e_S \, \phi_S$ where $\phi_S \in M_N^+$ and e_S is a basis for $\mathcal{A}$.

PROOF   The Fourier transform of ϕ is

$$\mathcal{F}_+(\phi) = b = b\chi_+ \in H_N^+ \quad \text{so}$$

$$\mathcal{F}_+(\phi) = \sum e_S \, b_S \, \chi_+ \quad \text{where } b_S \text{ are complex–valued}$$

$$= \sum e_S \, b_S^+$$

where $b_S^+ = b_S \chi_+ \in H_N^+$. Now $b_S^+(\xi)\,\xi = \xi\,b_S^+(\xi)$. That is, $b_S^+ \in A_N^+$, so there exists $\phi_S \in M_N^+$ such that $\mathcal{F}_+(\phi_S) = b_S^+$. Therefore

$$\phi = \sum e_S \, \phi_S$$

as required.   ∎

Although this proof using Fourier transforms is simple, the result itself is not obvious. Terry Tao independently observed and derived it by direct methods in his thesis [T].

5.6   Lecture 5

In this lecture, $\mathcal{A}$ is a finite dimensional complex algebra in which (I) is satisfied, and $\mathcal{X}$ is a finite dimensional left $\mathcal{A}$ module. The surface Σ is the Lipschitz surface consisting of all points

$$x = g(\mathbf{x})\,e_0 + \mathbf{x} \in \mathbb{R}^{m+1}$$

where $\mathbf{x} \in \mathbb{R}^m$, and g is a real–valued Lipschitz function which satisfies $\|\mathbf{D}g\|_\infty \leq \tan\omega < \infty$ where $0 \leq \omega < \pi/2$. The open set Ω_+ is the region above Σ, namely $\Omega_+ = \{X \in \mathbb{R}^{m+1} : X_0 > g(\mathbf{X})\}$. The set N is a compact set of unit vectors in $\mathbb{R}_+^{m+1}$ which is starlike about e_0, contains almost all the unit normals to Σ, and satisfies $\angle(n, e_0) \leq \omega$ for all $n \in N$.

5.6.1   (J) Algebras of singular integrals on Lipschitz surfaces, and functional calculi of Dirac operators on these surfaces

Let us first consider the case when $\Sigma = \mathbb{R}^m$ and $p = 2$. In this case we developed the functional calculus of the Dirac operator $\mathbf{D}$ in $L_2(\mathbb{R}^m)$ in section (D).

We now show that the bounded linear operator $b(\frac{1}{i}\mathbf{D})$ on $L_2(\mathbb{R}^m)$ can be represented as a singular integral operator in the case when $b \in H^+_{\{e_0\}}$. Indeed, for $\phi = \mathcal{G}_+(b) \in K^+_{\{e_0\}}$, then

$$b(\frac{1}{i}\mathbf{D})u(\mathbf{x}) = T_\phi u(\mathbf{x})$$

$$= \lim_{\delta \to 0+} \int_{\mathbb{R}^m} \phi(\mathbf{x} + \delta e_0 - \mathbf{y}) u(\mathbf{y}) \, d\mathbf{y}$$

for all $u \in L_2(\mathbb{R}^m)$ and almost all $\mathbf{x} \in \mathbb{R}^m$. Let us prove this.

Recall from section (D) that, for all $u \in L_2(\mathbb{R}^m)$,

$$b(\frac{1}{i}\mathbf{D})u = b(\frac{1}{i}\mathbf{D})\chi_+(\frac{1}{i}\mathbf{D})u$$

$$= \lim_{\delta \to 0+} b_\delta(\frac{1}{i}\mathbf{D})\chi_+(\frac{1}{i}\mathbf{D})u$$

$$= \lim_{\delta \to 0+} b_\delta(\frac{1}{i}\mathbf{D})u$$

where $b_\delta(\xi) = b(\xi)e^{-\delta|\xi|} = b(\xi)e^{-\delta\{i\xi\}}$ for $\delta > 0$.

Recall from section (F) that

$$T_\phi u = \lim_{\delta \to 0+} T_{\phi_\delta} u$$

in the sense of L_2 convergence, where $\phi_\delta(x) = \phi(x+\delta e_0)$, and, from section (H) that $\phi_\delta = \mathcal{G}_+(b_\delta)$. So it suffices to prove that

$$b_\delta(\frac{1}{i}\mathbf{D})u = T_{\phi_\delta} u$$

for all $\delta > 0$.

The operators T_{ϕ_δ} are not singular, so it is easy to verify that

$$T_{\phi_\delta} u(x) = \int_{\mathbb{R}^m} \phi_\delta(x-y) u(y) \, dy \quad \text{implies}$$

$$(T_{\phi_\delta} u)^\wedge(\xi) = b_\delta(\xi)\,\hat{u}(\xi) \quad \text{and hence}$$

$$T_{\phi_\delta} u = b_\delta(\frac{1}{i}\mathbf{D})u$$

as required.

In particular, if $b = \chi_+$, then $\phi(x) = k(x) = \frac{\bar{x}}{\sigma_m|x|^{m+1}}$, so that

$$\chi_+(\frac{1}{i}\mathbf{D}) = T_k = P_+$$

where P_+ is the projection onto the Hardy space $L_2^+(\mathbb{R}^m)$ defined in section

(F). It is given by

$$P_+u(\mathbf{x}) = \lim_{\delta \to 0+} \int_{\mathbb{R}^m} k(\mathbf{x} + \delta e_0 - \mathbf{y}) \, u(\mathbf{y}) \, d\mathbf{y}$$

$$= \lim_{\delta \to 0+} \frac{1}{\sigma_m} \int_{\mathbb{R}^m} \frac{-\mathbf{x} + \delta e_0 + \mathbf{y}}{|\mathbf{x} + \delta e_0 - \mathbf{y}|^{m+1}} \, u(\mathbf{y}) \, d\mathbf{y}$$

$$= \lim_{\varepsilon \to 0} \frac{1}{\sigma_m} \int_{|\mathbf{x} - \mathbf{y}| > \varepsilon} \frac{-\mathbf{x} + \mathbf{y}}{|\mathbf{x} - \mathbf{y}|^{m+1}} \, u(\mathbf{y}) \, d\mathbf{y} \; + \; \frac{1}{2} u(\mathbf{x})$$

for all $u \in L_2(\mathbb{R}^m)$ and almost all $\mathbf{x} \in \mathbb{R}^m$.

This is in agreement with the definition of the Hardy projection P_+ in section (D).

Recall that the operators T_ϕ are bounded linear operators on $L_p(\mathbb{R}^m)$ for all $p \in (1, \infty)$. That is, $T_\phi \in \mathcal{L}(L_p(\mathbb{R}^m))$. So it is natural to write $T_\phi = b(\frac{1}{i}\mathbf{D})$ when $b = \mathcal{F}_+(\phi)$, where D_j are the closed operators in $L_p(\mathbb{R}^m)$ defined in section (D), and $\mathbf{D} = (D_1, D_2, \ldots, D_m)$. It is an exercise for the reader to show that this gives an $H^+_{\{e_0\}}$ functional calculus of $\frac{1}{i}\mathbf{D}$ in $L_p(\mathbb{R}^m)$ in the sense that (i) the linear mapping

$$\begin{array}{ccccc} H^+_{\{e_0\}} & \to & K^+_{\{e_0\}} & \to & \mathcal{L}(L_p(\mathbb{R}^m)) \\ b & \mapsto & \phi & \mapsto & T_\phi = b(\frac{1}{i}\mathbf{D}) \end{array}$$

has the property that if $f \in A^+_{\{e_0\}}$ corresponds to $\psi \in M^+_{\{e_0\}}$, then $bf \in H^+_{\{e_0\}}$ corresponds to $\phi * \psi \in K^+_{\{e_0\}}$ with associated operator

$$(bf)(\tfrac{1}{i}\mathbf{D}) = T_{\phi * \psi} = T_\phi T_\psi = b(\tfrac{1}{i}\mathbf{D}) f(\tfrac{1}{i}\mathbf{D}) \; ,$$

and (ii) there is agreement with the natural definition for polynomials of several variables.

These results all have analogues in the more general case of the surface Σ specified at the start of this lecture. That is, there is a closed linear operator $\mathbf{D}_\Sigma$ in $L_p(\Sigma)$ for which it is natural to write $T_\phi = b(\frac{1}{i}\mathbf{D}_\Sigma)$ when $b \in H^+_N$ and $\phi = \mathcal{G}_+(b) \in K^+_N$. Then, as above, $\frac{1}{i}\mathbf{D}_\Sigma$ has an H^+_N functional calculus in $L_p(\Sigma)$ in the sense that (i) the linear mapping

$$\begin{array}{ccccc} H^+_N & \to & K^+_N & \to & \mathcal{L}(L_p(\Sigma)) \\ b & \mapsto & \phi & \mapsto & T_\phi = b(\frac{1}{i}\mathbf{D}_\Sigma) \end{array}$$

has the property that if $f \in A^+_N$ corresponds to $\psi \in M^+_N$, then $bf \in H^+_N$ corresponds to $\phi * \psi \in K^+_N$ with associated operator

$$(bf)(\tfrac{1}{i}\mathbf{D}_\Sigma) = T_{\phi * \psi} = T_\phi T_\psi = b(\tfrac{1}{i}\mathbf{D}_\Sigma) f(\tfrac{1}{i}\mathbf{D}_\Sigma) \; ,$$

and (ii) there is agreement with the natural definition for polynomials of several variables.

Details can be found in [LMcQ]. See section (M) for some further material.

In particular, if $b = \chi_+$, then $\phi(x) = k(x) = \dfrac{\overline{x}}{\sigma_m |x|^{m+1}}$, so that

$$\chi_+\left(\frac{1}{i}\mathbf{D}_\Sigma\right) \ = \ T_k \ = \ P_+$$

where P_+ is the bounded projection onto the Hardy space $L_p^+(\Sigma)$ given by

$$
\begin{aligned}
(P_+u)(x) \ &= \ \lim_{\delta\to0+} \int_\Sigma k(x + \delta e_0 - y)\, n(y)\, u(y)\, dS_y \\
&= \ \lim_{\varepsilon\to0} \int_{\{y\in\Sigma\,:\,|x-y|>\varepsilon\}} k(x - y)\, n(y)\, u(y)\, dS_y \ + \ \tfrac{1}{2}u(x)
\end{aligned}
$$

for all $u \in L_2(\Sigma)$ and almost all $\mathbf{x} \in \Sigma$.

As we know, the operator obtained by taking the scalar part P_{+0} of P_+ is the singular double–layer potential operator. In order to use this operator in potential theory, it is important to know that it has a bounded inverse $(P_{+0})^{-1} \in \mathcal{L}(L_2^+(\Sigma))$. This fact was first proved by Greg Verchota in his PhD thesis by using the Rellich estimate which was stated in section (F) [V].

Note however, that although the boundedness of P_+ can be used to prove the boundedness of P_{+0}, this operator does not belong to any holomorphic functional calculus of $\frac{1}{i}\mathbf{D}_\Sigma$, and so our theory cannot be used to give a representation of $(P_{+0})^{-1}$ as a singular integral operator. It would be useful to have some kind of integral representation of $(P_{+0})^{-1}$.

What is the operator $\mathbf{D}_\Sigma$? If, for some $\varepsilon > 0$, there is a left–monogenic function U defined on the neighbourhood $\Sigma_\varepsilon = \{\Sigma + te_0 : |t| < \varepsilon\}$ of Σ such that $U|_\Sigma = u \in L_p(\Sigma)$, and if $DU|_\Sigma = -\frac{\partial U}{\partial x_0}|_\Sigma = v \in L_p(\Sigma)$, then define $\mathbf{D}_\Sigma u = v$. (It is an easy exercise to show that the set of such functions u is dense in $L_p(\Sigma)$.) Define the domain $\mathcal{D}(\mathbf{D}_\Sigma)$ of $\mathbf{D}_\Sigma$ to be the set of all $u \in L_p(\Sigma)$ for which there is a sequence of such functions U_n such that $U_n|_\Sigma \to u$ and $DU_n|_\Sigma \to v \in L_p(\Sigma)$. Then let $\mathbf{D}_\Sigma u = v$ for these u. For details, see [LMcQ].

It is an exercise for the reader to show that its complex spectrum $\sigma(\mathbf{D}_\Sigma)$ is a subset of the closed double sector $S_\omega(\mathbb{C}) = \{\zeta \in \mathbb{C} : \zeta = 0 \text{ or } |\arg\zeta| \leq \omega \text{ or } |\arg(-\zeta)| \leq \omega\}$. Moreover, $\mathbf{D}_\Sigma$ has a bounded H_∞ functional calculus defined by $B\{\mathbf{D}_\Sigma\} = b(\frac{1}{i}\mathbf{D}_\Sigma)$ for all $B \in H_\infty(S_\mu^0(\mathbb{C}))$, $\mu > \omega$ where b is the corresponding function of m–variables, namely $b(\xi) = B\{i\xi\}$.

In particular, the projections associated with the parts of the spectrum of $\mathbf{D}_\Sigma$ in the right half plane and in the left half plane are

$$
\begin{aligned}
\chi_{\mathbb{R}+}\{\mathbf{D}_\Sigma\} \ &= \ \chi_+(\imath\mathbf{D}_\Sigma) \ = \ P_+ \quad \text{and} \\
\chi_{\mathbb{R}-}\{\mathbf{D}_\Sigma\} \ &= \ \chi_-(\imath\mathbf{D}_\Sigma) \ = \ P_- \ .
\end{aligned}
$$

Recall that we already did this in section (D) in the case when $\Sigma = \mathbb{R}^m$, $\mathbf{D}_\Sigma = \mathbf{D}$ and $p = 2$. It is a further exercise to show that the remainder of that section also carries across to this context.

5.6.2 (K) Boundary value problems for harmonic functions

Let us recall that the Cauchy operator $\mathcal{C}_\Sigma^+$ and its scalar part, the double–layer potential operator $\mathcal{C}_{\Sigma 0}^+$, are defined by

$$(\mathcal{C}_\Sigma^+ v)(X) = \int_\Sigma k(X - y)\, n(y)\, v(y)\, dS_y$$

$$(\mathcal{C}_{\Sigma 0}^+ v)(X) = \int_\Sigma \langle k(X - y)\,,\, n(y) \rangle\, v(y)\, dS_y$$

for all $v \in L_2(\Sigma)$ and all $X \in \Omega_+$. The scalar part of the Hardy projection, namely P_{+0}, is the singular double–layer potential operator which, as we have seen, is an invertible bounded operator on $L_2(\Sigma, \mathbb{C})$.

In this section we assume that $m \geq 2$. (Slight modifications are needed if $m = 1$, owing to the slower decay of solutions U at ∞.)

The solution of the Dirichlet problem

(D) $\begin{cases} \Delta U &= 0 \quad \text{on } \Omega_+ \\ U|_\Sigma &= w \in L_2(\Sigma, \mathbb{C}) \end{cases}$

which decays at ∞ is

$$U = \mathcal{C}_{\Sigma 0}^+ v \quad \text{where}$$

$$v = (P_{+0})^{-1} w \in L_2(\Sigma, \mathbb{C}).$$

This is because

$$\Delta U(X) = 0 \quad \text{for all } X \in \Omega_+,$$

$$\lim_{t \to 0} U(x + te_0) = \lim_{t \to 0} \int_\Sigma \langle k(x + te_0 - y)\,,\, n(y) \rangle\, v(y)\, dS_y = P_{+0} v(x) = w(x)$$

for almost all $x \in \Sigma$, and

$$\lim_{t \to \infty} U(x + te_0) = 0 \quad \text{for all } x \in \Sigma.$$

The construction of the solution by this method of layer potentials depends on the boundedness [CMcM] and invertibility [V] of P_{+0}. See Kenig's book [Ke] for a treatment of other methods.

A related problem is

(R) $\begin{cases} \Delta U &= 0 \quad \text{on } \Omega_+ \\ \dfrac{\partial U}{\partial X_0}\Big|_\Sigma &= w \in L_2(\Sigma, \mathbb{C}). \end{cases}$

The solution of (R) which decays at ∞ is

$$U(X) = -\int_{X_0}^{\infty} (\mathcal{C}_{\Sigma 0}^+ v)(\mathbf{X} + te_0)\, dt \quad \text{where}$$

$$v = (P_{+0})^{-1} w \in L_2(\Sigma, \mathbb{C}).$$

This is because

$$\Delta U(X) = 0 \quad \text{for all } X \in \Omega_+ ,$$

$$\lim_{t \to 0} \frac{\partial U}{\partial X_0}(x + te_0) = w(x) \quad \text{for almost all } x \in \Sigma, \text{ and}$$

$$\lim_{t \to \infty} U(x + te_0) = 0 \quad \text{for all } x \in \Sigma .$$

Let us now consider the boundary value problem

(B)
$$\begin{cases} \Delta U &= 0 \quad \text{on } \Omega_+ \\ \beta_0 \frac{\partial U}{\partial X_0} + \sum_{k=1}^{m} \beta_k \frac{\partial U}{\partial X_k}\Big|_{\Sigma} &= w \in L_2(\Sigma, \mathbb{C}) \end{cases}$$

where β_0, $\beta_k \in \mathbb{C}$. Let

$$\beta = \sum_{j=0}^{m} \beta_j e_j = \beta_0 e_0 + \sum_{j=1}^{m} \beta_j e_j.$$

We have just solved this problem in the special case when $\beta = e_0$.

Note first that this problem is not well posed for all choices of β. For example, if $m = 1$, then

$$\Delta U = \left(\frac{\partial U}{\partial X_0} \mp i \frac{\partial U}{\partial X_1} \right) \left(\frac{\partial U}{\partial X_0} \pm i \frac{\partial U}{\partial X_1} \right),$$

so that, if $\beta_1 = \pm i\beta_0$, then every solution of the Cauchy–Riemann equation

$$\frac{\partial U}{\partial X_0} \pm i \frac{\partial U}{\partial X_1} = 0$$

is a solution of (B) when $w = 0$.

In higher dimensions, it is natural to impose the covering condition of Agmon, Douglis and Nirenberg with respect to those normal and tangent vectors which are defined. This is the condition

(#) $\langle \beta, n + it \rangle \neq 0$ for all $n \in N$ and all $t \in \mathbb{R}^{m+1}$

such that $mod t = 1$ and $\langle n, t \rangle = 0$ By the compactness of N, this condition is equivalent to requiring that $|\langle \beta, n + it \rangle| \geq \kappa$ for some $\kappa > 0$ and for all the n and t which are specified above. It is an exercise in algebra to verify that this is equivalent to

(##) $\qquad |\langle \beta, |\zeta|_* e_0 - i\zeta\rangle| \geq \kappa ||\zeta|_*|$ for all $\zeta \in N(\mathbb{C}^m)$,

possibly for a different choice of κ. (To derive (##) from (#), choose $\zeta \in N(\mathbb{C}^m)$, meaning that there exists $n \in N$ and $c > 0$ such that $Re |\zeta|_* e_0 + \eta = cn$. Apply (#) with this choice of n, and with $t = c^{-1}(Im |\zeta|_* e_0 - \xi)$.)

Therefore, for some $\mu > 0$, the function b defined by

$$b(\zeta) = \frac{|\zeta|_* \, \chi_+(\zeta)}{\langle \beta, |\zeta|_* - i\zeta\rangle}$$

belongs to $H_\infty^+(N_\mu(\mathbb{C}^m))$, and hence to H_N^+, so that $b(\frac{1}{i}\mathbf{D}_\Sigma) \in \mathcal{L}(L_2(\Sigma, \mathcal{A}))$. Actually $b \in A_N^+$ because $b(\zeta)\zeta = \zeta b(\zeta)$.

The solution of (B) which decays at ∞ is

$$U(X) = -\int_{X_0}^\infty (C_\Sigma^+ b(\frac{1}{i}\mathbf{D}_\Sigma)v)_0 \, (\mathbf{X} + te_0) \, dt \qquad \text{where}$$

$$v = (P_{+0})^{-1}w \in L_2(\Sigma, \mathbb{C}) \subset L_2(\Sigma, \mathcal{A}).$$

This is because

$$\Delta U(X) = 0 \quad \text{for all } X \in \Omega_+ ,$$

$$\lim_{t\to 0} \sum_{k=0}^m \beta_k \frac{\partial U}{\partial X_k}(x + te_0) = w(x) \quad \text{for almost all } x \in \Sigma, \text{ and}$$

$$\lim_{t\to\infty} U(x + te_0) = 0 \quad \text{for all } x \in \Sigma .$$

It is an exercise to verify that the boundary limit is w. (Hint: first check that the identity

$$\left(\beta_0 \xi - \sum_{k=1}^m \beta_k \xi_k\right) b(\xi) = \xi \chi_+(\xi)$$

holds for all non–zero $\xi \in \mathbb{R}^m$, and then apply the functional calculus.)

Further, $(C_\Sigma^+ b(\frac{1}{i}\mathbf{D}_\Sigma)v)_0 \, (x + te_0) = (T_{\phi_t} v)_0(x)$ when $x \in \Sigma$ and $t > 0$, where $\phi = \mathcal{G}_+(b) \in M_N^+$. So the integrand can be expressed as

$$(C_\Sigma^+ b(\frac{1}{i}\mathbf{D}_\Sigma)v)_0 \, (X) = \int_\Sigma \overline{\langle \phi(X - y)}, n(y)\rangle \, v(y) \, dS_y , \quad X \in \Omega_+.$$

It may be of interest to compute the function ϕ explicitly. What the Fourier theory has shown is that $\phi \in M_N^+$ and hence that $T_\phi \in \mathcal{L}(L_2(\Sigma, \mathcal{A}))$.

We conclude with the remark that these methods solve (B) in $L_p(\Sigma)$ provided P_{+0} is invertible in $L_p(\Sigma)$. This is always true when $p \geq 2$, but is not true for every Lipschitz surface when $1 < p < 2$.

5.7   Additional Material

5.7.1   (L) Singular integrals on the boundary of a strongly Lipschitz domain

We could also have embedded $\mathbb{R}^{m+1}$ in an algebra $\mathcal{A}$ by using

$$\{e_1, e_2, \ldots, e_m, e_{m+1}\}$$

as a basis for $\mathbb{R}^{m+1}$, and identifying $x = \sum_{k=1}^{m+1} x_k e_k = \mathbf{x} + x_{m+1} e_{m+1}$ with a corresponding element of $\mathcal{A}$. We would then replace the condition (I) on $\mathcal{A}$ by

(I)
$$x^2 = -|x|^2 e_0 = -\left\{\sum_{k=1}^{m+1} x_k{}^2\right\}^{\frac{1}{2}}.$$

For example, $\mathcal{A}$ could be the real Clifford algebra $\mathbb{R}_{(m+1)}$ or the complex Clifford algebra $\mathbb{C}_{(m+1)}$.

The conjugate $\bar{x}$ of $x = \mathbf{x} + x_{m+1} e_{m+1} \in \mathbb{R}^{m+1}$ is then $\bar{x} = -x \in \mathbb{R}^{m+1}$.

In this context, the differential operator $D = \mathbf{D} + \frac{\partial}{\partial x_{m+1}} e_{m+1}$ acts from the left on a C^1 function ϕ defined on an open subset Ω of $\mathbb{R}^{m+1}$ to give

$$D\phi = \sum_{k=1}^{m} e_k \frac{\partial \phi}{\partial x_k} + e_{m+1} \frac{\partial \phi}{\partial x_{m+1}} = \sum_{k=1}^{m+1} e_k \frac{\partial \phi}{\partial x_k}$$

and also from the right to give

$$\phi D = \sum_{k=1}^{m+1} \frac{\partial \phi}{\partial x_k} e_k.$$

If $D\phi = 0$ then ϕ is called *left–monogenic*, while if $\phi D = 0$ then ϕ is *right–monogenic*.

With these conventions, all the theorems in section (E) remain valid.

We can also introduce the analogous sets and function spaces to those which were defined at the end of that section, provided N is a compact set of unit vectors in $\mathbb{R}_+^{m+1} = \{x \in \mathbb{R}^{m+1} : x_{m+1} > 0\}$ which contains e_{m+1}.

The results of section (F) go through unscathed in the case when $\mathbb{R}^{m+1}$ is embedded in $\mathcal{A}$ as specified above, provided e_0 is replaced by e_{m+1} every time it appears. The proofs are the same.

One reason for doing it this way is that the results are now invariant under Euclidean transformations. This does not help much in the general case of kernels $\phi \in K_N^+$, for the conditions on N must then be transformed appropriately, too. But it does mean that the L_p boundedness results for the singular Cauchy integral operator C_Σ and the singular double–layer potential operator $C_{\Sigma 0}$ are more geometrical. They can then be localised easily as follows.

Recall that an open subset Ω of $\mathbb{R}^{m+1}$ is called strongly Lipschitz if its boundary $b\Omega$ is contained in finitely many open sets W such that each set $\Omega \cap W$ is represented by the inequality $X_{m+1} > g(X_1, X_2, \ldots, X_m)$ in some Cartesian coordinate system, where g is a Lipschitz function. (If $b\Omega$ is unbounded, we make the additional assumption that all except one of the sets W are bounded.)

THEOREM 5.7.1
Let Ω be a strongly Lipschitz open subset Ω of $\mathbb{R}^{m+1}$ with boundary $b\Omega$. Let $1 < p < \infty$. The singular Cauchy integral operator $C_{b\Omega}$, defined for all $u \in L_p(b\Omega)$ and almost all $x \in b\Omega$ by

$$(C_{b\Omega} u)(x) = 2 \lim_{\varepsilon \to 0} \int\limits_{\{y \in b\Omega \,:\, |x-y| > \varepsilon\}} k(x-y)\, n(y)\, u(y)\, dS_y$$

is bounded on $L_p(b\Omega)$ with

$$\|C_{b\Omega}\, u\|_p \leq c_p \|u\|_p$$

for all $u \in L_p(b\Omega)$ and some constant c_p which depends only on p and $b\Omega$.

Therefore its scalar part, the singular double–layer potential operator $C_{b\Omega 0}$, defined for all $u \in L_p(b\Omega)$ and almost all $x \in b\Omega$ by

$$(C_{b\Omega 0} u)(x) = \frac{2}{\sigma_m} \lim_{\varepsilon \to 0} \int\limits_{\{y \in b\Omega \,:\, |x-y| > \varepsilon\}} \frac{\langle x - y, \, n(y) \rangle}{|x-y|^{m+1}}\, u(y)\, dS_y$$

is also bounded on $L_p(b\Omega)$ with the same bound.

PROOF   Cover $b\Omega$ with finitely many open sets $W^1, W^2, \ldots, W^M$ which have the properties specified prior to the theorem. Let χ_j be continuous real–valued functions defined on $b\Omega$ such that $0 \leq \chi_j \leq 1$ and $\mathrm{sppt}(\chi_j) \subset W_j$ for each k and $\sum_{j=1}^{M} \chi_j = 1$. Let θ_j be a second collection of continuous real–valued functions on $b\Omega$ with $\mathrm{sppt}(\theta_j) \subset W_j$ for each j, such that $0 \leq \theta_j \leq 1$ and $\mathrm{dist}(\mathrm{sppt}(1-\theta_j), \mathrm{sppt}(\chi_j)) \geq \delta$ for some $\delta > 0$.

Then

$$C_{b\Omega} u(x) = \sum_{j=1}^{M} \theta_j(x)\, C_{b\Omega}(\chi_j u)(x)$$

$$+ \sum_{j=1}^{M} \int\limits_{\mathrm{sppt}(\chi_j)} (1-\theta_j)(x)\, k(x-y)\, n(y)\, \chi_j(y)\, u(y)\, dS_y.$$

Each term in the first sum satisfies

$$\| \theta_j \, C_{b\Omega}(\chi_j u) \|_p \; \leq \; c \| \chi_j u \|_p \; \leq \; c \| u \|_p$$

because, under a Euclidean transformation, $b\Omega \cup W^j$ can be expressed as a piece of the graph of a Lipschitz function g_j. In the terms of the second sum, there are no singularities, so it is easy to verify the L_p estimates. Actually it is trivial if $b\Omega$ is bounded, and not much harder otherwise. (Note that we are assuming that either sppt$(1 - \theta_j)$ or sppt(χ_j) is bounded.) ∎

We remark that this result has been known since the boundedness of the Cauchy integral on a Lipschitz graph in the complex plane was proved in 1981 [CMcM], and it was used by Verchota to study boundary value problems for harmonic functions on strongly Lipschitz domains shortly thereafter [V].

It would be interesting to include the operators $C_{b\Omega}$ in the functional calculus of some operator in $L_p(b\Omega)$. Some results of this nature are contained in the paper by Tao Qian in this volume [Q], for the case when $b\Omega$ is the graph of a radial Lipschitz function. Further cases could probably be treated using the methods of John Ryan in his paper on conformal covariance [R2]. However it is not clear how to handle the general situation.

It would also be interesting to investigate the boundary value problem (B) in this context. The solution presented in section (K) can be expressed as a convolution operator of $v = (P_{+0})^{-1}w$. What is the precise expression in a strongly Lipschitz domain? Can it be expressed directly in terms of some functional calculus?

5.7.2 (M) More about singular convolution integrals on Lipschitz surfaces

Let us return to the convention used in the lectures of identifying $\mathbb{R}^{m+1}$ with the subspace of $\mathcal{A}$ spanned by $\{e_0, e_1, e_2, \ldots, e_m\}$. As before, let N be a compact set of unit vectors in $\mathbb{R}^{m+1}_+$ which is starlike about e_0, contains almost all the unit normals to Σ, and satisfies $\angle(n, e_0) \leq \omega$ for all $n \in N$.

When $0 < \mu < \frac{\pi}{2} - \omega$, we previously defined N_μ and $C^+_{N_\mu}$. Now define $C^-_{N_\mu} = -C^+_{N_\mu}$ and $S_{N_\mu} = C^+_{N_\mu} \cap C^-_{N_\mu}$. Also define the Banach space $K(C^-_{N_\mu})$ in a similar way to $K(C^+_{N_\mu})$.

Define $K(S_{N_\mu})$ to be the Banach space of pairs $(\phi, \underline{\phi})$ of functions with ϕ right–monogenic from S_{N_μ} to $\mathcal{A}$ and with $\underline{\phi}$ a differentiable function on $(0, \infty)$ such that

$$\underline{\phi}'(r) \; = \; \int\limits_{\{\mathbf{x} \in \mathbb{R}^m \,:\, |\mathbf{x}| = r\}} \phi(\mathbf{x}) \, d\mathbf{x}$$

and

$$\| (\phi, \underline{\phi}) \|_{K(S_{N_\mu})} = \frac{\sigma_m}{2} \sup \left\{ |x|^m \| \phi(x) \| : x \in S_{N_\mu} \right\}$$
$$+ \sup \left\{ \| \underline{\phi}(r) \| : r > 0 \right\} < \infty.$$

Note that $\underline{\phi}$ is determined by ϕ up to an additive constant, and that

$$\underline{\phi}(R) - \underline{\phi}(r) = \int_{r \leq |\mathbf{x}| \leq R} \phi(\mathbf{x}) \, d\mathbf{x}$$

when $0 < r < R$. Moreover $\underline{\phi}$ has a unique extension to the open cone $\mathbb{R}^+ N_\mu = \{ rn : r > 0, n \in N_\mu \}$ which satisfies

$$\underline{\phi}(y) - \underline{\phi}(z) = \int_{A(y,z)} \phi(x) \, n(x) \, dS_x$$

where $A(y, z)$ is a smooth oriented m–manifold in $\mathbb{R}^+ N_\mu$ joining the $(m - 1)$–sphere $S_y = \{ x \in \mathbb{R}^{m+1} : \langle x, y \rangle = 0$ and $|x| = |y| \}$ to the $(m - 1)$–sphere S_z, in which case $\| \underline{\phi}(y) \| \leq \| (\phi, \underline{\phi}) \|_{K(S_{N_\mu})}$ for all $y \in \mathbb{R}^+ N_\mu$.

There is an isomorphism

$$\left\{ \begin{array}{rcl} K(C_{N_\mu}^+) \oplus K(C_{N_\mu}^-) & \approx & K(C_{N_\mu}) \\ (\phi_+, \phi_-) & \leftrightarrow & (\phi, \underline{\phi}) \end{array} \right.$$

defined by

$$\phi = \phi_+ + \phi_- \quad \text{and}$$
$$\underline{\phi}(r) = \int_{S_r^+} \phi_+(x) n(x) \, dS_x - \int_{S_r^-} \phi_-(x) n(x) \, dS_x, \quad r > 0,$$

where $S_r^\pm = \{ x \in \mathbb{R}_\pm^{m+1} : |x| = r \}$ and $n(x) = x/r$. See [LMcS] and [LMcQ].

Define K_N^- and K_N in a similar way to K_N^+, so that

$$K_N^+ \oplus K_N^- = K_N.$$

Using this decomposition, it is straightforward to generalise the results in section (F). In particular, we have the following result.

THEOREM 5.7.2
Let $1 < p < \infty$. Given $(\phi, \underline{\phi}) \in K_N$, there is a bounded linear operator $T_{(\phi,\underline{\phi})}$ on $L_p(\Sigma)$, defined for all $u \in L_p(\Sigma)$ and almost all $x \in \Sigma$ by

$$(T_{(\phi,\underline{\phi})} u)(x) = \lim_{\varepsilon \to 0} \left\{ \int_{\{y \in \Sigma : |x-y| \geq \varepsilon\}} \phi(x - y) n(y) u(y) dS_y + \underline{\phi}(\varepsilon n(x)) u(x) \right\}.$$

Moreover, if $(\phi, \underline{\phi}) \in K(S_{N_\mu})$ for $0 < \mu < \frac{1}{2}\pi - \omega$, then

$$\left\| T_{(\phi, \underline{\phi})} u \right\|_p \leq C_{\omega, \mu, p} \left\| \phi \right\|_{K(S_{N_\mu})} \left\| u \right\|_p$$

for some constant $C_{\omega, \mu, p}$ which depends only on ω, μ and p (and the dimension m).

If $(\phi, \underline{\phi})$ corresponds to $(\phi_+, 0)$ (that is, if $\phi_- = 0$), then $T_{(\phi, \underline{\phi})} = T_{\phi_+}$. In particular,

$$P_+ = T_k = T_{(k, \frac{1}{2})} \,, \quad P_- = T_{(-k, \frac{1}{2})} \,, \quad C_\Sigma = T_{(2k, 0)} \,, \quad I = T_{(0, 1)}.$$

If $\phi(x)$ is an odd function which is right–monogenic and satifies $\| \phi(x) \| \leq C |x|^{-m}$ for all $x \in S_{N_\mu}$, then $T_{(\phi, 0)}$ is a principal value convolution operator, which, as we have seen, is bounded on $L_p(\Sigma)$.

The space K_N is not a convolution algebra. But if we define

$$M_N = \{ (\phi, \underline{\phi}) \in K_N \, : \, \phi \text{ is left–monogenic and (*) holds} \}$$

where

$$(*) \qquad \int_{\{ \mathbf{y} \in \mathbb{R}^m \, : \, |\mathbf{y}| = 1 \}} \langle \mathbf{y}, \, \mathbf{x} \rangle \, (\phi(\mathbf{y})\mathbf{y} - \mathbf{y}\phi(\mathbf{y})) \, dS_{\mathbf{y}} = \underline{\phi}(1)\,\mathbf{x} - \mathbf{x}\,\underline{\phi}(1) \,,$$

then $M_N^+ \oplus M_N^- \approx M_N$.

Convolution is defined as follows. Suppose that $(\phi, \underline{\phi}) \in K_N$ and $(\psi, \underline{\psi}) \in M_N$. If $(\phi, \underline{\phi})$ corresponds to $(\phi_+, \phi_-) \in K_N^+ \oplus K_N^-$ and $(\psi, \underline{\psi})$ corresponds to $(\psi_+, \psi_-) \in M_N^+ \oplus M_N^-$, then $(\phi, \underline{\phi}) * (\psi, \underline{\psi})$ corresponds to $(\phi_+ * \psi_+, \phi_- * \psi_-)$. Hence

$$T_{(\phi, \underline{\phi})} \, T_{(\psi, \underline{\psi})} = T_{(\phi, \underline{\phi}) * (\psi, \underline{\psi})}.$$

Let us see how this relates to the material in section (H).

The algebra H_N^- is defined to be the algebra of all those functions b on $\mathbb{R}^m \setminus \{0\}$ which have holomorphic extensions to $b \in H_\infty^-(\overline{N}_\mu(\mathbb{C}^m))$ for some $\mu > 0$, where $\overline{N} = \{ \overline{n} \in \mathbb{R}_+^{m+1} \, : \, n \in N \}$.

Let $H_N = H_N^+ \oplus H_N^-$. Also let A_N^- and A_N be the subspaces of H_N^- and H_N which consist of all functions b which satisfy $b(\xi)\xi = \xi\,b(\xi)$, $\xi \in \mathbb{R}^m$.

Then the Fourier transform $\mathcal{F}_-$ defined by Parseval's identity maps K_N^- onto H_N^- and M_N^- onto A_N^-, and the main theorem in section (H) remains valid once the appropriate changes have been made. Parseval's identity is the same as the one there with $\delta \to 0+$ replaced by $\delta \to 0-$. In both cases, it can be reexpressed as

$$\frac{1}{(2\pi)^m} \int_{\mathbb{R}^m} b(\xi)\hat{u}(-\xi)d\xi = \lim_{\varepsilon \to 0} \left\{ \int_{|\mathbf{x}| \geq \varepsilon} \phi(\mathbf{x})u(\mathbf{x})d\mathbf{x} + \underline{\phi}(\varepsilon)u(0) \right\}$$

for all u in the Schwartz space $\mathcal{S}(\mathbb{R}^m)$, where $b = \mathcal{F}_\pm(\phi)$.

It is now an easy matter to use this form of Parseval's identity to define the Fourier transform $\mathcal{F}$ from K_N onto H_N, and to adapt the main theorem of section (H) to this case. Under the isomorphism

$$\begin{cases} K_n^+ \oplus K_N^- & \approx K_N \\ (\ \phi_+\ ,\ \phi_-\ ) & \leftrightarrow (\phi, \underline{\phi}) \end{cases}$$

we have

$$b = \mathcal{F}(\phi, \underline{\phi}) = \mathcal{F}_+(\phi_+) + \mathcal{F}_-(\phi_-) = b\chi_+ + b\chi_-$$

and

$$b(\frac{1}{i}\mathbf{D}_\Sigma) = T_{(\phi, \underline{\phi})} = T_{\phi_+} + T_{\phi_-} = (b\chi_+)(\frac{1}{i}\mathbf{D}_\Sigma) + (b\chi_-)(\frac{1}{i}\mathbf{D}_\Sigma).$$

This time $\frac{1}{i}\mathbf{D}_\Sigma$ has an H_N functional calculus in $L_p(\Sigma)$ in the sense that (i) the linear mapping

$$\begin{array}{ccccc} H_N & \to & K_N & \to & \mathcal{L}(L_p(\Sigma)) \\ b & \mapsto & (\phi, \underline{\phi}) & \mapsto & T_{(\phi, \underline{\phi})} = b(\frac{1}{i}\mathbf{D}_\Sigma) \end{array}$$

has the property that if $f \in A_N$ corresponds to $(\psi, \underline{\psi}) \in M_N$, then $bf \in H_N$ corresponds to $(\phi, \underline{\phi}) * (\psi, \underline{\psi}) \in K_N$ with associated operator

$$(bf)(\frac{1}{i}\mathbf{D}_\Sigma) = T_{(\phi, \underline{\phi})*(\psi, \underline{\psi})} = T_{(\phi, \underline{\phi})} T_{(\psi, \underline{\psi})} = b(\frac{1}{i}\mathbf{D}_\Sigma) f(\frac{1}{i}\mathbf{D}_\Sigma) ,$$

and (ii) there is agreement with the natural definition for polynomials of several variables.

A full treatment of this material is developed in my paper with Chun Li and Tao Qian [LMcQ].

Let me summarise what we have done. Our aim is to characterise the Fourier transforms of those right–monogenic functions ϕ which arise naturally as convolution kernels of singular convolution integrals on Lipschitz surfaces Σ in $\mathbb{R}^m$. We have succeeded in characterising them by those functions b on $\mathbb{R}^m$ which have holomorphic extensions to an appropriate cone in $\mathbb{C}^{m+1}$. This informs us that there is an interesting duality between monogenic functions of $m+1$ real variables and holomorphic functions of m complex variables.

Using this Fourier transform, we have represented the singular integrals T_ϕ with convolution kernel ϕ in terms of operators $b(\frac{1}{i}\mathbf{D}_\Sigma)$ in the functional calculus of a Dirac operator $\mathbf{D}_\Sigma$ on the surface Σ.

Bibliography

[AT]   P. Auscher and Ph. Tchamitchian, *Bases d'ondelettes sur les courbes corde-arc, noyau de Cauchy et espaces de Hardy associés*, Revista Matemática Iberoamericana **5**, (1989), 139–170.

[BDS]   F. Brackx, R. Delanghe and F. Sommen, *Clifford Analysis*, Research Notes in Mathematics **76**, Pitman Advanced Publishing Program, 1982.

[Ca]   A.P. Calderón, *Cauchy integrals on Lipschitz curves and related operators*, Proc. Natl. Acad. Sci. U.S.A. **74**, (1977), 1324–1327.

[C]   W.K. Clifford, *Applications of Grassman's extensive algebra*, American Journal of Mathematics **1**, (1878), 350–358.

[CJS]   R.R. Coifman, P. Jones and S. Semmes, *Two elementary proofs of the L^2 boundedness of Cauchy integrals on Lipschitz curves*, Journal of the American Mathematical Society **2**, (1989), 553–564.

[CM]   R.R. Coifman and Y. Meyer, *Au-delà des opérateurs pseudo-différentiels*, Astérisque **57**, Société Mathématique de France, 1978.

[CM1]   R.R. Coifman and Y. Meyer, *Fourier analysis of multilinear convolutions, Calderón's theorem, and analysis on Lipschitz curves*, Lecture Notes in Mathematics, Springer-Verlag **779**, (1980), 104–122.

[CMcM]   R.R. Coifman, A. McIntosh and Y. Meyer, *L'intégrale de Cauchy définit un opérateur borné sur L^2 pour les courbes lipschitziennes*, Annals of Mathematics **116** (1982), 361–387.

[D1]   B.E.J. Dahlberg, *Estimates of harmonic measure*, Arch. Rat. Mech. Anal. **65**, (1977), 275–288.

[D2]   B.E.J. Dahlberg, *Poisson semigroups and singular integrals*, Proceedings of the American Mathematical Society **97**, (1986), 41–48.

[Dix]   A.C. Dixon, *On the Newtonian Potential* Quarterly Journal of Mathematics **35**, (1904), 283–296.

[DJS]   G. David, J.-L. Journé and S. Semmes, *Opérateurs de Calderón-Zygmund, fonctions para–accrétives et interpolation*, Revista Matemática Iberoamericana **1**, (1985), 1–57.

[DSS]   R. Delanghe, F. Sommen and V. Souček, *Clifford Algebra and Spinor–Valued Functions*, Kluwer, Academic, Norwell, MA, 1992.

[GLQ]   G.I. Gaudry, R.L. Long and T. Qian, *A martingale proof of L^2–boundedness of Clifford–valued singular integrals*, Annali di Matematica, Pura Appl. **165**, (1993), 369–394.

[GM]   J. Gilbert and M. Murray, *Clifford Algebras and Dirac Operators in Harmonic Analysis*, Cambridge, University Press, Cambridge, 1991.

[GS]   K. Gürlebeck and W. Sprössig, *Quaternionic Analysis and Elliptic Boundary Value Problems*, Birkhäuser Verlag, Basel, 1990.

[I]   V. Iftimie, *Fonctions hypercomplexes*, Bull. Math. de la Soc. Sci. Math. de la R. S. de Roumanie **9**, (1965), 279–332.

[JK]   D.S. Jerison and C.E. Kenig, *The Neumann problem on Lipschitz domains*, Bulletin of the American Mathematical Society **4**, (1981), 203–207.

[K]   Tosio Kato, *Perturbation Theory for Linear Operators, 2nd ed.*, Springer–Verlag, New York, 1976.

[Ke]   Carlos Kenig, *Harmonic Analysis Techniques for Second Order Elliptic Boundary Value Problems, C.B.M.S.*, American Mathematical Society **83**, 1994.

[LM]   H. Blaine Lawson Jr and Marie-Louise Michelsohn *Spin Geometry*, Princeton University Press, Princeton, NJ, 1989.

[LMc]   Chun Li and Alan McIntosh, *Clifford algebras and H_∞ functional calculi of commuting operators*, in *Clifford Analysis*, (John Ryan, ed.), CRC Press, Boca Raton, FL, 1995.

[LMcQ] C. Li, A. McIntosh and T. Qian, *Clifford algebras, Fourier transforms, and singular convolution operators on Lipschitz surfaces* Revista Matemática Iberoamericana, (to appear), 10, (1994), 665–721.

[LMcS] C. Li, A. McIntosh and S. Semmes *Convolution singular integrals on Lipschitz surfaces*, Journal of the American Mathematical Society **5**, (1992), 455–481.

[M]   Yves Meyer (vol. III with R.R. Coifman), *Ondelettes et Opérateurs, I, II, III*, Hermann, Paris editeurs des sciences et des arts, 1990.

[M1]   Marius Mitrea, *Clifford Wavelets, Singular Integrals, and Hardy Spaces*, Lecture Notes in Mathematics, Springer-Verlag, New York, 1994, 1575.

[M2]   Marius Mitrea, *Hypercomplex variable techniques in harmonic analysis*, in *Clifford Analysis*, (John Ryan, ed.), CRC Press, Boca Raton, FL, 1995.

[Mc]   Alan McIntosh, *Clifford algebras and the higher dimensional Cauchy integral*, Approximation Theory and Function Spaces, Banach Center Publications **22**, (1989), 253–267.

[McM]   Alan McIntosh and Yves Meyer, *Algèbres d'opérateurs définis par des intégrales singulières*, Comptes Rendus Acad. Sci., Paris, Sér. I, Math **301**, (1985), 395–397.

[McP]   A. McIntosh and A. Pryde, *A functional calculus for several commuting operators*, Indiana Univ. Math. Journal **36**, (1987), 421–439.

[McPR] A. McIntosh, A. Pryde and W. Ricker , *Comparison of joint spectra for certain classes of commuting operators*, Studia Math. **88**, (1988), 23–36.

[McQ]   A. McIntosh and T. Qian , *Convolution singular integral operators on Lipschitz curves*, Proceedings of the Special Year on Harmonic Analysis at Nankai Institute of Mathematics, Tianjin, China Lecture Notes in Mathematics, Springer-Verlag, New York, **1494**, (1991), 142–162.

[Mu]   Margaret Murray, *The Cauchy integral, Calderón commutators and conjugations of singular integrals in* $\mathbb{R}^m$, Transactions of the American Mathematical Society **289**, (1985), 497–518.

[N]   J. Nečas, *Les Méthodes Directes en Théorie des Equations Elliptiques*, Masson, Paris and Academia, Prague, 1967.

[PS]   J. Peetre and P. Sjölin, *Three-line theorems and Clifford analysis*, Complex Variables **19**, (1992), 151-163.

[Q]   Tao Qian, *Singular integrals with monogenic kernels on the m−torus and its Lipschitz perturbations*, in *Clifford Analysis*, (John Ryan, ed.), CRC Press, Boca Raton, FL, 1995.

[R1]   John Ryan, *Plemelj formulae and transformations associated to plane wave decomposition in complex Clifford analysis*, Proc. London Math. Soc. **64**, (1992), 70-94.

[R2]   John Ryan, *Conformal covariance in Clifford analysis*. To appear in Zeutschrift für Analysis und die Anwendugen.

[S]   Stephen Semmes, *Square function estimates and the T(b) theorem*, Proceedings of the American Mathematical Society **110**, (1990), 721-726.

[S1]   Fraciscus Sommen, *Plane waves, biregular functions and hypercomplex Fourier analysis*, Supplemento ai Rendiconti del Circolo Matematico di Palermo, Serie II, **9**, (1985), 205–219.

[S2]   Fraciscus Sommen, *Microfunctions with values in a Clifford algebra II*, Scientific Papers of the College of Arts and Sciences, The University of Tokyo **36**, (1986), 15–37.

[SW]   E.M. Stein and G. Weiss, *On the theory of harmonic functions of several variables, I. The theory of H^p spaces*, Acta Math.**103**,

(1960), 25–62.

[St]   E.M. Stein, *Singular Integrals and Differentiability Properties of Functions*, P.U.P., Princeton, NJ, 1970.

[T]   Terry Tao, *Convolution operators generated by right–monogenic and harmonic kernels*, Masters thesis, Flinders University, Adelaide, South Australia, 1992.

[V]   Greg Verchota, *Layer potentials and regularity for the Dirichlet problem for Laplace's equation in Lipschitz domains*, Journal of Functional Analysis **59**, (1984), 572–611.

School of Mathematics, Physics, Computing and Electronics,
Macquarie University,
North Ryde, New South Wales 2109,
AUSTRALIA
E-mail: alan@macadam.mpce.mq.edu.au

6

Clifford Algebras and H_∞ Functional Calculi of Commuting Operators

Chun Li and Alan McIntosh*

6.1 Introduction

We wish to thank John Ryan and his colleagues at the University of Arkansas who have organised an excellent conference.

It is possible to use Clifford algebras to study the spectral theory of commuting n–tuples of operators in a Banach space. See, for example, [McP] and [McPR] for some results about operators with real spectrum.

In this paper we consider operators of type ω in a Hilbert space. A single operator of type ω has a bounded H_∞ functional calculus if and only if T and its adjoint T^* satisfy square function estimates. Our aim is to generalise this result to the case of n–tuples $(T_1, T_2, \ldots, T_n)$ of operators.

First we define what it means for $(T_1, T_2, \ldots, T_n)$ to be of type ω in terms of whether it generates a monogenic semigroup in an appropriate sense. We then show that $(T_1, T_2, \ldots, T_n)$ has an H_∞ functional calculus on an appropriate cone in $\mathbb{C}^n$ if the single operator $iT = i\sum_{j=1}^{n} T_j e_j$ has an H_∞ functional calculus in the Cliffordised Hilbert space $\mathcal{H}_{(n)}$, or equivalently, if iT and its adjoint satisfy square function estimates.

6.2 The H_∞ functional calculus of a single operator

Throughout, $\mathcal{H}$ denotes a complex Hilbert space. By an *operator* in $\mathcal{H}$ we mean a linear mapping $T : \mathcal{D}(T) \to \mathcal{H}$ whose *domain* $\mathcal{D}(T)$ is a linear subspace of $\mathcal{H}$.

Let $0 \le \omega < \dfrac{\pi}{2}$. Let us say that an operator T in $\mathcal{H}$ is of *type* ω_+ if its

---

*The authors were supported by the Australian Government through the Australian Research Council.

1991 *Mathematics Subject Classification.* 47A60, 42B10

spectrum

$$\sigma(T) \subset S_{\omega+}(\mathbb{C}) = \{\, z \in \mathbb{C} \mid \arg z \mid \le \omega \,\} \cup \{\, 0 \,\}$$

and, for each $\mu > \omega$,

$$\| (T - zI)^{-1} \| \le C_\mu |z|^{-1}, \quad z \notin S_{\mu+}(\mathbb{C}).$$

This is equivalent [K] to the statement that $-T$ generates a bounded holomorphic semigroup e^{-zT} for

$$z \in S^0_{(\frac{\pi}{2}-\mu)+}(\mathbb{C}) = \{\, z \in \mathbb{C} : z \ne 0, \mid \arg z \mid < \frac{\pi}{2} - \mu \,\}.$$

Assume that T is one–one. We remark that every one–one operator of type ω_+ in $\mathcal{H}$ is a closed operator with dense domain and dense range [CDMcY].

Of concern to us is the question whether T has a bounded H_∞ functional calculus, that is whether there is a definition of $f(T)$ as a bounded linear operator on $\mathcal{H}$ for all $f \in H_\infty(S^0_{\mu+})$ such that $\|f(T)\| \le c_\mu \|f\|_\infty$. This definition should be natural in the sense that the mapping $f \mapsto f(T)$ is an algebra homomorphism which agrees with the usual definition of polynomials of T and resolvent operators.

Here $H_\infty(S^0_{\mu+})$ denotes the Banach space of bounded holomorphic functions defined on $S^0_{\mu+}(\mathbb{C})$ for $\mu > \omega$. Actually the above definition is independent of the choice of μ. This is proved in [Mc1], along with the following result.

THEOREM 6.2.1
Suppose that T is a one–one operator of type ω_+ in $\mathcal{H}$. Then T has a bounded H_∞ functional calculus if and only if T and its adjoint T^* satisfy the square function estimates

$$\int_0^\infty \| \psi_t(T) u \|^2 \, \frac{dt}{t} \le c \| u \|^2$$

$$\int_0^\infty \| \psi_t(T^*) u \|^2 \, \frac{dt}{t} \le c \| u \|^2$$

for all $u \in \mathcal{H}$ and for some function $\psi \in H_\infty(S^0_{\mu+})$ which satisfies

$$\int_0^\infty \psi^3(t) \, \frac{dt}{t} = 1 \quad and \quad |\psi(z)| \le C \, \frac{|z|^s}{1 + |z|^{2s}}$$

for some $s > 0$ and all $z \in S^0_{\mu+}(\mathbb{C})$, $\mu > \omega$.

Here and elsewhere, ψ_t is defined by $\psi_t(z) = \psi(tz)$.

There are results about holomorphic functional calculi of such operators and related quadratic estimates in the papers [Mc1, McY].

Let us also define the closed sectors and closed double sectors

$$S_{\omega-}(\mathbb{C}) = -S_{\omega+}(\mathbb{C}) \quad \text{and} \quad S_\omega(\mathbb{C}) = S_{\omega+}(\mathbb{C}) \cup S_{\omega-}(\mathbb{C}),$$

and the open sectors and open double sectors

$$S^0_{\mu-}(\mathbb{C}) = -S^0_{\mu+}(\mathbb{C}) \quad \text{and} \quad S^0_\mu(\mathbb{C}) = S^0_{\mu+}(\mathbb{C}) \cup S^0_{\mu-}(\mathbb{C}).$$

We now say that an operator T in $\mathcal{H}$ is of *type* ω if $\sigma(T) \subset S_\omega(\mathbb{C})$ and, for each $\mu > \omega$,

$$\left\| (T - zI)^{-1} \right\| \leq C_\mu |z|^{-1}, \quad z \notin S_\mu(\mathbb{C}).$$

The same technique that is used to prove theorem 2.1 now gives the following result.

THEOREM 6.2.2
Suppose that T is a one–one operator of type ω in $\mathcal{H}$. Then T has a bounded H_∞ functional calculus if and only if T and its adjoint T^* satisfy the square function estimates

$$\int_0^\infty \| \psi_t(T)u \|^2 \, \frac{dt}{t} \leq c \| u \|^2$$

$$\int_0^\infty \| \psi_t(T^*)u \|^2 \, \frac{dt}{t} \leq c \| u \|^2$$

for all $u \in \mathcal{H}$ and for some function $\psi \in H_\infty(S^0_\mu)$ which satisfies

$$\int_0^\infty \psi^3(-t) \, \frac{dt}{t} = \int_0^\infty \psi^3(t) \, \frac{dt}{t} = 1 \quad \text{and} \quad |\psi(z)| \leq C \, \frac{|z|^s}{1 + |z|^{2s}}$$

for some $s > 0$ and all $z \in S^0_\mu(\mathbb{C})$, $\mu > \omega$.

Define

$$\chi_+(z) = \begin{cases} 1, & Re\, z > 0 \\ 0, & Re\, z < 0 \end{cases}$$

and let $\chi_-(z) = 1 - \chi_+(z)$.

We remark that if T is a one–one operator of *type* ω with a bounded H_∞ functional calculus, then $E_\pm = \chi_\pm(T)$ are bounded projections associated with the parts of the spectrum in $S_{\omega+}(\mathbb{C})$ and $S_{\omega-}(\mathbb{C})$. They give rise to the spectral decomposition

$$\begin{cases} \mathcal{H} &= \mathcal{H}_+ \oplus \mathcal{H}_- \\ u &= E_+ u + E_- u \end{cases}$$

where $\mathcal{H}_\pm = E_\pm(\mathcal{H})$.

6.3   Fourier transform between holomorphic functions and monogenic functions

For a bounded measurable function b defined on $\mathbb{R}^n$, the measurable function ϕ is called its (generalised) inverse Fourier transform if it satisfies, for any $u \in \mathcal{S}(\mathbb{R}^n)$,

$$\frac{1}{(2\pi)^n} \int_{\mathbb{R}^n} b(\xi)\hat{u}(-\xi)\, d\xi = \int_{\mathbb{R}^n} \phi(x)u(x)\, dx,$$

where $\hat{u}$ is the classical Fourier transform of u.

For $\zeta = \xi + i\eta \in \mathbb{C}^n$, denote

$$|\zeta|_*^2 = |\xi|^2 - |\eta|^2 + 2i(\xi, \eta).$$

If $|\zeta|_*^2 \notin (-\infty, 0]$, let $|\zeta|_*$ denote its square root with positive real part. For $\mu > 0$, define

$$S_\mu^0(\mathbb{C}^n) = \{\, \zeta = \xi + i\eta \in \mathbb{C}^n \mid |\zeta|_*^2 \notin (-\infty, 0]\,,\ |\eta| < Re|\zeta|_* \tan\mu \,\}.$$

It is easy to verify that if $\zeta \in S_\mu^0(\mathbb{C}^n)$, then $|\zeta|_* \in S_{\mu+}^0(\mathbb{C})$, and $-|\zeta|_* \in S_{\mu-}^0(\mathbb{C})$.

Let $H_\infty(S_\mu^0(\mathbb{C}^n))$ denote the Banach space of bounded holomorphic functions of n–variables defined on the open cone $S_\mu^0(\mathbb{C}^n)$. We shall show that, for a function in $H_\infty(S_\mu^0(\mathbb{C}^n))$, the inverse Fourier transform of its restriction to $\mathbb{R}^n$ can be extended to a monogenic function on a cone $S_\mu^0(\mathbb{R}^{n+1})$ containing $\mathbb{R}^n$.

We first define the real and complex 2^n–dimensional Clifford algebras $\mathbb{R}_{(n)}$ and $\mathbb{C}_{(n)}$ as follows.

For each subset S of $\{\, 1, 2, \ldots, n \,\}$, let $\{\, e_S \,\}$ denote a basis element of the vector space $\mathbb{R}_{(n)}$ (or $\mathbb{C}_{(n)}$). The algebraic structure is defined by

$$\begin{aligned}
e_0 &= e_\varphi = 1 \quad \text{the identity} \\
e_j &= e_{\{j\}}, \quad \text{and} \quad e_j^2 = -1 \quad \text{if } 1 \le j \le n \\
e_i e_j &= -e_j e_i = e_{\{i,j\}} \quad \text{for } 1 \le i < j \le n \\
e_{j_1} e_{j_2} \cdots e_{j_s} &= e_{\{j_1, j_2, \ldots, j_s\}} \quad \text{if } 1 \le j_1 < j_2 < \ldots < j_s \le n \,.
\end{aligned}$$

For $u \in \mathbb{R}_{(n)}$ (or $\mathbb{C}_{(n)}$), define $|u| = \sqrt{\sum_S |u_S|^2}$. Then $|uv| \le C|u|\,|v|$ for some constant C. We embed the vector space $\mathbb{R}^{n+1}$ ($\mathbb{C}^{n+1}$) into $\mathbb{R}_{(n)}$ ($\mathbb{C}_{(n)}$) by identifying the standard basis $e_0, e_1, \ldots, e_n$ of $\mathbb{R}^n$ ($\mathbb{C}^n$) with its counterpart in $\mathbb{R}_{(n)}$ ($\mathbb{C}_{(n)}$). So from now on we denote $x = (x_0, x_1, \ldots, x_n) \in \mathbb{R}^{n+1}$ ($\mathbb{C}^{n+1}$) by $x = x_0 + x_1 e_1 + \ldots + x_n e_n$. On letting $\bar{x} = x_0 - x_1 e_1 - \ldots - x_n e_n$ denote the Clifford conjugate of x, we have $\bar{x}x = x\bar{x} = |x|^2$.

The Dirac operator $D = \dfrac{\partial}{\partial x_0} e_0 + \dfrac{\partial}{\partial x_1} e_1 + \ldots + \dfrac{\partial}{\partial x_n} e_n$ acts on Clifford–

valued functions ϕ of $n + 1$ variables in two ways:

$$(D\phi)(x) = \sum_{j=0}^{n} e_j \frac{\partial \phi}{\partial x_j}(x)$$
$$(\phi D)(x) = \sum_{j=0}^{n} \frac{\partial \phi}{\partial x_j}(x) e_j$$

Since Clifford algebras are noncommutative when $n \geq 2$, $D\phi$ is usually different from ϕD.

We say that a function ϕ is *left–monogenic* (or *right–monogenic*) on a domain Ω if $(D\phi)(x) = 0$ (or $(\phi D)(x) = 0$) on Ω.

Let $k(x) = \dfrac{1}{\sigma_n} \dfrac{\bar{x}}{|x|^{n+1}}$, where σ_n is the volume of the unit n–sphere in $\mathbb{R}^{n+1}$. Then k is both left– and right–monogenic for $x \neq 0$. The following theorem is the higher-dimensional version of the Cauchy Integral Formula. For the proof, please see [BDS].

THEOREM 6.3.1
Suppose that Ω is a bounded subset of $\mathbb{R}^{n+1}$ with piecewise smooth boundary, and that ϕ is a right–monogenic function on a neighbourhood of Ω. Then, for all $x \in \Omega$,

$$\phi(x) = \int_{\partial\Omega} \phi(y) n(y) k(y - x) \, dS_y$$

where $n(y)$ is the outward-pointing unit normal at $y \in \partial\Omega$.

For $0 < \mu < \pi/2$, let

$$C_\mu^+(\mathbb{R}^{n+1}) = \{\, x = x_0 + \mathbf{x} \in \mathbb{R}^{n+1} : x_0 > -\tan\mu \, |\mathbf{x}| \,\}$$
$$C_\mu^-(\mathbb{R}^{n+1}) = -C_\mu^+(\mathbb{R}^{n+1})$$

and

$$S_\mu^0(\mathbb{R}^{n+1}) = \{\, x = x_0 + \mathbf{x} \in \mathbb{R}^{n+1} : |x_0| < \tan\mu \, |\mathbf{x}| \,\}$$
$$= C_\mu^+(\mathbb{R}^{n+1}) \cap C_\mu^-(\mathbb{R}^{n+1}).$$

The following result is due to Li, McIntosh and Qian [LMcQ]. Also see the lectures of McIntosh in this volume [Mc2].

THEOREM 6.3.2
Suppose that $b \in H_\infty(S_\mu^0(\mathbb{C}^n))$ and that

$$|b(\zeta)| \leq C \frac{|\zeta|^s}{1 + |\zeta|^{2s}}$$

for some $s \in (0, n)$. Then the inverse Fourier transform of its restriction to $\mathbb{R}^n$ exists and can be extended to a right–monogenic function ϕ on

$S_\mu^0(\mathbb{R}^{n+1})$, *which satisfies the estimate*

$$|\phi(x)| \le C_\nu \frac{|x|^s}{|x|^n(1 + |x|^{2s})}$$

for $x \in S_\nu^0(\mathbb{R}^{n+1})$ and $\nu < \mu$.

PROOF   For $x \in \mathbb{R}^{n+1}$, $\zeta \in \mathbb{C}^n$, let

$$e(x, \zeta) = e^{\{i \sum_{j=1}^n x_j \zeta_j - ix_0 \sum_{j=1}^n \zeta_j e_j\}} = e^{i(\mathbf{X},\zeta) - ix_0 \zeta}.$$

Then $e(x, \zeta)$ is monogenic in x and holomorphic in ζ. Let

$$\chi_\pm(\zeta) = \frac{1}{2}(1 \pm \frac{i \sum_{j=1}^n \zeta_j e_j}{|\zeta|_*}) \qquad \text{for all } \zeta \text{ such that } |\zeta|_*^2 \notin (-\infty, 0].$$

Then $\chi_\pm(\zeta)$ are bounded holomorphic functions on $S_\mu^0(\mathbb{C}^n)$ for all $\mu < \pi/2$. Furthermore,

$$\begin{aligned}
\chi_\pm{}^2(\zeta) &= \chi_\pm(\zeta), \\
\chi_+(\zeta) + \chi_-(\zeta) &= 1 \\
\chi_-(\zeta)\chi_+(\zeta) &= \chi_+(\zeta)\chi_-(\zeta) = 0 \\
i\zeta\chi_+(\zeta) &= |\zeta|_*\chi_+(\zeta), \quad i\zeta\chi_-(\zeta) = -|\zeta|_*\chi_-(\zeta).
\end{aligned}$$

For $x \in C_\mu^+(\mathbb{R}^{n+1})$, i.e., $x_0 > -\tan\mu|\mathbf{x}|$, choose a unit vector $v \in \mathbb{R}^{n+1}$ such that $v_0 > \cos\mu$, $(x, v) > 0$. Let

$$v(\mathbb{C}^n) = \{\zeta = \xi + i\eta \in \mathbb{C}^n : |\zeta|_*^2 \notin (-\infty, 0], \eta + Re|\zeta|_* = \kappa v \text{ for some } \kappa > 0\}.$$

Then $v(\mathbb{C}^n) \subset S_\mu^0(\mathbb{C}^n)$. Define

$$\phi_+(x) = \frac{1}{(2\pi)^n} \int_{v(\mathbb{C}^n)} b(\zeta)\chi_+(\zeta)e(x, \zeta)\, d\zeta.$$

The convergence of the integral follows from the estimate

$$|\chi_+(\zeta)e(x, \zeta)| \le B_\mu e^{-(\mathbf{X},\eta) - x_0 Re|\zeta|_*} \le B_\mu e^{-\rho(x,v)|\zeta|}$$

for some positive numbers s, ρ and B_μ. In fact, we have

$$|\phi_+(x)| \le B_\mu \frac{(x, v)^s}{(x, v)^n(1 + (x, v)^{2s})}.$$

By the holomorphic property of b, the value of $\phi_+(x)$ is independent of the choice of v. Also ϕ is right–monogenic. For $\nu < \mu$ and $x \in C_\nu^+(\mathbb{R}^{n+1})$, v can be chosen so that $(x, v) \ge B_\nu|x|$. Thus, we have

$$|\phi_+(x)| \le B_{\mu,\nu} \frac{|x|^s}{|x|^n(1 + |x|^{2s})}, \qquad x \in C_\nu^+(\mathbb{R}^{n+1}).$$

Similarly, for $x \in C_\mu^-(\mathbb{R}^{n+1})$, i.e., $x_0 < \tan \mu |\mathbf{x}|$, choose a unit vector $v \in \mathbb{R}^{n+1}$ such that $v_0 > \cos \mu$, $(x, v) < 0$. Let

$$\bar{v}(\mathbb{C}^n) = \{\varsigma = \xi + i\eta \in \mathbb{C}^n : |\varsigma|_*^2 \notin (-\infty, 0], \ -\eta + Re|\varsigma|_*$$
$$= \kappa v \ \text{ for some } \ \kappa > 0\}.$$

Then $\bar{v}(\mathbb{C}^n) \subset S_\mu^0(\mathbb{C}^n)$. Define

$$\phi_-(x) = \frac{1}{(2\pi)^n} \int_{\bar{v}(\mathbb{C}^n)} b(\varsigma) \chi_-(\varsigma) e(x, \varsigma) d\varsigma.$$

The convergence of the integral follows from the estimate

$$|\chi_-(\varsigma) e(x, \varsigma)| \le B_\mu e^{-(\mathbf{X}, \eta) + x_0 Re|\varsigma|_*} \le B_\mu e^{-e(x, v)|\varsigma|}$$

and we have

$$|\phi_-(x)| \le B_\mu \frac{|(x, v)|^s}{|(x, v)|^n (1 + |(x, v)|^{2s})}.$$

The definition is also independent of the choice of v, and ϕ_- is right–monogenic.

Let $\phi = \phi_+ + \phi_-$. Then ϕ is right–monogenic on $S_\mu^0(\mathbb{R}^{n+1})$, and satisfies the estimate

$$|\phi(x)| \le B_{\mu, \nu} \frac{|x|^s}{|x|^n (1 + |x|^{2s})} \qquad \text{for } \nu < \mu \text{ and } x \in S_\nu^0(\mathbb{R}^{n+1}).$$

Now we need to prove that the restriction of ϕ on $\mathbb{R}^n$ is the inverse Fourier transform of b.

For $u \in S(\mathbb{R}^n)$, we have

$$\int_{\mathbb{R}^n} \phi(x) u(x) \, dx = \int_{\mathbb{R}^n} \phi_+(x) u(x) \, dx + \int_{\mathbb{R}^n} \phi_-(x) u(x) \, dx.$$

Define u_+, u_- by $\hat{u}_\pm(\xi) = \chi_\pm(\xi) \hat{u}(\xi)$. Then $u = u_+ + u_-$ and u_+ can be extended left–monogenically to $\mathbb{R}_+^{n+1}$ and u_- to $\mathbb{R}_-^{n+1}$ by

$$u_\pm(x) = \frac{1}{(2\pi)^n} \int_{\mathbb{R}^n} e(x, \xi) \chi_\pm(\xi) \hat{u}(\xi) d\xi.$$

Then by using the Cauchy Formula, and considering the decay of ϕ at zero and infinity,

$$\int_{\mathbb{R}^n} \phi_+(x) u_+(x) \, dx = \lim_{\delta \to 0} \int_{\mathbb{R}^n} \delta e_0 + x) u(x) \, dx.$$

For $x \in \mathbb{R}^n$,

$$\phi_+(\delta e_0 + x) = \frac{1}{(2\pi)^n} \int_{\mathbb{R}^n} b(\xi) \chi_+(\xi) e(x, \xi) e^{-\delta |\xi|} \, d\xi.$$

By Fubini's theorem, and letting δ tend to 0,

$$\int_{\mathbb{R}^n} \phi_+(x)u(x)\, dx = \frac{1}{(2\pi)^n} \int_{\mathbb{R}^n} b(\xi)\chi_+(\xi)\hat{u}_-(-\xi)d\xi$$

$$= \frac{1}{(2\pi)^n} \int_{\mathbb{R}^n} b(\xi)\hat{u}_-(-\xi)d\xi.$$

That is,

$$\int_{\mathbb{R}^n} \phi_+(x)u(x)\, dx = \frac{1}{(2\pi)^n} \int_{\mathbb{R}^n} b(\xi)\hat{u}_-(-\xi)d\xi.$$

Similarly, we have

$$\int_{\mathbb{R}^n} \phi_-(x)u(x)\, dx = \frac{1}{(2\pi)^n} \int_{\mathbb{R}^n} b(\xi)\hat{u}_+(-\xi)d\xi.$$

These give us

$$\int_{\mathbb{R}^n} \phi(x)u(x)\, dx = \frac{1}{(2\pi)^n} \int_{\mathbb{R}^n} b(\xi)\hat{u}(-\xi)d\xi,$$

which proves that ϕ is the inverse Fourier transform of b. ∎

Let $H_\theta = \{\, x = x_0 + \mathbf{x} \in \mathbb{R}_+^{n+1} : x_0/|\mathbf{x}| = \tan\theta \,\}$ for some $0 < \theta < \mu$.

We have proved that ϕ is the inverse Fourier transform of the restriction of b to $\mathbb{R}^n$. Since $\phi \in L^1(\mathbb{R}^n)$, we in fact have,

$$
\begin{aligned}
b(\xi) &= \int_{\mathbb{R}^n} \phi(x)e(x, -\xi)\, dx \\
&= \int_{\mathbb{R}^n} \phi_+(x)e(x, -\xi)\, dx + \int_{\mathbb{R}^n} \phi_-(x)e(x, -\xi)\, dx \\
&= \int_{\mathbb{R}^n} \phi_+(x)e(x, -\xi)\chi_+(\xi)\, dx + \int_{\mathbb{R}^n} \phi_-(x)e(x, -\xi)\chi_-(\xi)\, dx \\
&= \int_{-H_\theta} \phi_+(x)n(x)e(x, -\xi)\chi_+(\xi)\, dx \\
&\quad + \int_{H_\theta} \phi_-(x)n(x)e(x, -\xi)\chi_-(\xi)\, dx \qquad \text{by Cauchy's Formula.}
\end{aligned}
$$

Therefore for $\zeta \in S_\theta(\mathbb{C}^n)$,

$$
\begin{aligned}
b(\zeta) &= \int_{-H_\theta} \phi_+(x)n(x)e(x, -\zeta)\chi_+(\zeta)\, dx \\
&\quad + \int_{H_\theta} \phi_-(x)n(x)e(x, -\zeta)\chi_-(\zeta)\, dx
\end{aligned}
$$

since both sides are holomorphic on $S_\theta(\mathbb{C}^n)$, and coincide on $\mathbb{R}^n$.

6.4 H_∞ functional calculi of n–tuples of commuting operators

Suppose that $(T_1, T_2, \ldots, T_n)$ is an n-tuple of commuting operators on a Hilbert space $\mathcal{H}$, and $\cap_{j=1}^n \mathcal{D}(T_j)$ is dense in $\mathcal{H}$. Let

$$\mathcal{H}_{(n)} = \{\, u = \sum_S u_S e_S, \quad u_S \in \mathcal{H} \,\}$$

be the Cliffordised Hilbert space of $\mathcal{H}$. Then $T = T_1 e_1 + T_2 e_2 + \ldots + T_n e_n$ is the densely defined operator on $\mathcal{H}_{(n)}$ defined by

$$T(\sum_S u_S e_S) = \sum_{k=1}^n \sum_S T_k u_S e_k e_S.$$

We know that if iT is of type ω, and satisfies square function estimates, then iT has a bounded H_∞ functional calculus on the sector $S_\mu^0(\mathbb{C})$ for all $\mu > \omega$. In this case, what we are interested in is under what conditions $(T_1, T_2, \ldots, T_n)$ has a bounded H_∞ functional calculus on some cone in $\mathbb{C}^n$ containing the joint spectrum of $(T_1, T_2, \ldots, T_n)$.

In the single operator case, an operator is of type ω_+ if and only if it generates an analytic semigroup on a sector related to ω. We are going to generalise this concept to the case of n–tuples of operators . First, let

$$N_\mu^+ = \{\, x = (x_0, \mathbf{x}) \in \mathbb{R}^{n+1} : x_0 \geq \tan \mu |\mathbf{x}| \,\}$$

and

$$N_\mu^- = \{\, x = (x_0, \mathbf{x}) \in \mathbb{R}^{n+1} : -x_0 \geq \tan \mu |\mathbf{x}| \,\} = -N_\mu^+.$$

DEFINITION 6.4.1 *We say that $(T_1, T_2, \ldots, T_n)$ generates a bounded monogenic semigroup on $N_\mu = N_\mu^+ \cup N_\mu^-$, if, for some constant C,*

$$\left\| e^{-i(\mathbf{X}, T) + i x_0 T} \chi_+(T) \right\| \leq C, \quad \forall x \in N_\mu^-$$

and

$$\left\| e^{-i(\mathbf{X}, T) + i x_0 T} \chi_-(T) \right\| \leq C, \quad \forall x \in N_\mu^+.$$

REMARK Notice that both operator–valued functions are monogenic in x on the corresponding cones.

DEFINITION 6.4.2 *We say that $(T_1, T_2, \ldots, T_n)$ is of type ω, if for any $\mu > \omega$, $(T_1, T_2, \ldots, T_n)$ generates a bounded monogenic semigroup on N_μ.*

REMARK It is not so hard to prove that iT is of type ω as a single operator whenever $(T_1, T_2, \ldots, T_n)$ is of type ω.

We have the following theorem.

THEOREM 6.4.3
Suppose that iT is a one-one operator of type ω which has a bounded H_∞ functional calculus on $S^0_\mu(\mathbb{C})$ for any $\mu > \omega$. If, furthermore, $(T_1, T_2, \ldots, T_n)$ is of type ω, then $(T_1, T_2, \ldots, T_n)$ has a bounded H_∞ functional calculus on $S^0_\mu(\mathbb{C}^n)$ for any $\mu > \omega$, i.e., for $b \in H_\infty(S^0_\mu(\mathbb{C}^n))$, $b(T_1, \ldots, T_n)$ is a bounded operator on $\mathcal{H}_{(n)}$, and

$$\| b(T_1, \ldots, T_n) \| \leq C_\mu \| b \|_\infty .$$

PROOF   First we define $b(T)$ for functions $b \in H_\infty(S^0_\mu(\mathbb{C}^n))$ which satisfy

$$|b(\zeta)| \leq C \frac{|\zeta|^s}{1 + |\zeta|^{2s}} \quad \text{for some } 0 < s < n.$$

From Theorem 3.2, there exist right–monogenic functions ϕ_+, ϕ_- on $C^+_\mu(\mathbb{R}^{n+1})$ and $C^-_\mu(\mathbb{R}^{n+1})$, respectively, such that for $\theta < \mu$,

$$
\begin{aligned}
b(\zeta) \quad &= \quad \int_{-H_\theta} \phi_+(x)n(x)e(x, -\zeta)\chi_+(\zeta)\, dx \\
&+ \int_{H_\theta} \phi_-(x)n(x)e(x, -\zeta)\chi_-(\zeta)\, dx
\end{aligned}
$$

for $\zeta \in S_\theta(\mathbb{C}^n)$.
   Now we choose $\omega < \theta < \mu$ fixed, and define

$$
\begin{aligned}
b(T) \quad &= \quad \int_{-H_\theta} \phi_+(x)n(x)e(x, -T)\chi_+(T)\, dx \\
&+ \int_{H_\theta} \phi_-(x)n(x)e(x, -T)\chi_-(T)\, dx
\end{aligned}
$$

where

$$e(x, -T) = e^{-i(\mathbf{X}, T) + ix_0 T} \quad \text{for all } x \in \mp H_\theta.$$

The convergence of the integrals is ensured by the assumption that $(T_1, T_2, \ldots, T_n)$ is of type ω, which means,

$$\| e(x, -T)\chi_\pm(T) \| \leq C_\theta.$$

Thus, $b(T)$ is well defined.
   Suppose that $\psi \in H_\infty(S^0_\mu(\mathbb{C}))$ is a function of one–variable satisfying

$$|\psi(z)| \leq C \frac{|z|^s}{1 + |z|^{2s}} \quad \text{for some } 0 < s < n \text{ and for all } z \in S^0_\mu(\mathbb{C}).$$

There is an associated holomorphic function $\widetilde{\psi}$ of n-complex variables defined by

$$\widetilde{\psi}(\zeta) = \psi\{i\zeta\} = \psi(|\zeta|_*)\chi_+(\zeta) + \psi(-|\zeta|_*)\chi_-(\zeta), \qquad \zeta \in S_\mu^o(\mathbb{C}^n).$$

(See [LMcQ] or [Mc2].) It is easy to see that $\chi_\pm\{i\zeta\} = \chi_\pm(\zeta)$.

Then, $\widetilde{\psi}(\zeta) = \psi\{i\zeta\}$ satisfies

$$|\widetilde{\psi}(\zeta)| \leq C\frac{|\zeta|^s}{1+|\zeta|^{2s}} \quad \text{for all } \zeta \in S_\mu^0(\mathbb{C}^n).$$

For $t > 0$, let $\widetilde{\psi}_t(\zeta) = \widetilde{\psi}(t\zeta)$. Then $\widetilde{\psi}_t$ has the same properties as $\widetilde{\psi}$.

Now, we can define $b(T)$ for $b \in H_\infty(S_\mu^0(\mathbb{C}^n))$. Since for such b, $b\widetilde{\psi}_t$ satisfies

$$|(b\widetilde{\psi}_t)(\zeta)| \leq C\,\|b\|_\infty \frac{t^s|\zeta|^s}{1+t^{2s}|\zeta|^{2s}} \quad \text{for all } \zeta \in S_\mu^0(\mathbb{C}^n),$$

with C independent of t, then $b\widetilde{\psi}_t(T)$ is well defined and

$$\left\| b\widetilde{\psi}_t(T) \right\| \leq C\,\|b\|_\infty \qquad \text{for all } t > 0.$$

We choose ψ such that

$$\int_0^\infty \psi^3(-t)\,\frac{dt}{t} = \int_0^\infty \psi^3(t)\,\frac{dt}{t} = 1.$$

Then we define $b(T)$ by

$$(b(T)u, v) = \int_0^\infty \left((b\widetilde{\psi}_t)(T)\psi_t(iT)u, \psi_t(iT)^*v \right)\frac{dt}{t}$$

for $u, v \in \mathcal{H}_{(n)}$. Since iT satisfies square function estimates, $b(T)$ is well defined, and satisfies

$$|(b(T)u,v)| \leq \sup_{t>0}\|(b\widetilde{\psi}_t)(T)\| \left\{ \int_0^\infty \|\psi_t(iT)u\|^2\frac{dt}{t} \right\}^{\frac{1}{2}} \left\{ \int_0^\infty \|\psi_t(iT)^*v\|^2\frac{dt}{t} \right\}^{\frac{1}{2}}$$

$$\leq C\,\|b\|_\infty\,\|u\|\,\|v\|,$$

which means $\|b(T)\| \leq C\,\|b\|_\infty$. ∎

The following corollary is easier to use in practice.

COROLLARY 6.4.4
Suppose that $(T_1, \ldots, T_n)$ is an n–tuple of commuting operators on a Hilbert space $\mathcal{H}$, and $\bigcap_{j=1}^n \mathcal{D}(T_j)$ is dense in $\mathcal{H}$. Suppose further that

(i) $T_1^2 + \ldots + T_n^2$ is one-one, and of type $2\omega_+$, therefore, $|T| = (T_1^2 + \ldots + T_n^2)^{1/2}$ exists and is of type ω_+;

(ii) $|T|$ has a bounded H_∞ functional calculus on $S_\mu^0(\mathbb{C})$ for any $\mu > \omega$;

(iii) $T_j|T|^{-1}$ is bounded on $\mathcal{H}$;

(iv)

$$\left\| e^{i(x_1 T_1 + \ldots + x_n T_n) + x_0|T|} \right\| \leq C_\mu, \quad \text{for any } \frac{-x_0}{|\mathbf{x}|} \geq \tan\mu, \ \mu > \omega;$$

Then $(T_1, \ldots, T_n)$ has a bounded H_∞ functional calculus on $S_\mu^0(\mathbb{C}^n)$ for $\mu > \omega$.

Bibliography

[BDS]   F. Brackx, R. Delanghe and F. Sommen, *Clifford Analysis*, Research Notes in mathematics 76, Pitman Advanced Publishing Program, 1982.

[CDMcY]   M. Cowling, I. Doust, A. McIntosh and A. Yagi, *Banach space operators with a bounded H^∞ functional calculus*, Journal of the Australian Mathematical Society, Series A, (to appear).

[K]   Tosio Kato, *Perturbation Theory for Linear Operators*, 2nd edition, Springer–Verlag, New York, 1976.

[LMcQ]   C. Li, A. McIntosh and T. Qian, *Clifford algebras, Fourier transforms, and singular convolution operators on Lipschitz surfaces*, Revista Mathematica Iberoamericana, 10(1994), 665–671.

[Mc1]   Alan McIntosh, *Operators which have an H_∞ functional calculus*, Miniconference on Operator Theory and Partial Differential Equations, 1986, Proceedings of the Centre for Mathematical Analysis, ANU, Canberra, Vol. 14, 1986, 210–231.

[Mc2]   Alan McIntosh *Clifford algebras, Fourier theory, singular integrals, and harmonic functions on Lipschitz domains*, in *Clifford Analysis*, (John Ryan, ed.), CRC Press, Boca Raton, FL, 1995.

[McP]   A. McIntosh and A. Pryde, *A functional calculus for several commuting operators*, Ind. Univ. Math. Journal **36** (1987), 421–439.

[McPR]   A. McIntosh, A. Pryde and W. Ricker, *Comparison of joint spectra for certain classes of commuting operators*, Studia Math. **88**, (1988), 23–36.

[McY]   A. McIntosh and A. Yagi *Operators of type ω without a bounded H∞ functional calculus*, Miniconference on Operators in Analysis, 1989, Proceedings of the Centre for Mathematical Analysis, ANU, Canberra, **24**, (1989), 159–172.

School of Mathematics, Physics, Computing and Electronics,
Macquarie University,
North Ryde, New South Wales 2109,
AUSTRALIA
E-mail: chun@macadam.mpce.mq.edu.au
E-mail: alan@macadam.mpce.mq.edu.au

7

Hypercomplex Variable Techniques in Harmonic Analysis

Marius Mitrea

7.1 Introduction

Taking "square roots" of quadratic forms is one of the primary reasons for working in the Clifford algebra context. In particular, one can consider the square root of the Laplacian and still be within the class of differential operators. Originating in the pioneering work of Moisil [Mo], [MT], Teodorescu [Te] and Fueter [Fu] among others, the study of the resulting elliptic first-order differential operator, much in the spirit of the Cauchy-Riemann $\bar{\partial}$ operator, has become by now a well-established, active area of research (see for instance the monographs [BDS], [HS], [GM2], [M3], and the references therein).

More recently, Coifman, McIntosh, Murray, Semmes and their collaborators ([Mc], [Mu], [Se1, Se2]) have considered Clifford algebra-valued singular integral operators of Cauchy type. A basic motivation is that the double-layer potential operator, which solves the Dirichlet problem for the Laplacian, is one of the components (actually the "real part") of the higher-dimensional Cauchy integral.

Hardy spaces in the Clifford algebra framework arise quite naturally in connection with the qualitative analysis of the boundary behavior of the solutions of the generalized Cauchy-Riemann equations. Attempts to develop such a $\mathcal{H}^p$ space theory, much in the spirit of the results of Kenig [Ke], are due to McIntosh, Meyer, Gilbert and Murray [LMS], [Me], [GM1, GM2]; cf. also [M1, M3].

In this paper we present a survey of some recent results in this direction. The layout of the paper is as follows. Sections 2-4 contain some background material. The Clifford algebra version of the T(b) theorem is recalled in section 5. As for the $\mathcal{H}^p$ theory in Lipschitz domains, the departure point is Theorem 7.6.1 containing several alternative characterizations for these Hardy spaces. In turn, this result is used to prove a Clifford algebra version

0-8493-8481-8/96/$0.00+$.50

of the Burkholder-Gundy-Silverstein theorem (see section 8). Another essential ingredient used here is a certain Rellich type identity for monogenic functions which we devise in section 7.

Applications to the theory of harmonic functions in non-smooth domains are consider in section 9. Among other things, we discuss the connection with the classical boundary value problems for the Laplacian, and give a short proof of a theorem of Dahlberg [D] regarding the norm equivalence of the area-function and the nontangential maximal function for a function harmonic in a Lipschitz domain.

I would like to express my sincere thanks to Professor Alan McIntosh for several enriching discussions, as well as to the organizers of this workshop for their generous hospitality and support.

7.2   Clifford Algebra Rudiments

In this section we shall briefly review the construction and the basic properties of the Clifford algebra associated with $\mathbb{R}^m$. For a more detailed account on these matters, see [GM2].

The Clifford algebra associated with $\mathbb{R}^m$ endowed with the Euclidean metric is the enlargement of $\mathbb{R}^m$ to a unitary algebra $\mathbb{R}_{(m)}$ for which

$$x^2 = -|x|^2, \qquad (7.2.1)$$

for any $x \in \mathbb{R}^m$, and such that $\mathbb{R}_{(m)}$ is not generated (as an algebra) by any proper subspace of $\mathbb{R}^m$. Note that (7.2.1) becomes by polarization

$$xy + yx = -2\langle x, y \rangle, \qquad (7.2.2)$$

for any $x, y \in \mathbb{R}^m$. In particular, if $\{e_j\}_{j=1}^m$ is the standard basis for $\mathbb{R}^m$, we must have

$$e_j e_k + e_k e_j = -2\,\delta_{jk}, \qquad (7.2.3)$$

in other words $e_j^2 = -1$ and $e_j e_k = -e_k e_j$ for any $j \neq k$. Consequently, any element $A \in \mathbb{R}_{(m)}$ has a representation of the form

$$A = \sum_I{}' A_I e_I, \qquad A_I \in \mathbb{R}. \qquad (7.2.4)$$

Here $\sum'$ indicates that the sum is performed only on strictly increasing multi-indices I, i.e., ordered l-tuples of the form $I = (i_1, i_2, \ldots, i_l)$, with $1 \leq i_1 < i_2 < \ldots < i_l \leq m$, where $0 \leq l \leq m$. Furthermore, e_I stands for the product $e_{i_1} \cdot e_{i_2} \cdot \ldots \cdot e_{i_l}$ (by convention $e_\varnothing := e_0 := 1$). We also set $e^I := (-1)^l e_{i_l} \cdot e_{i_{l-1}} \cdot \ldots \cdot e_{i_1}$, so that $e_I e^I = e^I e_I = 1$.

THEOREM 7.2.1
The Clifford algebra $\mathbb{R}_{(m)}$ exists and is uniquely determined up to an isomorphism.

PROOF Let us first sketch the proof of the uniqueness part. Obviously, it suffices to show that if A from (7.2.4) is zero, then necessarily $A_I = 0$ for all I. To this effect, we note the following identity

$$2^{-m} \sideset{}{'}\sum_{I} e_I A e^I = \begin{cases} A_\varnothing, & \text{if } m \text{ is even,} \\ A_\varnothing + A_{(1,2,\ldots,m)} e_{(1,2,\ldots,m)}, & \text{if } m \text{ is odd,} \end{cases} \qquad (7.2.5)$$

(see [M2] for a proof). The minimality of $\mathbb{R}_{(m)}$ ensures that 1 and $e_{(1,2,\ldots,m)}$ are linearly independent so that, at any rate, (7.2.5) implies that $A_\varnothing = 0$. Applying this result to $e^I A$, for arbitrary I, the conclusion follows.

As far as the existence of such an algebra is concerned, we shall produce an example in the matrix algebra $\mathcal{M}_{2^m \times 2^m}(\mathbb{R})$. Consider the matrices $e_j := E_j^m$, $j = 1, 2, \ldots, m$, where, for each $1 \le k \le m$, $\{E_j^k\}_{j=1}^k$ are inductively defined by

$$E_1^1 := \begin{pmatrix} 0 & -1 \\ 1 & 0 \end{pmatrix}$$

and, in general, for $1 \le k \le n-1$, $1 \le j \le k$,

$$E_j^{k+1} := \begin{pmatrix} E_j^k & 0 \\ 0 & -E_j^k \end{pmatrix}, \text{ and } E_{k+1}^{k+1} := \begin{pmatrix} 0 & -I_{2^k} \\ I_{2^k} & 0 \end{pmatrix}.$$

It is easy to see that $\{e_j\}_j$ satisfy (7.2.3). Also, as the trace of e_I is zero for all $I \ne \varnothing$, we infer that $\{e_I\}_I$ are linearly independent over $\mathbb{R}$. Thus, we may take $\mathbb{R}_{(m)}$ to be the sub-algebra of $\mathcal{M}_{2^m \times 2^m}(\mathbb{R})$ consisting of all matrices A of the form (7.2.4). ∎

Note that this proof actually gives more. First, $\dim \mathbb{R}_{(m)} = 2^m$ and we have a natural conjugation on $\mathbb{R}_{(m)}$, denoted by $^-$, given by the usual transposition of matrices in $\mathcal{M}_{2^m \times 2^m}(\mathbb{R})$.

We also embed $\mathbb{R}^{m+1}$ into $\mathbb{R}_{(m)}$ by identifying $(x_0, x') \in \mathbb{R}^{m+1} = \mathbb{R} \oplus \mathbb{R}^m$ with $x_0 + x' \in \mathbb{R}_{(m)}$ (note that by (7.2.1), $1 \notin \mathbb{R}^m$), and call these elements *Clifford vectors*. An important observation is that any Clifford vector X has a multiplicative inverse, given by $X^{-1} = \bar{X}/|X|^2$. We define the real part of $A \in \mathbb{R}_{(m)}$ to be $\operatorname{Re} A := A_\varnothing$, if A is as in (7.2.4), and endow $\mathbb{R}_{(m)}$ with the natural Euclidean metric $|A|^2 := \operatorname{Re}(A\bar{A})$.

Finally, let us note that a similar construction works for the complex case.

7.3   Elements of Clifford Analysis

We shall work in the Euclidean space $\mathbb{R}^{m+1}$ assumed to be embedded in the Clifford algebra $\mathbb{R}_{(m)}$. Let f, g be two continuous, $\mathbb{R}_{(m)}$-valued functions defined in a domain $\Omega \subseteq \mathbb{R}^{m+1}$. We introduce the Clifford algebra derivative of f and g at a point $X \in \Omega$ by

$$D(f|g)(X) := \lim_{\mathcal{O} \downarrow X} \frac{\int_{\partial \mathcal{O}} f\, n\, g\, d\sigma}{\int\int_{\mathcal{O}} d\,\mathrm{Vol}}$$

(cf. [P2], [MT]). Here n is the outward unit normal to the boundary of the domain $\mathcal{O}$ which is shrinking to the point X, $d\sigma$ is the surface measure on $\partial \mathcal{O}$, and the top integrand must be interpreted in the sense of point-wise Clifford algebra multiplication.

It is easy to check the Leibnitz rule $D(f|g) = D(f|1)\,g + f\,D(1|g)$, and we shall simply set $Dg := D(1|g)$ and $fD := D(f|1)$. Using the Taylor series expansion for f, it is not difficult to see that, at any point of differentiability for $f = \sum'_I f_I e_I$, we have

$$Df = \sum_I{}' \sum_{j=0}^{m} \frac{\partial_j f_I}{\partial x_j} e_j e_I.$$

We also set $\overline{D}f := \overline{D(\overline{f}|1)}$ and $f\overline{D} := \overline{D(1|\overline{f})}$. Simple calculations give that the Laplace operator $\triangle$ in $\mathbb{R}^{m+1}$ has the factorizations

$$\triangle = D\overline{D} = \overline{D}D. \tag{7.3.1}$$

The next lemma, which in the complex case goes back to Pompeiu [P1, P3], (cf. also [Te]), can be thought of as the higher-dimensional analogue of the classical Leibnitz-Newton formula on the real line.

LEMMA 7.3.1
If $D(f|g)$ is continuous in Ω, then, for any relatively compact $\omega \subset \Omega$,

$$\int_{\partial \emptyset} f\, n\, g\, d\sigma = \int\int_{\emptyset} D(f|g).$$

In particular, if f, g are Lipschitz continuous, then

$$\int_{\partial \omega} f\, n\, g\, d\sigma = \int\int_{\omega} (fD)g + f(Dg). \tag{7.3.2}$$

Following Moisil and Teodorescu [MT], we shall call f (left) monogenic ((right) monogenic, or two-sided monogenic, respectively) if $Df = 0$ ($fD = 0$, or $Df = fD = 0$, respectively). Note that by (7.3.1), any monogenic function is harmonic.

Our basic example of a two-sided monogenic function, the so-called *Cauchy kernel*, is the fundamental solution of the operator D given by

$$E_0(X) := \frac{1}{\sigma_m} \frac{\overline{X}}{|X|^{m+1}}, \quad X \in \mathbb{R}^{m+1} \setminus \{0\}, \qquad (7.3.3)$$

where σ_m stands for the area of the unit sphere in $\mathbb{R}^{m+1}$. Note that $E_0(X) = \overline{D}\Gamma_0(X)$, where $\Gamma_0(X) := \frac{1}{(1-m)\sigma_m} \frac{1}{|X|^{m-1}}$, $X \neq 0$, $m \geq 2$, is the canonical fundamental solution for the Laplace operator in $\mathbb{R}^{m+1}$.

Lemma 7.3.1 applied to f and $g := E_0(X - \cdot)$ in $\Omega \setminus B_\epsilon(X)$ yields, after letting ϵ go to zero, the Pompeiu integral representation formula ([Mo], [Te])

$$f(X) = \int_{\partial\Omega} f(Y)\, n(Y)\, E_0(Y - X)\, d\sigma(Y)$$

$$- \iint_\Omega (fD)(Y) E_0(Y - X)\, d\text{Vol}(Y),$$

for $X \in \Omega$. In particular, if f is, e.g., right monogenic, we have the Cauchy type reproducing formula ([Di])

$$f(X) = \int_{\partial\Omega} f(Y)\, n(Y)\, E_0(Y - X)\, d\sigma(Y). \qquad (7.3.4)$$

For f right monogenic and g left monogenic, we also have from (7.3.2) that

$$\int_{\partial\Omega} f(X)\, n(X)\, g(X)\, d\sigma(X) = 0. \qquad (7.3.5)$$

The following simple observation will also be useful in the sequel.

LEMMA 7.3.2
Given the $(m+1)-$tuple of real valued functions $U = (u_j)_{j=0}^m$, then $F := u_0 - u_1 e_1 - \ldots - u_m e_m$ is a two-sided monogenic function if and only if U is a system of conjugate harmonic functions in the sense of Moisil-Teodorescu [Mo], [MT], and Stein-Weiss [SW], i.e. U satisfies the generalized Cauchy-Riemann equations $div\, U = 0$, $curl\, U = 0$.

Let us now recall the higher dimensional Cauchy integral on Lipschitz hyper-surfaces. An unbounded Lipschitz domain $\Omega \subset \mathbb{R}^{m+1}$ is the domain above the graph of a Lipschitz function $\varphi : \mathbb{R}^{m+1} \to \mathbb{R}$, i.e., $\Omega := \{(t, x); \varphi(x) < t\}$.

If $d\sigma$ denotes the canonical surface measure on $\Sigma := \partial\Omega$, then the outward unit normal $n(\varphi(x), x) := (-1, \nabla\varphi(x))(1 + |\nabla \varphi(x)|^2)^{-1/2}$ is defined $d\sigma-$almost everywhere on Σ.

The higher-dimensional Cauchy integral operator $\mathcal{C}$ is formally defined by

$$\mathcal{C}f(X) := \int_\Sigma f(Y)\,n(Y)\,E_0(Y-X)\,d\sigma(Y), \quad X \in \mathbb{R}^{m+1} \setminus \Sigma. \quad (7.3.6)$$

Also, its singular version, the so called *Hilbert transform* on Σ is the principal value integral operator

$$Hf(X) := \lim_{\epsilon \to 0} H_\epsilon f(X), \quad X \in \Sigma, \quad (7.3.7)$$

where

$$H_\epsilon f(X) := \int_{\substack{Y \in \Sigma \\ |X-Y|>\epsilon}} f(Y)\,n(Y)\,E_0(Y-X)\,d\sigma(Y), \quad X \in \Sigma.$$

The deep result of Coifman, McIntosh, Meyer [CMM] asserts that for all $1 < p < \infty$ the Hilbert transform is a bounded mapping of $L^p(\Sigma, d\sigma)$ (see also [Mu], [Mc], [LMS]).

Of course, there are also "right-handed" versions for these operators (i.e., f appears in the rightmost part of the integrand). When necessary, we shall indicate which one is used by employing the superscripts l, r (e.g., the operators (7.3.6), (7.3.7) will be denoted by $\mathcal{C}^l$ and H^l, respectively).

7.4 Non-Homogeneous Dirac Operators

We embed $\mathbb{R}^{m+1}$ into the Clifford algebra $\mathbb{R}_{(M)}$, $M \geq m+1$, by identifying $(x_0, x_1, \cdots, x_m)$ from $\mathbb{R}^{m+1}$ with $\sum_{j=0}^m x_j e_j$ from $\mathbb{R}_{(M)}$.

For $k \in \mathbb{C}$, $\operatorname{Im} k > 0$, set

$$D_k := \partial_0 e_0 + \partial_1 e_1 + \cdots + \partial_m e_m + k e_{m+1},$$

and

$$\overline{D_k} := \partial_0 e_0 - \partial_1 e_1 - \cdots - \partial_m e_m - k e_{m+1}.$$

Note that $D_k \overline{D_k} = \overline{D_k} D_k = \triangle + k^2$.

Next, let Γ_k be the canonical fundamental solution for the Helmholtz operator $\triangle + k^2$ in $\mathbb{R}^{m+1}$, i.e.,

$$\Gamma_k(X) := -\frac{1}{\sigma_m}\frac{(-ik)^{m-2}}{(m-2)!}\int_1^\infty (t^2-1)^{\frac{m-3}{2}} e^{ik|X|t}dt, \quad |X| \neq 0, \quad (7.4.1)$$

for $m \geq 2$ (see e.g., [Ga]). Then $E_k(X) := \overline{D_k}\Gamma_k(X)$ is a fundamental solution for the non-homogeneous Dirac operator D_k, i.e., $D_k E_k(X) = \delta(X)$ in $\mathbb{R}^{m+1}$. Furthermore, it is easy to check that $D_{-k}E_k(-X) = -\delta(X)$, and that

$$|E_k(X) + E_k(-X)| \leq C|\Gamma_k(X)|, \quad X \in \mathbb{R}^{m+1}. \quad (7.4.2)$$

We shall also need the fact that, for $m \geq 2$,

$$|\Gamma_k(X)| + |X||E_k(X)| + |X|^2|(\nabla E_k)(X)|$$
$$= \begin{cases} \mathcal{O}(|X|^{-m+1}) \text{ as } |X| \to 0, \\ \mathcal{O}(\exp\{-\operatorname{Im} k\,|X|\}) \text{ as } |X| \to +\infty. \end{cases}$$

Many of the results of the previous section continue to hold in this more general non-homogeneous context. For example, (7.3.2) becomes

$$\int_{\partial\Omega} f\,n\,g\,d\sigma = \iint_\Omega (fD_k)g + f(D_{-k}g), \qquad (7.4.3)$$

for Lipschitz continuous functions f, g on $\overline{\Omega}$, say. Also, by replacing the kernel E_0 with E_k, for $k \in \mathbb{C}$, we obtain some versions C_k, H_k of the Cauchy integral operator and the Hilbert transform, respectively. When suitably interpreted, the Cauchy reproducing formula and the Cauchy vanishing theorem are still valid in this case (cf. also [Ry]).

Before closing this section, let us introduce a notational convention many times used in the sequel. The estimate $F \lesssim G$, for two quantities depending on some parameter $s \in S$, will signify that there exists a positive constant, Const, such that $F(s) \leq \operatorname{Const} G(s)$, for all $s \in S$. Furthermore, $F \approx G$ stands for $F \lesssim G$ and $G \lesssim F$.

7.5 Clifford Algebra-Valued Singular Integral Operators

Recall that a Clifford algebra-valued, Lipschitz continuous function $k(x, y)$ defined for $x \neq y$, $x, y \in \mathbb{R}^m$, is said to be a *standard kernel* if

$$|k(x, y)| + |x - y||\nabla_x\, k(x, y)| + |x - y||\nabla_y\, k(x, y)| \lesssim |x - y|^{-m}, \quad (7.5.1)$$

uniformly for all $x \neq y$, and that the operator T is *associated with the kernel* $k(x, y)$ if

$$\langle \psi, T\varphi \rangle = \iint_{\mathbb{R}^m \times \mathbb{R}^m} \psi(x)k(x, y)\varphi(y)\,dx dy,$$

for φ, ψ integrable, compactly supported with disjoint supports.

We are now ready to state the Clifford algebra version of the T(b) theorem. Although valid in more generality, we restrict ourselves to the version relevant here.

THEOREM 7.5.1

[DJ], [MM], [DJS] Suppose that the essentially bounded, Clifford vector-valued functions b_i, $i = 1, 2$, satisfy $\operatorname{Re} b_i \geq c_0 > 0$ on $\mathbb{R}^m$. If the operator

T is associated with a Clifford algebra-valued standard kernel $k(x,y)$, then T is bounded on $L^2(\mathbb{R}^m)$ if and only if

$$\int_B \left| \int_B k(x,y)b_1(y)dy \right| dy \lesssim |B|, \qquad (7.5.2)$$

$$\int_B \left| \int_B b_2(x)k(x,y)dx \right| dx \lesssim |B|, \qquad (7.5.3)$$

uniformly for all balls $B \subset \mathbb{R}^m$ (here $|B|$ stands for the euclidean volume of the ball).

A more compact, equivalent way of expressing the conditions (7.5.2)-(7.5.3) in the above theorem is to say that $T(b_1)$, $T^t(b_2) \in BMO$ and $M_{b_2}^r T M_{b_1}^l$ has the weak boundedness property (where M_b^l, or M_b^r, stands for the operator of left, or right respectively, multiplication with the Clifford algebra-valued function b). See, e.g., [Me]. There are several proofs of this result which fully employ the Clifford algebra machinery. See [AT], [Da], [GLQ], [Se1, Se2], [AJM].

It must also be mentioned here that this theorem admits various extensions, some of which we shall use in the sequel. Hence, for further reference, we note that any operator associated with a standard kernel which is bounded on $L^2(\mathbb{R}^m)$ is in fact bounded on any $L^p(\mathbb{R}^m)$ and even on $L^p(\mathbb{R}^m, \omega dx)$, where the weight ω is in the Muckenhoupt class A_p, for $1 < p < \infty$. Moreover, when suitably interpreted, these results continue to hold for, e.g., $\mathcal{H} \otimes \mathbb{R}_{(M)}$-valued standard kernels where $\mathcal{H}$ is an arbitrary Hilbert space.

We illustrate these ideas by considering a generalization of the Hilbert transform discussed in the previous section. Let Γ be a fixed, upright, closed, circular cone in the upper-half space, whose vertex is at the origin and which has a sufficiently small overture such that $(X + \Gamma) \setminus \{X\} \subset \Omega$, for all $X \in \Sigma$.

Now let $\mathcal{H}$ be a Hilbert space which is also a two-sided module over $\mathbb{R}_{(M)}$, and let $K(X,Y)$ be a $\mathcal{H}$-valued function defined for $Y - X \notin \Gamma$, such that, for some $k \in \mathbb{C}$ with $\operatorname{Im} k > 0$,

$$|X - Y|^m \|K(X,Y)\|_{\mathcal{H}} \lesssim \exp\left(-\operatorname{Im} k \, |X - Y|\right), \quad \text{for} \quad Y - X \notin \Gamma, \quad (7.5.4)$$

and having the property that for any $h \in \mathcal{H}$ there exists $\epsilon = \epsilon(h) > 0$ such that, for any $X \in \mathbb{R}^m$,

$$D_k \langle K(\cdot, X), h \rangle = 0 \quad \text{on } \mathbb{R}^{m+1} \setminus (-\Gamma - \epsilon + X), \qquad (7.5.5)$$

$$\langle K(\cdot, X), h \rangle D_k = 0 \quad \text{on } \mathbb{R}^{m+1} \setminus (\Gamma + \epsilon + X). \qquad (7.5.6)$$

Let us make the notation $L_{\mathcal{H}}^p(\Sigma, d\sigma)$ for the Banach space of (classes of) measurable functions on Σ which are $\mathcal{H}$−valued and L^p−integrable (with the natural norm), $1 < p < \infty$.

THEOREM 7.5.2
If the kernel $K(X, Y)$ of the integral operator

$$T f(X) := p.v. \int_\Sigma f(Y) K(X, Y) \, d\sigma(Y), \quad X \in \Sigma,$$

satisfies (7.5.4), (7.5.5) and (7.5.6), then T is bounded on $L^p_\mathcal{H}(\Sigma, \, d\sigma)$ for any $1 < p < \infty$.

In fact, as it will become more apparent in the next section, the continuity of the operators of the type described above can be nicely expressed in terms of some Hardy spaces $\mathcal{H}^p(\Omega)$ naturally associated to Ω. The following holds.

THEOREM 7.5.3
With the above hypotheses, for any $h \in \mathcal{H}$, the operator

$$T f(X) := \int_\Sigma \langle h, f(Y) K(X, Y) \rangle \, d\sigma(Y), \quad X \in \Omega,$$

maps $L^p_\mathcal{H}(\Sigma, \, d\sigma)$ into $\mathcal{H}^p(\Omega)$.

For kernels of the form $K(X, Y) = \Phi(X - Y)$, with $\Phi(X)$ right monogenic and satisfying $|\Phi(X)| \lesssim |X|^{-m}$ in $\mathbb{R}^{m+1} \setminus \Gamma$, direct proofs of these results can be found in [LMS]. We shall present here an alternative argument based on an idea of Meyer [Me] which utilizes Theorem 7.5.1.

PROOF [**Proof of Theorem 7.5.2.**] To see that $K((\varphi(x), x), (\varphi(y), y))$ is a standard kernel, it suffices to show that for all $h \in \mathcal{H}$, $\|h\|_\mathcal{H} = 1$,

$$|\nabla_X \langle K(X, Y), h \rangle| \lesssim |X - Y|^{-m-1},$$

uniformly for $X, Y \in \Sigma$ (the estimate of $\nabla_Y K$ is completely similar). Fix $h \in \mathcal{H}$ with $\|h\|_\mathcal{H} = 1$, two disjoint points X, Y in Σ, and set $d := \frac{1}{2} \, \text{dist} \, (Y - X, \Gamma)$. By Cauchy's integral representation formula

$$|\nabla_X \langle K(X, Y), h \rangle| \lesssim \int_{|Y - X - Z| = d} |K(Y - Z, Y)|$$

$$|(\nabla E_k)(Y - X - Z)| \, d\sigma(Z) \lesssim |X - Y|^{-m-1},$$

by (7.5.4), since $|\nabla E_k| \lesssim d^{-m-1}$ and $|Z| \approx |X - Y| \approx d$ on the contour of integration.

Let us now prove that for any surface ball B, which without any loss of generality is assumed to be of the form $B_r := \Sigma \cap B((\varphi(0), 0); r)$, one has

$$\int_{B_r} \left| \int_{B_r} n(Y) \langle K(X, Y), h \rangle \, d\sigma(Y) \right| d\sigma(X) \lesssim d\sigma(B_r), \qquad (7.5.7)$$

uniformly for $h \in \mathcal{H}$ with $\|h\|_{\mathcal{H}} = 1$. To this end, using (7.4.3), we can deform the contour of integration in the innermost integral to $\partial C \backslash \Sigma$, where C is the cylinder $C := \{X + t; X \in B_r, 0 < t < r\}$. Therefore the left-hand side of (7.5.7) can be majorized by

$$\int_{B_r} \int_{r + B_r} |\langle K(X, Y), h \rangle| \, d\sigma(X) \, d\sigma(Y)$$

$$+ \int_{B_r} \int_V |\langle K(X, Y), h \rangle| \, d\sigma(X) \, d\sigma(Y)$$

$$+ |k| \int_{B_r} \int\!\!\int_C \|K(X, Y)\|_{\mathcal{H}} dY \, d\sigma(X)$$

$$=: I + II + III,$$

where V is the "vertical" part of ∂C and $d\sigma$ its canonical surface measure.

Using $|\langle K(X, Y), h \rangle| \lesssim \|K(X, Y)\|_{\mathcal{H}} \lesssim |X - Y|^{-m}$, we have that the integrand in I is $\lesssim r^{-m} \approx d\sigma(B_r)^{-1}$, hence $I \lesssim d\sigma(B_r)$. Also, by an appropriate change of variables,

$$II \lesssim \int_{B_r} \int_{B_{2r} \backslash B_r} \frac{1}{|X - Y|^m} \, dx dy \lesssim r^m \left(\int_{\substack{|y| < 1 \\ y \in \mathbb{R}^m}} \int_{\substack{1 < |x| < 2 \\ x \in \mathbb{R}^m}} \frac{1}{|x - y|^m} dx dy \right),$$

where the last estimate was obtained by projecting Σ onto $\mathbb{R}^m$. Since the rightmost integral is finite, we also get $II \leq d\sigma(B_r)$. Finally, as

$$\int\!\!\int_{\mathbb{R}^{m+1}} \|K(X, Y)\|_{\mathcal{H}} dY \leq \text{constant} < +\infty,$$

uniformly in X, we see that III also has the right order. Hence, the conclusion follows. ∎

7.6 Hardy Spaces on Lipschitz Domains

Let Ω be an unbounded Lipschitz domain in $\mathbb{R}^{m+1}$. For $0 < p < \infty$ and $k \in \mathbb{C}$, we introduce the Hardy spaces $\mathcal{H}_k^p(\Omega)$ by

$$\mathcal{H}_k^p(\Omega) :=$$

$$\left\{ F : \Omega \to \mathbb{R}_{(M)} \, ; FD_k = 0, \, \|F\|_{\mathcal{H}_k^p} := \sup_{\epsilon > 0} \left(\int_\Sigma |F(\cdot + \epsilon)|^p \, d\sigma \right)^{\frac{1}{p}} < \infty \right\}.$$

In the commutative case, i.e., for $m = N = 1$ and $k = 0$, these spaces have been considered in Kenig's thesis [Ke], where a complete theory has been developed. However, his approach relies on conformal mapping techniques and factorization theorems for holomorphic functions—tools that break down in higher dimensions.

To state the main result of this section we need to introduce some more notation. For a $\mathbb{R}_{(M)}$-valued function F in Ω, we define *the nontangential maximal function* of F by $\mathcal{N}(F)(X) := \sup\{|F(Y)|; Y \in X + \Gamma\}$, $X \in \Sigma$, where Γ denotes a fixed sufficiently sharp upright cone in $\mathbb{R}^{m+1}$ centered at the origin.

Corresponding to the case when Γ from the definition of $\mathcal{N}$ is degenerate, *the maximal radial function* of F is given by $F_{\mathrm{rad}}(X) := \sup\{|F(X + \epsilon)|; \epsilon > 0\}$, $X \in \Sigma$. Also, recall *the Lusin area-function*

$$\mathcal{A}(F)(X) := \left(\iint_{X+\Gamma} |\partial_0 F(Y)|^2 |X - Y|^{1-m} d\mathrm{Vol}(Y) \right)^{1/2}, \quad X \in \overline{\Omega},$$

and its radial analogue, *the Littlewood-Paley g−function*

$$g(F)(X) := \left(\int_0^\infty |\partial_0 F(X + t)|^2 t\, dt \right)^{1/2}, \quad X \in \overline{\Omega}.$$

The following theorem from [M1, M3], extending some earlier work from [GM1, GM2], can be thought of as the higher-dimensional analogue of certain results of Kenig [Ke].

THEOREM 7.6.1
If $k \in \mathbb{C}$, Im $k > 0$, and $1 < p < \infty$, then for a Clifford algebra-valued function F satisfying $FD_k = 0$ in Ω, the following are equivalent:

1. $F \in \mathcal{H}_k^p(\Omega)$;
2. $F = \mathcal{C}_k f$ *in Ω for some $f \in L^p(\Sigma, d\sigma)$.*
3. $\mathcal{N}(F) \in L^p(\Sigma, d\sigma)$;
4. $F_{rad} \in L^p(\Sigma, d\sigma)$;
5. $\mathcal{A}(F) \in L^p(\Sigma, d\sigma)$ *and $\lim_{t\to\infty} F(X + t) = 0$ for some $X \in \Sigma$;*
6. $g(F) \in L^p(\Sigma, d\sigma)$ *and $\lim_{t\to\infty} F(X + t) = 0$ for some $X \in \Sigma$.*

In addition, F has the nontangential boundary limit

$$F(X) := \lim_{\substack{Y \to X \\ Y \in \Gamma + X}} F(Y) = \frac{1}{2}(f + H_k f)(X)$$

at almost any point $X \in \Sigma$, F has the reproduction property $F = \mathcal{C}_k F$ in Ω, and

$$\|F\|_{\mathcal{H}_k^p} \approx \|\mathcal{N}(F)\|_{L^p} \approx \|F_{rad}\|_{L^p} \approx \|\mathcal{A}(F)\|_{L^p} \approx \|g(F)\|_{L^p} \approx \|F\|_{L^p}.$$
$$(7.6.1)$$

The result remains true even in the weighted case, with a weight $\omega \in A_p$ (the Muckenhoupt class), and the analogue version for right monogenic functions in Ω holds as well.

In showing that $(1) \Rightarrow (2)$, the main idea is to use the Cauchy reproducing formula for F in the cylinder

$$C_{\epsilon,N,r} := \{(t,x)\,;\,\epsilon \leq t - \varphi(x) \leq N,\,|x| \leq r\}$$

and then to use a limiting argument (cf. [H]). The full details of this have been worked out in [M1, M3], hence we omit them here. The existence of the nontangential boundary limit a.e. on Σ also follows from (2) and, using once again a limiting argument, it is not difficult to infer the reproducing formula $F = \mathcal{C}_k F$. Furthermore, standard reasonings show that

$$\mathcal{N}(\mathcal{C}_k f)(X) \lesssim \sup_{\epsilon>0} |H_{k,\epsilon} f(X)| + f^*(X),$$

uniformly in $X \in \Sigma$, for any $f \in L^p(\Sigma, d\sigma)$ (recall that $*$ stands for the usual Hardy-Littlewood maximal operator). This and a Cotlar type inequality (see, e.g., [Me]) give in turn that (2) implies (3). That $(3) \Rightarrow (4) \Rightarrow (1)$ is obvious.

The rest of this section is devoted to proving the equivalence of (5) and (6) with (1)-(4). First we shall deal with $(1) - (4) \Rightarrow (5) - (6)$. The key estimates in doing so are formulated in the next lemma.

LEMMA 7.6.2
For any $1 < p < \infty$ one has

$$\|f + H_k f\|_{L^p(\Sigma, d\sigma)} \lesssim \|g(\mathcal{C}_k f)\|_{L^p(\Sigma, d\sigma)} \lesssim \|f\|_{L^p(\Sigma, d\sigma)},$$

$$\|f + H_k f\|_{L^p(\Sigma, d\sigma)} \lesssim \|\mathcal{A}(\mathcal{C}_k f)\|_{L^p(\Sigma, d\sigma)} \lesssim \|f\|_{L^p(\Sigma, d\sigma)},$$

uniformly for $f \in L^p(\Sigma, d\sigma)$.

Accepting this lemma for the moment, these estimates written for $f := F|_\Sigma \in L^p(\Sigma, d\sigma)$, where F is an arbitrary function satisfying $FD_k = 0$ in Ω and (1)-(4) of Theorem 7.6.1, give that $g(F)$, $\mathcal{A}(F)$ belong to $L^p(\Sigma, d\sigma)$ and have equivalent norms to $\|F\|_{L^p}$.

The proof of the above lemma is accomplished in several steps. The idea is to use the Hilbert space-valued version of the Clifford T(b) theorem. Let us deal with the area function first. To this end, we introduce the space $\mathcal{K}$ of (classes of) measurable functions $h : \Gamma \to \mathbb{C}$ such that

$$\|h\|_{\mathcal{K}} := \left(\int\!\!\int_\Gamma |h(Z)|^2 |Z|^{1-m} dZ \right)^{1/2} < +\infty,$$

and note that if

$$\mathcal{S}f(X)(Z) := \int_{\Sigma} f(Y)(\partial_0 E_k)(X - Y + Z)\, d\sigma(Y), \quad X \in \Sigma,\ Z \in \Gamma,$$

then $\mathcal{A}(f)(X) = \|\mathcal{S}(f\, n)(X)\|_{\mathcal{K}}$, $X \in \Sigma$. Hence, the estimate from above for the area function will be a consequence of the boundedness of the operator $\mathcal{S}$, from $L^p(\Sigma, d\sigma)$ into $L^p_{\mathcal{K}}(\Sigma, d\sigma)$, the space of $\mathcal{K}$-valued, L^p-integrable functions on Σ. This in turn, by Theorem 7.5.2, comes down to checking that $K(X, Y)(Z) := (\partial_0 E_k)(X - Y + Z)$ satisfies the conditions (7.5.4)-(7.5.6). Since (7.5.5) and (7.5.6) are simple consequences of $D_k E_k = E_k D_k = 0$, we are left with (7.5.4). However, by definition,

$$\|K(X, Y)\|_{\mathcal{K}}^2 = \int\int_{\Gamma} |(\partial_0 E_k)(X - Y + Z)|^2 |Z|^{1-m} dZ, \quad X, Y \in \Sigma,$$

and since for $Z \in \Gamma$ we have $\max\{|Z|, |X - Y|\} \lesssim |X - Y + Z|$, (7.5.4) follows easily by employing the decay of ∇E_k. This concludes the proof of the boundedness of $\mathcal{S}$.

The treatment of the estimate from above for the g-function is essentially the same. More specifically, this time we take $\mathcal{K} := L^2((0, \infty), t\, dt)$ and consider

$$\mathcal{T}_{\pm} f(X)(t) := \partial_0(\mathcal{C}_k f)(X \pm t), \quad X \in \Sigma,\ t > 0, \tag{7.6.2}$$

such that $g(f)(X) = \|\mathcal{T}_+ f(X)\|_{\mathcal{K}}$. Since the kernels associated with the integral operators $\mathcal{T}_{\pm}$ can be seen to satisfy (7.5.4)-(7.5.6), the conclusion follows by using again Theorem 7.5.2.

To obtain the bound from below we use a Littlewood-Paley type identity (here we adapt an argument from [DJS]).

First we claim that for any $F \in \mathcal{H}^p_k(\Omega)$ and any fixed $t > 0$, $\partial_0 F(\cdot + t)$ belongs to $\mathcal{H}^p_k(\Omega)$, and that, moreover,

$$t(\partial_0 F)(X + t) \to 0, \text{ as } t \to \infty \text{ or } t \to 0, \text{ for a.e. } X \in \Sigma. \tag{7.6.3}$$

To justify this, observe that, by differentiating the Cauchy integral reproducing formula, we get

$$t(\partial_0 F)(X + t) = \int_{\Sigma} F(Y) n(Y)\, t\, (\partial_0 E_k)(Y - X - t)\, d\sigma(Y).$$

Now since $t(\partial_0 E_k)(Y - X - t)$ decays on Ω, the well-known argument gives us that $|t(\partial_0 F)(X + t)| \lesssim F^*(X)$, uniformly in $X \in \Sigma$ and $t > 0$. Consequently,

$$\|\sup_{t>0} |t(\partial_0 F)(\cdot + t)|\|_{L^p(\Sigma, d\sigma)} \lesssim \|F\|_{L^p(\Sigma, d\sigma)} \tag{7.6.4}$$

hence, in particular, $(\partial_0 F)(\cdot + t) \in \mathcal{H}^p_k(\Omega)$.

The first convergence in (7.6.3) is easily seen by using, e.g., the Poisson-like decay of $t(\partial_0 E_k)(X + t)$ on Ω. However, the limit for $t \to 0$ is a bit more subtle. First we note that

$$\lim_{t \to 0} \int_{\substack{|Y| \leq C \\ Y \in \Sigma}} n(Y)(\partial_0 E_k)(Y - X - t) \, d\sigma(Y) = 0, \quad X \in \Sigma.$$

Using (7.4.3), this can be readily reduced to verifying that

$$\lim_{t \to 0} \int \int_{\substack{|Y| \leq C \\ Y \in \Omega}} (\partial_0 E_k)(Y + t) dY = 0,$$

which, in turn, follows from the decay properties of ∇E_k. With this at hand, one can easily check that for all $X \in \Sigma$

$$\lim_{t \to 0} \int_\Sigma f(Y) n(Y) t (\partial_0 E_k)(Y - X - t) \, d\sigma(Y) = 0,$$

if, e.g., f is Lipschitz continuous, compactly supported on Σ. Moreover, once again due to the Poisson-like behavior of $t(\partial_0 E_k)(X + t)$ on Ω,

$$\sup_{t > 0} \left| \int_\Sigma f(Y) n(Y) t (\partial_0 E_k)(Y - X - t) \, d\sigma(Y) \right| \lesssim f^*(X),$$

therefore the maximal operator canonically associated to the type of convergence in question is bounded on $L^p(\Sigma, d\sigma)$. Thus, the usual argument completes the proof of the claim.

Our next claim is that for any $f \in L^p(\Sigma, d\sigma)$ we have

$$\left\| \sup_{\epsilon, N > 0} \left| \int_\epsilon^N t \partial_0^2 (C_k f)(X + 2t) \, dt \right| \right\|_{L^p(\Sigma, d\sigma)} \lesssim \|f\|_{L^p(\Sigma, d\sigma)}, \tag{7.6.5}$$

and, at almost every $X \in \Sigma$,

$$\lim_{\substack{\epsilon \to +0 \\ N \to +\infty}} \int_\epsilon^N t \partial_0^2 (C_k f)(X + 2t) \, dt = -\frac{1}{8}(I + H_k) f(X). \tag{7.6.6}$$

To prove this, we integrate by parts twice

$$\int_\epsilon^N t \partial_0^2 (C_k f)(X + 2t) \, dt = \frac{1}{2} t \partial_0 (C_k f)(X + 2t) \Big|_\epsilon^N - \frac{1}{4}(C_k f)(X + 2t) \Big|_\epsilon^N.$$

Thus (7.6.5) is a consequence of (7.6.4) and (2) $\Rightarrow$ (4) in Theorem 7.6.1, while (7.6.6) follows from (7.6.3) and the trace formula for the Cauchy integral.

Let us now remark that if $F \in \mathcal{H}_k^p(\Omega)$, then for any $X \in \Sigma$ and any $t > 0$,

$$(\partial_0^2 F)(X + 2t) = -\int_\Sigma (\partial_0 F)(Y + t) \, n(Y) \, (\partial_0 E_k)(Y - X - t) \, d\sigma(Y). \tag{7.6.7}$$

In fact, this is simply obtained by differentiating

$$(\partial_0 F)((X + s) + t) = \int_\Sigma (\partial_0 F)(Y + t)\, n(Y)\, E_k(Y - X - s)\, d\sigma(Y)$$

with respect to s, and then making $s = t$.

In what follows, we shall use the superscripts l and r to be able to distinguish between the two versions of the the operators in (7.6.2) corresponding to $\mathcal{C}_k^l$ and $\mathcal{C}_k^r$, respectively (here we employ the convention made at the end of section 3).

For arbitrary $f \in L^p(\Sigma, d\sigma)$ and $f' \in L^q(\Sigma, d\sigma)$, we write the identity (7.6.7) for $F := \mathcal{C}_k^l f$, multiply both sides by $n(X)\, f'(X)$ on the right, and then integrate the resulting identity on Σ against $d\sigma(X)$. If we introduce the Clifford algebra-bilinear form $\langle \cdot, \cdot \rangle$ by setting

$$\langle f, f' \rangle := \int_\Sigma f(X)\, n(X)\, f'(X)\, d\sigma(X),$$

then the final equality reads

$$\int_\Sigma \partial_0^2 (\mathcal{C}^l f)(X + 2t)\, n(X)\, f'(X)\, d\sigma(X) = \langle T_+^l f, T_-^r f' \rangle + \langle T_+^l f, \mathcal{R}_t f' \rangle.$$

Here, for any t, the kernel $R_t(X, Y)$ of the "residual" integral operator $\mathcal{R}_t$ satisfies

$$|R_t(X, Y)| \lesssim |E_k(Y - X - t) + E_k(X - Y + t)| = \mathcal{O}(\Gamma_k(X - Y + t)),$$

uniformly in t. Using this, Schur's test gives that the operator $\mathcal{R} f(X)(t) := \mathcal{R}_t f(X)$ maps $L^p(\Sigma, d\sigma)$ boundedly into $L_{\mathcal{K}}^\infty(\Sigma, d\sigma)$, for any $1 < p < \infty$.

If we now integrate this identity against $\int_0^\infty t\, dt$, then, by permuting the integrals in the left-hand side and using (7.6.6), we get

$$-\frac{1}{8}\langle (I + H_k^l) f, f' \rangle = \int_0^\infty \langle T_+^l f, T_-^r f' + \mathcal{R}_t f' \rangle\, t\, dt.$$

To see how this can be used to conclude the proof of Lemma 7.6.2, for $1 < p, q < \infty$ conjugate exponents, and for $f \in L^p(\Sigma, d\sigma)$, $f' \in L^q(\Sigma, d\sigma)$, we write

$$|\langle (I + H_k^l) f, f' \rangle|$$

$$\lesssim \int_0^\infty \int_\Sigma |T_+^l f(X)(t)|(|T_-^r f'(X)(t)| + |\mathcal{R}_t f'(X)|)\, t\, dt\, d\sigma(X)$$

$$\lesssim \|T_+^l f\|_{L_{\mathcal{K}}^p(\Sigma, d\sigma)}(\|T_-^r f'\|_{L_{\mathcal{K}}^q(\Sigma, d\sigma)} + \|\mathcal{R} f'\|_{L_{\mathcal{K}}^q(\Sigma, d\sigma)})$$

$$\lesssim \|T_+^l f\|_{L_{\mathcal{K}}^p(\Sigma, d\sigma)} \|f'\|_{L^q(\Sigma, d\sigma)},$$

hence the proof is completed by a standard duality argument. To conclude, we note that the area-function can be handled similarly.

At this point we have completed the proof of the implication $(1) - (4) \Rightarrow$ $(5) - (6)$ together with the corresponding part in the estimate $(7.6.1)$. Consequently, we turn our attention to converse implications, i.e., $(5) \Rightarrow$ $(1) - (4)$ and $(6) \Rightarrow (1) - (4)$.

Let F be as in (5) (the reasoning for F as in (6) is completely similar). Consider the Hilbert space $\mathcal{K} := L^2((0, \infty), t\,dt)$, and the $\mathcal{K}$-valued function U defined on Ω by

$$U(X)(t) := \partial_0 F(X + t), \quad X \in \Omega, \ t > 0.$$

Note that $U_{\mathrm{rad}}(X) = g(F)(X)$, hence $U_{\mathrm{rad}} \in L^p(\Sigma, d\sigma)$. Also, $UD_k = 0$ in Ω. According to the part already proved in Theorem 7.6.1, U has a nontangential boundary trace on Σ, denoted by the same symbol U, $U \in L^p_{\mathcal{K}}(\Sigma, d\sigma)$, which is readily seen to be

$$U(Y)(t) = \partial_0 F(Y + t), \quad \text{for a.e. } Y \in \Sigma \text{ and } t > 0.$$

Next we claim that

$$t\,|\partial_0 F(X + t)| \lesssim U^*(X), \tag{7.6.8}$$

uniformly for $t > 0$ and $X \in \Sigma$. To see this, note that there exists a constant $0 < \lambda < 1$ depending only on Ω such that $B_{\lambda t}(X + t) \subset \Omega$ so that, using the mean-value theorem for subharmonic functions, we have

$$|\partial_0 F(X + t)| \lesssim \frac{1}{|B_{\lambda t}(X + t)|} \iint_{B_{\lambda t}(X+t)} |\partial_0 F(W)|\, dW$$

(writing $W := Y + s$, with $Y \in \Sigma$ and $s > 0$)

$$\lesssim t^{-m-1} \int_{\Sigma \cap B_{\lambda t}(X)} \left(\int_{(1-\lambda)t}^{(1+\lambda)t} |\partial_0 F(Y + s)|\, ds \right) d\sigma(Y)$$

(using Hölder's inequality in the innermost integral)

$$\lesssim t^{-m-1} \int_{\Sigma \cap B_{\lambda t}(X)} \left(\int_{(1-\lambda)t}^{(1+\lambda)t} |\partial_0 F(Y + s)|^2 s\, ds \right)^{1/2} d\sigma(Y)$$

$$\lesssim t^{-m-1} \int_{\Sigma \cap B_{\lambda t}(X)} \|U(Y)\|_{\mathcal{K}}\, d\sigma(Y)$$

$$\lesssim t^{-1} U^*(X).$$

In particular, $\partial_0 F(\cdot + t) \in \mathcal{H}^p_k(\Omega)$ for any fixed $t > 0$. Now take $0 < \delta < N < \infty$, arbitrary otherwise. If we can prove that

$$\|F(\cdot + \delta) - F(\cdot + N)\|_{L^p(\Sigma, d\sigma)} \leq \text{const} < +\infty \tag{7.6.9}$$

uniformly in δ, N, and that $\lim_{t\to\infty} F(\cdot + t) = 0$, then Fatou's lemma will give

$$\|F(\cdot + \delta)\|_{L^p(\Sigma,\, d\sigma)} \lesssim \liminf_{N\to\infty} \|F(\cdot + \delta) - F(\cdot + N)\|_{L^p(\Sigma,\, d\sigma)} \lesssim 1,$$

i.e. $F \in \mathcal{H}_k^p(\Omega)$ and we are done. To this end, for a fixed $X \in \Sigma$, we write

$$F(X + N) - F(X + \delta) = \int_\delta^N \partial_0 F(X + t)\,dt$$

$$= t\,\partial_0 F(X + t)|_\delta^N - \int_\delta^N t\,\partial_0^2 F(X + t)\,dt.$$

In virtue of (7.6.8), the first term in the right-hand side of the last equality above is in $L^p(\Sigma,\, d\sigma)$ uniformly in δ and N, so we only have to control the second term in a similar fashion. The idea is to use the fact that $\partial_0 F(\cdot + t)$ belongs to $\mathcal{H}_k^p(\Omega)$ and therefore one can still use the identity (7.6.7). Integrating both sides of this identity against $\int_\delta^N t\,dt$ we get

$$\int_\delta^N t\,\partial_0^2 F(X + t)\,dt =$$

$$4 \int_\Sigma \int_{\delta/2}^{N/2} \partial_0 F(Y + t)\, n(Y)\, (\partial_0 E_k)(Y - X - t)\, t\,dt\,d\sigma(Y)$$

and, by introducing $G(Y)(t) := \overline{\partial_0 F(Y + t)\, n(Y)}\, \chi_{(\delta/2,N/2)}(t)$, $Y \in \Sigma$, $t > 0$, and the kernel $K(X,Y)(t) := \overline{(\partial_0 E_k)(Y - X - t)}$

$$= \int_\Sigma \langle G(Y), K(X,Y) \rangle\, d\sigma(Y)$$

$$=: SG(X),$$

where the pairing $\langle \cdot, \cdot \rangle$ refers to the Hilbert space $\mathcal{K} \otimes \mathbb{R}_{(M)}$. With the aid of Theorem 7.5.2, the integral operator S is easily checked to be bounded so that

$$\left\| \int_\delta^N t\,\partial_0^2 F(X + t)\,dt \right\|_{L^p(\Sigma,\, d\sigma)} \lesssim \|SG\|_{L^p(\Sigma,\, d\sigma)}$$

$$\lesssim \|G\|_{L_{\mathcal{K}}^p(\Sigma,\, d\sigma)} \lesssim \|g(F)\|_{L^p(\Sigma,\, d\sigma)}.$$

From this, (7.6.9) follows.

Finally, standard arguments show that $\lim_{t\to\infty} F(X + t)$ exists and it is independent of $X \in \Sigma$, hence, by hypothesis, this limit is zero. The proof of Theorem 7.6.1 is therefore complete.

7.7   Rellich Type Formulas for Monogenic Functions

For the rest of the paper, we shall specialize some of our previous results for $k = 0$, in which case we shall omit to write it as a subscript, i.e., we shall use $\mathcal{H}^p(\Omega)$ in place of $\mathcal{H}_0^p(\Omega)$, etc.

First we note a boundary cancellation property for two-sided monogenic functions in $\mathcal{H}^2(\Omega)$. In $\mathbb{R}^2 \equiv \mathbb{C}$, this statement corresponds precisely to the usual Cauchy theorem for the holomorphic function F^2.

LEMMA 7.7.1
For any two-sided monogenic function F in $\mathcal{H}^2(\Omega)$ one has

$$\int_\Sigma F\,n\,F\,d\sigma = 0.$$

The proof is readily seen from (7.2.5) plus a limiting argument. In fact, one can show that (any component of) FnF belongs to the atomic Hardy space $H_{\mathrm{at}}^1(\Sigma)$, but we shall not pursue this here.

For a Clifford algebra-valued function F we set $F^\pm := \frac{1}{2}(F \pm \overline{F})$. Next, we shall establish the relevant Rellich type identities for monogenic functions.

LEMMA 7.7.2
In the above hypotheses, one has

$$\mathrm{Re}\int_\Sigma F(nF)^\pm\,d\sigma = \mathrm{Re}\int_\Sigma FnF^\pm\,d\sigma = \pm\frac{1}{2}\int_\Sigma |F|^2\,\mathrm{Re}\,n\,d\sigma.$$

Note that the classical Rellich identities for harmonic functions (see [DK]) are obtained for the particular case in which F is the gradient of a harmonic function u in Ω (see also section 9).

PROOF   As $2(Fn)^\pm = Fn \pm \overline{Fn} = Fn \pm \overline{n}\,\overline{F}$, and since $n + \overline{n} = 2\mathrm{Re}\,n$,

$$\mathrm{Re}\,(F\overline{n}\,\overline{F}) = \frac{1}{2}\mathrm{Re}\,(F\overline{n}\,\overline{F} + \overline{F\overline{n}\,\overline{F}}) = \frac{1}{2}\mathrm{Re}\,\{F(\overline{n} + n)\overline{F}\}$$

$$= \frac{1}{2}\mathrm{Re}\,(F\overline{F})(2\,\mathrm{Re}\,n) = |F|^2\mathrm{Re}\,n,$$

everything is readily seen from Lemma 7.7.1.   ∎

Finally, the main result of this section is the following.

THEOREM 7.7.3
For any two-sided monogenic function F in $\mathcal{H}^2(\Omega)$ one has

$$\|F\|_{L^2(\Sigma)} \approx \|F^\pm\|_{L^2(\Sigma)} \approx \|Fn\|_{L^2(\Sigma)} \approx \|(Fn)^\pm\|_{L^2(\Sigma)} \approx \|(nF)^\pm\|_{L^2(\Sigma)}.$$

The proof is an immediate consequence of Schwarz inequality, the fact that $\mathrm{Re}\, n \le C < 0$ almost everywhere on Σ and the identities deduced in the previous lemma.

7.8 A Burkholder-Gundy-Silverstein Theorem for Monogenic Functions in Lipschitz Domains

In the classical setting of one complex variable, the theorem of Burkholder-Gundy-Silverstein asserts that a holomorphic function belongs to the Hardy space $\mathcal{H}^p(\mathbb{R}^2_+)$ if and only if the nontangential maximal function of its real part belongs to $L^p(\mathbb{R})$, $0 < p < \infty$ (see [BGS]).

In this section we shall state and outline the proof of a Burkholder-Gundy-Silverstein type result in the Clifford algebra framework. Complete details can be found in [M3].

THEOREM 7.8.1

If F is two-sided monogenic in Ω such that $\lim_{t \to \infty} F(X + t) = 0$ for some $X \in \Sigma$ then, for $0 < p < \infty$, the following are equivalent:

1. $\mathcal{N}(F) \in L^p(\Sigma, d\sigma)$;
2. $\mathcal{A}(F) \in L^p(\Sigma, d\sigma)$;
3. $g(F) \in L^p(\Sigma, d\sigma)$;
4. $\mathcal{N}(F^{\pm}) \in L^p(\Sigma, d\sigma)$;
5. $\mathcal{A}(F^{\pm}) \in L^p(\Sigma, d\sigma)$;
6. $g(F^{\pm}) \in L^p(\Sigma, d\sigma)$.

In addition, if all these conditions are fulfilled, then also

$$\|\mathcal{N}(F)\|_{L^p} \approx \|\mathcal{A}(F)\|_{L^p} \approx \|g(F)\|_{L^p}$$

$$\approx \|\mathcal{N}(F^{\pm})\|_{L^p} \approx \|g(F^{\pm})\|_{L^p} \approx \|\mathcal{A}(F^{\pm})\|_{L^p}.$$

PROOF [Sketch of proof] We shall only explain the case $0 < p \le 2$. The idea of proof (cf. also [K1, K2]) is to use Theorem 7.7.3 and Theorem 7.6.1 to treat first the case $p = 2$. Then, the fact that this L^2 theory is valid in arbitrary Lipschitz domains will allow us to extend the results in the range $0 < p < 2$ via some "good λ" inequalities.

Let us first assume that $\mathcal{N}(F^{\pm}) \in L^p(\Sigma, d\sigma)$. Since

$$F(X) = \int_0^{\infty} \overline{D} F^{\pm}(X + t)dt,$$

using Theorem 7.6.1 and some well-known estimates for harmonic functions we infer that $F_{\epsilon} := F(\cdot + \epsilon) \in \mathcal{H}^2(\Omega)$ for any $\epsilon > 0$.

Fix $\epsilon > 0$, $\lambda > 0$, introduce $\mathcal{O}_\lambda := \{X \in \Sigma; \mathcal{N}(F_\epsilon^\pm)(X) > \lambda\}$ and consider the "sawtooth region" $D_\lambda := \cup_X (\Gamma + X)$, $X \in \Sigma \backslash \mathcal{O}_\lambda$. By Theorem 7.6.1 and Theorem 7.7.3 in which we take $\mathcal{N}'$ to be the nontangential maximal operator corresponding to the Lipschitz domain D_λ, we have

$$\int_{\Sigma \backslash \mathcal{O}_\lambda} |\mathcal{N}'(F_\epsilon)|^2 \, d\sigma \leq \int_{\partial D_\lambda} |F_\epsilon|^2 \, d\sigma \lesssim \int_{\partial D_\lambda} |F_\epsilon^\pm|^2 \, d\sigma$$

$$\lesssim \int_{\Sigma \backslash \mathcal{O}_\lambda} |F_\epsilon^\pm|^2 \, d\sigma + \int_{\partial D_\lambda \backslash \Sigma} |F_\epsilon^\pm|^2 \, d\sigma.$$

Note that $|F_\epsilon^\pm| \leq \mathcal{N}(F_\epsilon^\pm) \leq \lambda$ on $\partial D_\lambda \backslash \Sigma$ by construction, and that

$$\int_{\Sigma \backslash \mathcal{O}_\lambda} |F_\epsilon^\pm|^2 \, d\sigma \leq \int_{\Sigma \backslash \mathcal{O}_\lambda} |\mathcal{N}(F_\epsilon^\pm)|^2 \, d\sigma \leq 2 \int_0^\lambda t \, d\sigma(\mathcal{O}_t) \, dt.$$

Since $d\sigma(\partial D_\lambda) \approx d\sigma(\Sigma \backslash \mathcal{O}_\lambda)$, the above estimates amount to

$$\int_{\Sigma \backslash \mathcal{O}_\lambda} |\mathcal{N}'(F_\epsilon)|^2 \, d\sigma \lesssim \int_0^\lambda t \, d\sigma(\mathcal{O}_t) \, dt + \lambda^2 \, d\sigma(\mathcal{O}_\lambda)$$

which in turn implies that

$$d\sigma(\{X \in \Sigma; |\mathcal{N}'(F_\epsilon)(X)| > \lambda\}) \lesssim d\sigma(\mathcal{O}_\lambda) + \lambda^{-2} \int_0^\lambda t \, d\sigma(\mathcal{O}_t) \, dt$$

After multiplying by λ^{p-1} and integrating against $\int_0^\infty d\lambda$, this yields

$$\|\mathcal{N}(F_\epsilon)\|_{L^p(\Sigma)} \lesssim \|\mathcal{N}'(F_\epsilon)\|_{L^p(\Sigma)} \lesssim \|\mathcal{N}(F_\epsilon^\pm)\|_{L^p(\Sigma)} \lesssim \|\mathcal{N}(F^\pm)\|_{L^p(\Sigma)}.$$

Finally, letting $\epsilon \to 0$ we obtain the equivalence

$$\|\mathcal{N}(F)\|_{L^p(\Sigma)} \approx \|\mathcal{N}(F^\pm)\|_{L^p(\Sigma)}.$$

Next we turn our attention to the area- and g-function. First, with $\mathcal{A}(F_\epsilon)$ instead of F_ϵ and $\mathcal{N}(F_\epsilon)$ in place of $\mathcal{N}(F_\epsilon^\pm)$, the same arguments as before give that

$$\|g(F)\|_{L^p(\Sigma)} \lesssim \|\mathcal{A}(F)\|_{L^p(\Sigma)} \lesssim \|\mathcal{N}(F)\|_{L^p(\Sigma)}.$$

For the converse inequalities we follow the same line, taking this time $g(F_\epsilon)$ to stand for $F_\epsilon^\pm$, so that we arrive at $\|\mathcal{N}(F_\epsilon)\|_{L^p} \lesssim \|\mathcal{N}(g(F_\epsilon))\|_{L^p}$. By Lebesgue's monotone convergence theorem it suffices to show that

$$\|\mathcal{N}(g(F_\epsilon))\|_{L^p(\Sigma, d\sigma)} \lesssim \|g(F_\epsilon)\|_{L^p(\Sigma, d\sigma)}.$$

To this effect, consider the Hilbert space $\mathcal{H} := L^2((0, \infty), t \, dt)$, and the $\mathcal{H}$-valued harmonic function U in Ω defined by $U(X)(t) := \partial_0 F_\epsilon(X + t)$. It is easy to see that the argument in [FS, page 170] extends to this situation so that

$$[\mathcal{N}(U)]^{p/2}(X) \lesssim [(U_{\mathrm{rad}})^{p/2}]^*(X).$$

Now $\|U(X)\|_{\mathcal{H}} = g(F_\epsilon)(X)$, hence $U_{\mathrm{rad}}(X) \lesssim g(F_\epsilon)(X)$. Consequently,

$$\|\mathcal{N}(g(F_\epsilon))\|_{L^p(\Sigma)} \lesssim \|U_{\mathrm{rad}}\|_{L^p(\Sigma)} \lesssim \|g(F_\epsilon)\|_{L^p(\Sigma)},$$

by the L^2-boundedness of the maximal operator.

Similar arguments complete the proof of this theorem. $\blacksquare$

An useful observation is that any $\mathbb{R}^{m+1}$-valued, left (or right) monogenic function is automatically two-sided monogenic (cf. Lemma 3.2) so the conclusions of Theorem 7.8.1 are valid for such functions.

7.9 Some Applications to the Theory of Harmonic Functions in Lipschitz Domains

Recall the Dirichlet problem for the Laplace operator in Ω

$$(D) \quad \left\{ \begin{array}{l} \triangle u = 0 \text{ in } \Omega, \\ \mathcal{N}(u) \in L^2(\Sigma, d\sigma), \\ u|_\Sigma = f \in L^2(\Sigma, d\sigma), \end{array} \right.$$

and the Neumann problem for the Laplacian in Ω

$$(N) \quad \left\{ \begin{array}{l} \triangle u = 0 \text{ in } \Omega, \\ \mathcal{N}(\triangledown u) \in L^2(\Sigma, d\sigma), \\ \frac{\partial u}{\partial n}|_\Sigma = f \in L^2(\Sigma, d\sigma). \end{array} \right.$$

The traces appearing in (D) and (N) should be understood in the sense of the nontangential limit to the boundary.

THEOREM 7.9.1

If u is harmonic in Ω and $\mathcal{N}(\triangledown u) \in L^2(\Sigma, d\sigma)$, then $\triangledown u$ has a nontangential boundary limit $\triangledown u(X)$ at almost every point $X \in \Sigma$ and

$$\|\mathcal{N}(\triangledown u)\|_{L^2} \approx \|\partial_0 u\|_{L^2} \approx \sum_{j=1}^m \|\partial_j u\|_{L^2}$$

$$\approx \| \triangledown u\|_{L^2} \approx \|\partial u/\partial n\|_{L^2} \approx \| \triangledown_T u\|_{L^2}, \tag{7.9.1}$$

where $\triangledown_T := \triangledown - n\frac{\partial}{\partial n}$ is the usual tangential gradient.

The proof is readily seen by using Theorem 7.7.3 for $F := \overline{D}u$.

It is a well-known fact that the estimates (7.9.1) can be used to prove the existence part in (D) and (N) ([Ve]). The uniqueness part for (N) is already contained in (7.9.1), while the uniqueness part in (D) is seen from the *a priori* estimate

$$\|u\|_{L^2(\Sigma,\, d\sigma)} \approx \|\mathcal{N}(u)\|_{L^2(\Sigma,\, d\sigma)},$$

implied by Theorem 7.7.3, Theorem 7.8.1 and the next lemma.

LEMMA 7.9.2
Let u be a harmonic function in Ω such that $\mathcal{N}(u) \in L^p(\Sigma,\, d\sigma)$ (or $u_{rad} \in L^p(\Sigma,\, d\sigma)$, $\lim_{t\to\infty} u(X+t) = 0$ and $\mathcal{A}(u) \in L^p(\Sigma,\, d\sigma)$, $\lim_{t\to\infty} u(X+t) = 0$ and $g(u) \in L^p(\Sigma,\, d\sigma)$, respectively), $0 < p < \infty$.
Then there exists a unique $\mathbb{R}^{m+1}$−valued monogenic function F in Ω which dies at infinity and such that $\operatorname{Re} F = u$.

We can construct F in Ω by setting (cf. [FS])

$$F(X + t) := -\int_t^\infty (\overline{D}u)(X + s)\, ds,$$

where $X \in \Sigma$ and $t > 0$.

By Theorem 7.6.1, the important thing is that F has roughly the same "size" as u, i.e., $\|\mathcal{N}(F)\|_{L^p} \approx \|\mathcal{N}(u)\|_{L^p}$ (or $\|F_{rad}\|_{L^p} \approx \|u_{rad}\|_{L^p}$, $\|\mathcal{A}(F)\| L^p \approx \|\mathcal{A}(u)\|_{L^p}$, $\|g(F)\|_{L^p} \approx \|g(u)\|_{L^p}$, respectively).

An immediate corollary of Theorem 7.8.1 and Lemma 7.9.2 is the following graph version of a theorem of Dahlberg [D] concerning the norm equivalence between the area- and the nontangential maximal function of a harmonic function.

COROLLARY 7.9.3
Let $0 < p < \infty$. If the function u is harmonic in Ω and has $\lim_{t\to\infty} u(X + t) = 0$ for some $X \in \Sigma$, then the following are equivalent:

1. *$\mathcal{N}(u) \in L^p(\Sigma,\, d\sigma)$;*
2. *$u_{rad} \in L^p(\Sigma,\, d\sigma)$;*
3. *$\mathcal{A}(u) \in L^p(\Sigma,\, d\sigma)$;*
4. *$g(u) \in L^p(\Sigma,\, d\sigma)$.*

In addition, if all these conditions are satisfied, then

$$\|u_{rad}\|_{L^p(\Sigma)} \approx \|\mathcal{N}(u)\|_{L^p(\Sigma)} \approx \|\mathcal{A}(u)\|_{L^p(\Sigma)} \approx \|g(u)\|_{L^p(\Sigma)}.$$

Finally, we discuss the following corollary (cf. [D]).

COROLLARY 7.9.4

For any harmonic function u which dies at infinity in Ω*, one has*

$$\int_{\Sigma} |u|^2 \, d\sigma \approx \int\!\!\int_{\Omega} | \bigtriangledown u|^2 \, dist\{X, \Sigma\} \, dVol.$$

PROOF   Clearly, this is an equivalent formulation of $\|u\|_{L^2} \approx \|g(u)\|_{L^2}$, which is provided by Theorem 7.6.1 and Theorem 7.7.3.   ∎

Bibliography

[AJM]   L. Adolfsson, B. Jawerth and M. Mitrea, *The Cauchy singular integral operator and Clifford wavelets*, (J. Benedetto and M. Frazier Eds.), in Wavelets: Mathematics and Applications, CRC Press, Boca Raton, FL, 519–540.

[AT]   P. Auscher and Ph. Tchamitchian, *Bases d'ondelettes sur des courbes corde-arc, noyau de Cauchy et espaces de Hardy associés*, Rev. Matemática Iberoamericana **5**, (1989), 139–170.

[BDS]   F. Brackx, R. Delanghe and F. Sommen, *Clifford Analysis*, Pitman Adv. Pub. Program, 1982.

[BGS]   D. L. Burkholder, R. F. Gundy and M. L. Silverstein, *A maximal function characterization of the class H^p*, Trans. Amer. Math. Soc. **157**, (1971), 137–153.

[CMM]   R. R. Coifman, A. McIntosh and Y. Meyer, *L'intégrale de Cauchy définit un opérateur borné sur L^2 pour les courbes lipschitziennes*, Ann. of Math. **116**, (1982), 361–388.

[D]   B. E. J. Dahlberg, *Weighted norm inequalities for the Lusin area integral and the nontangential maximal function for functions harmonic in a Lipschitz domain*, Studia Math., **67**, (1980), 297–314.

[DK]   B. E. J. Dahlberg and C. E. Kenig, *Hardy spaces and the L^p-Neumann problem for Laplace's equation in a Lipschitz domain*, Ann. of Math., **125**, (1987), 437–465.

[Da]   G. David, *Wavelets and Singular Integrals on Curves and Surfaces*, Lecture Notes, vol. 1465, Springer Verlag, New York, 1991.

[DJ]   G. David, and J.-L. Journé, *A boundedness criterion for generalized Calderón-Zygmund operators* Annals of Mathematics, **120**, (1984), 371–397.

[DJS]   G. David, J.-L. Journé and S. Semmes *Operatéurs de Calderó-Zygmund, fonctiones para-accretives et interpolation*, Rev. Mat.

Iberoamericana, **1**, (1985), 1–56.

[Di]   A. C. Dixon, *On the Newtonian potential*, Quarterly J. of Math., **35**, (1904), 283–296.

[FS]   C. Fefferman and E. M. Stein, H^p *spaces of several variables* Acta Math., **129**, (1972), 137–193.

[Fu]   B. Fueter, *Analytische Funktionen einer Quaternionenvariablen* Comment. Math. Helv., **4**, (1932), 9–20.

[Ga]   H. G. Garnir, *Les problémes aux limites de la physique mathé matique*, Birkhäuser-Verlag, Basel, 1958.

[GLQ]   G. I. Gaudry, R. L. Long, and T. Qian, *A Martingale Proof of L^2 Boundedness of Clifford-Valued Singular Integrals*, Annali di Matematica Para ed Applicata, CLXV, (1993), 369–394.

[GM1]   J. Gilbert and M. A. Murray, H^p-*Theory on Euclidean space and the Dirac operator* Rev. Mat. Iberoamericana, **4**, (1988), 253–289.

[GM2]   J. Gilbert and M. A. Murray, *Clifford Algebras and Dirac Operators in Harmonic Analysis*, Cambridge Studies in Advanced Mathematics, **26**, (1991).

[HS]   D. Hestenes and G. Sobczyk, *Clifford Algebra to Geometric Calculus. A Unified Language for Mathematics and Physics* D. Reidel Publ. Comp., Dordrecht, Boston, 1984.

[H]   K. Hoffman, *Banach Spaces of Analytic Functions* Dover Publ., New York, 1988.

[Ke]   C. E. Kenig, *Weighted H^p spaces on Lipschitz domains*, Amer. J. Math., **102**, (1980), 129–163.

[K1]   P. Koosis, *Sommabilité de la fonction maximale et appartenance à H_1*, C. R. Acad. Sci. Paris, **286**, (1978), 1041–1044.

[K2]   P. Koosis, *Sommabilité de la fonction maximale et appartenance à H_1 Cas de plusieurs variables*, C. R. Acad. Sci. Paris, **288**, (1979), 489–492.

[LMS]   C. Li, A. McIntosh and S. Semmes, *Convolution singular integrals on Lipschitz surfaces*, J. Am. Math. Soc., **5**, (1992), 455–481.

[LMQ]   C. Li, A. McIntosh and T. Qian, *Clifford algebras, Fourier transforms, and singular convolution operators on Lipschitz surfaces*, Revista Matematica Iberoamericana, 10(1994), 665–721.

[Mc]   A. McIntosh, *Clifford algebras and the higher dimensional Cauchy integral*, Approximation Theory and Function Spaces, Banach Center Publications, **22**, (1989), 253–267.

[MM]   A. McIntosh and Y. Meyer, *Algebres d'opérateurs definis par des integrales singulieres* C.R. Acad. Sci. Paris, Ser. I Math., **301**, (1985), 395–397.

[Me]   Y. Meyer, *Ondelettes et opérateurs*, Hermann, Paris, 1990.

[M1]   M. Mitrea, *Hardy spaces and Clifford algebras*, 1991, preprint.

[M2]   M. Mitrea, *Constructions of Clifford Wavelets* Advances in Applied Clifford Algebras, **2**, (1992), 249–276.

[M3]   M. Mitrea, *Clifford wavelets, Singular integrals, and Hardy spaces*, Lecture Notes, Springer-Verlag, New York, 1994, 1575.

[M4]   M. Mitrea, *Harmonic and Clifford analytic functions in nonsmooth domains*, 1993, preprint.

[M5]   M. Mitrea, *Clifford algebras and boundary estimates for harmonic functions*, Clifford Algebras and Their Applications in Mathematical Physics, (F. Brackx, R. Delanghe and H. Serras eds.), Kluwer Academic Publ., Dordrecht, 1993, pp. 151–158.

[M6]   M. Mitrea, *The regularity of the Cauchy integral and related operators on Lipschitz hypersurfaces*, Complex Variables: Theory and Applications, (to appear).

[M7]   M. Mitrea, *Boundary value problems and Hardy spaces associated to the Helmholtz equation on Lipschitz domains*, 1992, preprint.

[M8]   M. Mitrea, *On Dahlberg's Lusin area integral theorem*, Proc. of Am. Math. Soc., (to appear).

[Mo]   G. Moisil, *Sur la généralisation des fonctions conjuguées*, Rendiconti della Reale Accad. Naz. dei Lincei, **14**, (1931), 401–408.

[MT]   G. C. Moisil and N. Teodorescu, *Fonctions holomorphes dans l'espace*, Mathematica Cluj, **5**, (1931), 142–150.

[Mu]   M. A. Murray, *The Cauchy integral, Calderón commutators and conjugations of singular integrals in $\mathbb{R}^m$*, Trans. Amer. Math. Soc., **289**, (1985), 497–518.

[P1]   D. Pompeiu, *Sur les singularités des fonctions analytiques uniformes*, C. R. Acad. Sci. Paris, **145**, (1909), 103–105.

[P2]   D. Pompeiu, *Sur une classe de fonctions d'une variable complexe*, Rendiconti del Circolo Matematico di Palermo, **33**, (1912), 108–113.

[P3]   D. Pompeiu, *Sur une définition des fonctions holomorphes*, C. R. Acad. Sci. Paris, **166**, (1918), 209–212.

[Ry]   J. Ryan, *Cauchy-Green type formulae in Clifford analysis*, Trans. of the Am. Math. Soc., 347, (1995), 1331–1341.

[Se1]   S. Semmes, *A criterion for the boundedness of singular integrals on hypersurfaces*, Trans. Amer. Math. Soc., **311**, (1989), 501–513.

[Se2]   S. Semmes, *Square function estimates and the $T(b)$ theorem*, Proc. of A.M.S., **110**, (1990), 721–776..

[St]   E. M. Stein, *Singular integrals and differentiability properties of functions*, Princeton University Press, Princeton, NJ, 1970.

[SW]   E. M. Stein and G. Weiss, *On the theory of harmonic functions of several variables, I. The theory of H^p spaces*, Acta Math., **103**, (1960), 25–62.

[Ta]   T. Tao, *Convolution operators generated by right-monogenic kernels*, 1992, preprint.

[Te]   N. Théodoresco, *La Dérivée Areolaire*, Ann. Roumanine des Math., Cahier 3, Bucharest, (1936).

[Ve]   G. Verchota, *Layer potentials and regularity for the Dirichlet problem for Laplace's equation in Lipschitz domains*, Journal of Functional Analysis, **59**, (1984), 572–611.

Department of Mathematics,
University of Minnesota,
Minneapolis, Minnesota 55455
E-mail: mitrea@math.umn.edu

8

Some Applications of Conformal Covariance in Clifford Analysis

John Ryan*

ABSTRACT Building on an earlier work, we show how the conformal group can be applied in both R^n and C^n to shed new light on results in Clifford analysis. This includes setting up a Bergman kernel over unbounded domains in C^n and exhibiting a conformal covariance for harmonic measure. Many of the results arise from the conformal covariance of cells of harmonicity and special real n-dimensional manifolds lying in C^n. These manifolds form natural generalizations of domains in R^n. We also use conformal covariance to adapt existing results to illustrate the L^2-boundedness of the double-layer potential operator over Lipschitz perturbations of the sphere, to describe a conformal covariance associated to mutually commuting operators over a real Banach algebra, and to set up boundary value problems for a particular inhomogeneous equation with Laplacian as the principal part.

8.1   Introduction

In an earlier work, [22], we show that the conformal group acting over $R^n \cup \{\infty\}$ plays a fundamental role in Clifford analysis in Euclidean space. In particular, we illustrate that the operators used by Gürlebeck and Sprössig in [8] to solve the Dirichlet problem, possess a conformal covariance. This enables us to transfer the Dirichlet problem over a bounded domain in R^n and solve a similar boundary value problem over an unbounded domain. This approach is made simple and unified by using a group of 2×2 matrices with entries in a Clifford algebra. This group gives a simple way of characterizing conformal transformations over $R^n \cup \{\infty\}$. It was first

---

*Research supported in part by a grant from the Arkansas Science and Technology Authority.

0-8493-8481-8/96/$0.00+$.50

introduced by Vahlen [28], and was reintroduced in a number of papers of Ahlfors; see, for instance, [1,2].

In [22], we use this group to describe intertwining operators for the Dirac operator in R^n, its iterates, and their fundamental solutions, seen as convolution operators. This builds upon ideas of Bojarski [5], and Peetre and Qian [16].

In this paper, we build upon results presented in [22] to illustrate further applications of the conformal group in Clifford analysis, particularly complex Clifford analysis. We begin by showing that the special manifolds described in [19] are preserved under conformal transformations over $\mathbf{C}^n$. From this, we are able to show that the cells of harmonicity associated to these manifolds are also preserved under these conformal transformations. As a consequence, it follows that the Huygens' principle described in [19,25] is preserved under Möbius transformations. Therefore, many of the existing formulae and results from complex Clifford analysis are preserved under these transformations. In particular, solutions to the equation $D_{\mathbf{C}}^k f = g$ possess a conformal covariance, as do the domains of holomorphy upon which these solutions live. Here, $D_{\mathbf{C}}$ is the Dirac operator in $\mathbf{C}^n$, k is a positive integer, and f, g are holomorphic functions. Also, the intrinsic Dirac operator D_M, introduced in [21] over a manifold $M, \subseteq \mathbf{C}^n$, with special properties, possesses a conformal covariance, as does the fundamental solution of D_M. Consequently, the methods used in [22] to establish conformal covariance for operators needed to set up and solve the Dirichlet problem in R^n carry over to the context described here.

The layout of the paper is as follows. After a preliminary section on the conformal group and Vahlen matrices over R^n and $\mathbf{C}^n$, we use geometric arguments to illustrate the conformal invariance of cells of harmonicity in several complex variables. We then return to the group of Vahlen matrices over R^n. First we illustrate that the Cayley map can be used to establish the L^2-boundedness of the singular Cauchy transform over Lipschitz perturbations of the sphere. We also use the Vahlen matrices to illustrate a conformal covariance of n-mutually commuting operators defined on a real Banach algebra. This builds upon ideas presented by McIntosh and Pryde in [12], see also [13]. We conclude this section by using the conformal group to study properties of solutions to the equation $\left(D + \frac{1}{\|c\underline{x}+d\|^2}\right) f(\underline{x}) = 0$, where $\begin{pmatrix} a & b \\ c & d \end{pmatrix}$ is a Vahlen matrix over R^n. This enables us to set the stage to carry over arguments used in [8] to study boundary value problems for the Laplacian and to set up and solve similar boundary problems for the operator $\left(D + \frac{1}{\|c\underline{x}+d\|^2}\right)^2$. This involves using Vahlen matrices to link results presented by X. Zhenyuan [29] with results presented in [8].

In the next section we return to the applications of conformal covariance in complex Clifford analysis. In particular, we introduce intertwining operators for the Dirac operator in $\mathbf{C}^n$, its iterates, the intrinsic Dirac operators introduced in [21], and their iterates. We also show that these intertwining operators are also intertwining operators for certain convolution operators. We then use results from this section to illustrate how the Bergman kernel for the complex Dirac operator can be set up, particularly over unbounded cells of harmonicity. This extends on results presented in [24]. Using homotopy arguments presented in [19], we are able to show that the Bergman kernel presented here is unique to the cell of harmonicity, and does not depend on the choice of representation over some underlying manifold. In this section, we also describe a conformal covariance for harmonic measure.

In the last section, we briefly illustrate how homotopy arguments can be used to describe intertwining operators for holomorphic continuations of the convolution operators arising from fundamental solutions to the Dirac operator, and its iterates, in $\mathbf{C}^n$. The arguments used enable one to see that the Huygens' principle described in [18,19], and [25], is also covariant, and leads to a generalization of some results presented in [18,19].

Acknowledgment

The author is grateful to Chun Li and Alan McIntosh for helpful comments while preparing this work, and to Josef Siciak for a brief conversation which helped to motivate some of the key ideas in this paper.

8.2   Preliminaries

In this section we introduce some basic properties that we need in the remainder of the paper. First, we introduce the real 2^n-dimensional Clifford algebra A_n with basis $1, e_1, \ldots, e_n, \ldots, e_1 \ldots e_n$, where $e_1, \ldots, e_n$ is an orthonormal basis for R^n and $e_i e_j + e_j e_i = -2\delta_{ij}$, the Kroneker delta function. So, we consider $R^n \subseteq A_n$. Each non-zero vector $\underline{x} \in R^n$ has a multiplicative inverse $\underline{x}^{-1} = \frac{-\underline{x}}{\|\underline{x}\|^2} \in R^n$.

This inverse is the Kelvin inverse of the vector x. It follows that the unit sphere S^{n-1} is the generator of a group

$$\{ a \in A_n : a = a_1 \ldots a_p : p \in N \text{ and } a_j \in S^{n-1} \text{ for } 1 \leq j \leq p \}$$

lying in A_n. This group is called the pin group, and it is denoted by $Pin(n)$.

There is an anti-automorphism $\tilde{\ } : A_n \rightarrow A_n : e_{j_1} \ldots e_{j_r} \mapsto e_{j_r} \ldots e_{j_1}$, where $j_1 < \cdots < j_r$ and $1 \leq r \leq n$. Moreover, by induction on p, it may be observed that for $\underline{x} \in R^n$ the action $a\underline{x}\tilde{a}$, with $a \in Pin(n)$ represents an

orthogonal transformation over R^n. In fact, as $y\,\underline{x}\,y$ for $y \in S^{n-1}$ represents
a reflection in the direction of y, each orthogonal transformation over R^n
can be expressed by $a\underline{x}\tilde{a}$ for some $a \in Pin(n)$. So there is a surjective
homomorphism $\theta : Pin(n) \rightarrow O(n)$, where $O(n)$ denotes the orthogonal
group over R^n. Moreover, $-a\underline{x}\,(-\tilde{a}) = a\underline{x}\tilde{a}$. So $\{\pm 1\} \subseteq \ker\theta$. Using direct
computation, it may be observed that $\ker\theta = \{\pm 1\}$.

We now consider the complex Clifford algebra $A_n(\mathbf{C})$. This is the com-
plexification of A_n. This algebra has the same basis as A_n, and the anti-
automorphism $\tilde{\;}: A_n \rightarrow A_n$ automatically extends to an anti-automorphism
$\tilde{\;}: A_n(\mathbf{C}) \rightarrow A_n(\mathbf{C})$. The complex space spanned by $e_1, \ldots, e_n$ can
easily be identified with $\mathbf{C}^n$. For each $\underline{z} = z_1 e_1 + \cdots + z_n e_n \in \mathbf{C}^n$ we
have that $\underline{z}^2 = -z_1^2 - \cdots - z_n^2$. So, not every non-zero vector in $\mathbf{C}^n$ has a
multiplicative inverse in $A_n(\mathbf{C})$; for instance, $(e_1 + ie_2)^2 = 0$.

We denote the set $\left\{\underline{z} \in \mathbf{C}^n : \underline{z}^2 = 0\right\}$ by $N(\underline{0})$, and for each $\underline{z_1} \in \mathbf{C}^n$
we denote the set $\left\{\underline{z} \in \mathbf{C}^n : (\underline{z} - \underline{z_1})^2 = 0\right\} = N(\underline{0}) + \underline{z_1}$ by $N(\underline{z_1})$. The
complex sphere in $\mathbf{C}^n$ is the set $\left\{\underline{z} \in \mathbf{C}^n : \underline{z}^2 = -1\right\}$, and we denote this
set by $S_{\mathbf{C}}^{n-1}$. Note that $S^{n-1} \subseteq S_{\mathbf{C}}^{n-1}$. Using $S_{\mathbf{C}}^{n-1}$, we can introduce a
complex extension of the group $Pin(n)$. We denote the group

$$\left\{a \in A_n(\mathbf{C}) : a = a_1 \ldots a_p : p \in \mathbf{N}, \text{ and } a_j \in S_{\mathbf{C}}^{n-1} \cup iS_{\mathbf{C}}^{n-1} \text{ for } 1 \le j \le p\right\}$$

by $Pin(\mathbf{C}^n)$. Moreover, $Pin(n) \subseteq Pin(\mathbf{C}^n)$.

Using induction on p, it may be observed that $a\underline{z}\tilde{a} \in \mathbf{C}^n$ for each $\underline{z} \in$
$\mathbf{C}^n$, and $(a\underline{z}\tilde{a})^2 = \underline{z}^2$. It follows that there is a group homomorphism
$\theta_{\mathbf{C}} : Pin(\mathbf{C}^n) \rightarrow O(\mathbf{C}^n)$, where $O(\mathbf{C}^n)$ is the complex orthogonal group
$\{O \in Hom_{\mathbf{C}}(\mathbf{C}^n, \mathbf{C}^n) : \langle O\underline{z}, O\underline{w}\rangle_{\mathbf{C}} = \langle \underline{z}, \underline{w}\rangle_{\mathbf{C}} \text{ with } \langle\underline{z}, \underline{w}\rangle_{\mathbf{C}} = \sum_{j=1}^n z_j w_j\}$.
Moreover, $\theta_{\mathbf{C}}|_{A_n} = \theta$. It may be deduced that $Pin(\mathbf{C}^n)$ is a covering group
of $O(\mathbf{C}^n)$, with finite kernel.

Besides the map $\tilde{\;}$, we also need the anti-automorphism $- : A_n(\mathbf{C}) \rightarrow$
$A_n(\mathbf{C}) : e_{j_1} \ldots e_{j_r} \mapsto (-1)^r e_{j_r} \ldots e_{j_1}$. We usually write $\bar{Z}$ for $-(Z)$, where
$Z \in A_n(\mathbf{C})$. Moreover, the identity component of $Z\bar{Z}$ is $z_0^2 + \cdots + z_{1\ldots n}^2$,
where $Z = z_0 + \cdots z_{1\ldots n} e_1 \ldots e_n$.

Let us now consider the following basic Möbius transformations within
$\mathbf{C}^n$. First, we have the translation

$$T_{\underline{w}} : \mathbf{C}^n \rightarrow \mathbf{C}^n : \underline{z} \mapsto \underline{z} + \underline{w}, \text{ with } \underline{w} \in \mathbf{C}^n,$$

second, there is the orthogonal transformation

$$O : \mathbf{C}^n \rightarrow \mathbf{C}^n : O \in O(\mathbf{C}^n),$$

then, there is the dilation

$$D_\lambda : \mathbf{C}^n \rightarrow \mathbf{C}^n : \underline{z} \mapsto \lambda\underline{z}, \text{ with } \lambda \in \mathbf{C}\backslash\{0\},$$

and last, there is the Kelvin inversion

$$Inv : \mathbf{C}^n \backslash N\left(\underline{0}\right) \to \mathbf{C}^n \backslash N\left(\underline{0}\right) : \underline{z} \mapsto \underline{z}^{-1}.$$

Associated to these transformations are the following matrices: $\begin{pmatrix} 1 & \underline{w} \\ 0 & 1 \end{pmatrix}$,

where $\underline{w} \in \mathbf{C}^n$, $\begin{pmatrix} a & 0 \\ 0 & \tilde{a}^{-1} \end{pmatrix}$, where $a \in Pin\left(\mathbf{C}^n\right)$, $\begin{pmatrix} \lambda^{\frac{1}{2}} & 0 \\ 0 & \lambda^{-\frac{1}{2}} \end{pmatrix}$, with

$\lambda \in \mathbf{C} \backslash \{0\}$, and $\begin{pmatrix} 0 & -1 \\ 1 & 0 \end{pmatrix}$. The group generated by these matrices is
called the complex Vahlen group, and it is denoted by $V\left(\mathbf{C}^n\right)$. In [20],
we use induction and density arguments to show that a general element of
$V\left(\mathbf{C}^n\right)$ can be expressed as $\begin{pmatrix} a & b \\ c & d \end{pmatrix}$, where $a, b, c, d \in A_n\left(\mathbf{C}\right)$, and have
the following properties:

(i)

$$\begin{aligned} a &= a_1 \ldots a_{p_1} \\ b &= b_1 \ldots b_{p_2} \\ c &= c_1 \ldots c_{p_3} \\ d &= d_1 \ldots d_{p_4}, \end{aligned}$$

with $p_1, p_2, p_3, p_4 \in \mathbf{N}$, $a_j, b_k, c_\ell, d_m \in \mathbf{C}^n$ for $1 \leq j \leq p_1$, $1 \leq k \leq p_2$,
$1 \leq \ell \leq p_3$, and $1 \leq m \leq p_4$.

(ii) $a\tilde{b}, b\tilde{c}, c\tilde{d}, d\tilde{a} \in \mathbf{C}^n$, and

(iii) $a\tilde{d} - b\tilde{c} = \pm 1$.

In [20] we establish:

LEMMA 8.2.1
Suppose $\begin{pmatrix} a & b \\ c & d \end{pmatrix} \in V\left(\mathbf{C}^n\right)$. *Then there is a sequence*

$$\left\{ \begin{pmatrix} a_m & b_m \\ c_m & d_m \end{pmatrix} \right\}_{m=1}^{\infty}, \subseteq V\left(\mathbf{C}^n\right)$$

such that $\lim_{n\to\infty} \begin{pmatrix} a_m & b_m \\ c_m & d_m \end{pmatrix} = \begin{pmatrix} a & b \\ c & d \end{pmatrix}$, *and the entries in each*

matrix $\begin{pmatrix} a_m & b_m \\ c_m & d_m \end{pmatrix}$ *are either zero or invertible, i.e.,* $a_m, b_m, c_m, d_m \in$
$Pin\left(\mathbf{C}^n\right) \times \mathbf{C}.$

The group $V\left(\mathbf{C}^n\right)$ contains as a subgroup the real Vahlen group $V(n) = \left\{ \begin{pmatrix} a & b \\ c & d \end{pmatrix} : a, b, c, d \in A_n, \text{ and } (i) \ a, b, c, d \in Pin(n) \times R, (ii) \ a\tilde{c}, \tilde{c}d, \tilde{d}b, \tilde{b}a \right.$
$\in R^n$, *and* $(iii) \ a\tilde{d} - b\tilde{c} = \pm 1 \Big\}$. Elements of this group are introduced by

Vahlen in [28] to describe Möbius transformations in R^n. This approach was rediscovered by Ahlfors, see, for instance [1,2], and has been used by a number of authors in recent years, see, for instance [5,16]. In particular, we have that for each $\begin{array}{cc} a & b \\ c & d \end{array} \in V(n)$, the expression $(a\underline{x} + b)(c\underline{x} + d)^{-1}$ is well defined for each $\underline{x} \in R^n \cup \{\infty\}$, and describes a Möbius transformation over $R^n \cup \{\infty\}$. Using Lemma 8.2.1, we are able to generalize this result in [20], and show that for each $\left(\begin{array}{cc} a & b \\ c & d \end{array} \right) \in V(\mathbf{C}^n)$, the expression $(a\underline{z} + b)(c\underline{z} + d))^{-1}$ is well defined in $\mathbf{C}^n$, except on some null cone, and gives a Möbius transformation within $\mathbf{C}^n$.

8.3   Conformal Invariance of Cells of Harmonicity

In [17,18,19] and elsewhere, we introduce the following special types of manifolds lying in $\mathbf{C}^n$.

DEFINITION 8.3.1   Suppose M is a connected, real, n-dimensional, C^1-submanifold of $\mathbf{C}^n$, with or without boundary, satisfying

(i) $M \cap N(\underline{z}) = \{\underline{z}\}$ and
(ii) $TM_{\underline{z}} \cap N(\underline{z}) = \{\underline{z}\}$

for each $\underline{z} \in M$. Then M is called a domain manifold. Here, $TM_{\underline{z}}$ is the tangent space to M at $\underline{z}$. If $M \subseteq R^n$, then $\overset{\circ}{M}$, the interior of M, corresponds to a domain. In [19], we use homotopy deformation arguments to show that there is an abundance of such manifolds. Also, in [17,18,19], and elsewhere, we show that much of classical function theory over domains in R^n extends to these domain manifolds.

We now establish the invariance of domain manifolds under conformal transformations.

THEOREM 8.3.2
Suppose that M is a domain submanifold, and $\left(\begin{array}{cc} a & b \\ c & d \end{array} \right) \in V(\mathbf{C}^n)$ is such that

$$\phi(\underline{z}) = (a\underline{z} + b)(c\underline{z} + d)^{-1}$$

is well defined for each $\underline{z} \in M$. Then the set $\phi(M)$ is also a domain manifold.

PROOF As ϕ is a holomorphic homeomorphism, it is easy to see that $\phi(M)$ is a C^1-manifold of real dimension n. So, we need to check conditions (i) and (ii) from Definition 1 to obtain the result.

Condition (i) follows if we can show that $(\phi(\underline{z}) - \phi(\underline{w}))^2 \neq 0$ for each distinct pair of vectors $\underline{z}$ and $\underline{w}$ in M. Let us first assume that c is invertible in $A_n(\mathbf{C})$. Then

$$
\begin{aligned}
(a\underline{z} + b) \, (c\underline{z} + d)^{-1} & \\
&= \left(ac^{-1}(cd + d) - ac^{-1}d + b\right) (c\underline{z} + d)^{-1} \\
&= ac^{-1} + \left(-ac^{-1}d + b\right)(c\underline{z} + d)^{-1} \\
&= ac^{-1} \pm (c\underline{z}\tilde{c} + d\tilde{c})^{-1}
\end{aligned}
$$

for each $\underline{z} \in M$, as $\phi(\underline{z})$ is well defined on M. So,

$$
\begin{aligned}
\phi(\underline{z}) - \phi(\underline{w}) &\\
&= \pm \left((c\underline{z}\tilde{c} + d\tilde{c})^{-1} - (c\underline{w}\tilde{c} + d\tilde{c})^{-1}\right) \\
&= \pm \left((c\underline{w}\tilde{c} + d\tilde{c})^{-1}(c\underline{w}\tilde{c} + d\tilde{c})(c\underline{z}\tilde{c} + d\tilde{c})^{-1}\right. \\
&\quad \left. - (c\underline{w}\tilde{c} + d\tilde{c})^{-1}(c\underline{z}\tilde{c} + d\tilde{c})(c\underline{z}\tilde{c} + d\tilde{c})^{-1}\right) \\
&= \pm (c\underline{w}\tilde{c} + d\tilde{c})^{-1}(c(\underline{w} - \underline{z})\tilde{c})(c\underline{z}\tilde{c} + d\tilde{c})^{-1} \\
&= \pm (\widetilde{c\underline{w} + d})^{-1}(\underline{w} - \underline{z})(c\underline{z} + d)^{-1}.
\end{aligned}
$$

It follows that in this case, if $(\underline{w} - \underline{z})^2 \neq 0$, then $(\phi(\underline{w}) - \phi(\underline{z}))^2 \neq 0$. The result is straightforward to obtain if $c = 0$.

Suppose now that $c \neq 0$, and that c is not invertible. Then by Lemma 8.2.1 there is a sequence of Möbius transformations

$$
\phi_n(\underline{z}) = (a_n\underline{z} + b_n)(c_n\underline{z} + d_n)^{-1}
$$

which converges locally uniformly to $\phi(\underline{z})$, and $c_n = 0$ or c_n is invertible. It follows that

$$
\phi(\underline{z}) - \phi(\underline{w}) = \lim_{n \to \infty} \phi_n(\underline{z}) - \phi_n(\underline{w}) = \pm(\widetilde{c\underline{w} + d})^{-1}(\underline{w} - \underline{z})(c\underline{z} + d)^{-1}.
$$

As $(c\underline{w} + d)^{-1}$ and $(c\underline{z} + d)^{-1}$ are well defined, and $(w - z)^2 \neq 0$, it follows that condition (i) in Definition 1 is satisfied.

Using similar arguments to those used to verify condition 1, it may be observed that $D\phi_{\underline{z}}u = \pm(\widetilde{c\underline{z} + d})^{-1}u(c\underline{z} + d)^{-1}$. It follows that condition (ii) from Definition 1 also holds. Consequently, $\phi(M)$ is a domain manifold.

∎

Suppose now that M is a closed domain manifold. Let us consider the component of $\mathbf{C}^n \setminus \bigcup_{\underline{z} \in M \setminus \overset{\circ}{M}} N(\underline{z})$, which contains the interior of M. We denote this set by M^+. It is straightforward to verify that this set is a domain in $\mathbf{C}^n$. We call the domain M^+ a cell of harmonicity. We have previously introduced these domains in [17,18,19]. When $M \subseteq R^n$, the domain M^+ corresponds to the cells of harmonicity described by Avanissian [4], and others.

We now show that cells of harmonicity are preserved by Möbius transformations.

THEOREM 8.3.3

Suppose that M is a closed domain manifold, and M^+ is its cell of harmonicity. Suppose also that $\begin{pmatrix} a & b \\ c & d \end{pmatrix} \in V(\mathbf{C}^n)$, and $\phi(\underline{z}) = (a\underline{z}+b)$ $(c\underline{z}+d)^{-1}$ is well defined on M. Then $\phi(M^+) = \phi(M)^+$.

PROOF   The case $c = 0$ is straightforward. So, let us proceed to the case where c is invertible. Suppose that $\underline{w} \in M^+ \backslash M$. Then there is a path $\psi : [0,1] \rightarrow M^+$ such that $\psi(0) = \underline{z} \in \overset{\circ}{M}$ and $\psi(1) = \underline{w}$. In [19], we show that for each $\psi(t) \in M^+ \backslash M$, we have that $M \cap N(\psi(t))$ is a manifold homeomorphic to S^{n-2}. If for some t we have that $\phi(\psi(t))$ is well defined, then $\phi(\psi(t)) - \phi(\underline{w}) = \pm(c\widetilde{\psi(t)+d})^{-1}(\underline{w} - \psi(t))(c\underline{w}+d)^{-1}$ for each $\underline{w} \in M$. By simple continuity and density arguments, it now follows that $\phi(\psi(t))$ is well defined for each $t \in [0,1]$, and for each $\phi(\psi(t)) \notin \phi(M)$ we have that $\phi(M) \cap N(\phi(\psi(t)))$ is a manifold homeomorphic to S^{n-2}. Furthermore,

$$\phi(M) \cap N(\phi(\psi(t))) = \phi(M \cap \psi(t)).$$

It follows that $\phi(\underline{z}) \in \phi(M)^+$ for each $\underline{z} \in M^+$. Consequently, $\phi(M^+) \subseteq \phi(M)^+$ whenever c is invertible.

The same density argument used to establish Theorem 8.3.2 now gives that $\phi(M^+) \subseteq \phi(M)^+$ for any $\phi(\underline{z}) = (a\underline{z}+b)(c\underline{z}+d)^{-1}$, which is well defined on M. We can now repeat the previous arguments, working with the Möbius transformation ϕ^{-1}, instead of ϕ, to obtain

$$\phi^{-1}(\phi(M)^+) \subseteq M^+.$$

Consequently, $\phi(M^+) = \phi(M)^+$.   ∎

Observation: It may easily be observed that Theorem 8.3.3 also holds, if we simply assume that M is a domain manifold and $c\ell M$, the closure of M, is also a domain manifold. In particular, if M is the unit disk in R^n, and ϕ is the Cayley transform, then $\phi(M)$ is upper half-space. In this case, M^+ is the Lie ball, $\phi(M)^+$ is the tube domain described in [27] and elsewhere, and the previous theorem shows us that the Cayley map transforms the Lie ball into the tube domain. The Lie ball is Cartan's classical domain of type four (see [9]), and this observation gives rise to the two realizations of this domain, one bounded and one unbounded, described in [27], and elsewhere.

THEOREM 8.3.4
Suppose that M is a domain manifold, and cℓM is also a domain manifold. Suppose also $\begin{pmatrix} a & b \\ c & d \end{pmatrix} \in V(\mathbf{C}^n)$, and $\phi(\underline{z}) = (a\underline{z} + b)(c\underline{z} + d)^{-1}$ is well defined on M. Then for each $\underline{w} \in M^+$, we have that

$$\phi(M) \cap N(\phi(\underline{w})) = \phi(M \cap N(\underline{w})).$$

In [10], a number of domains and real $(n-1)$-dimensional manifolds are set up in $\mathbf{C}^n$, and it is shown that they can be used to establish Cauchy integral formulae. Identical arguments to those used to establish Theorems 8.3.2, 8.3.3, and 8.3.4 can be used to show that these types of domains and manifolds are also conformally invariant.

8.4 $V(n)$ and Real Clifford Analysis

We begin this section by giving a brief description of the conformal covariance of the Dirac operator in R^n. The Dirac operator is the differential operator $D = \sum_{j=1}^{n} e_j \frac{\partial}{\partial x_j}$. We have that $D^2 = -\Delta_n$, where Δ_n is the Laplacian in R^n.

DEFINITION 8.4.1   Suppose U is a domain in R^n, and $f : U \to A_n$ is a C^1-function. Then f is said to be left-monogenic if $Df(\underline{x}) = 0$ for each $\underline{x} \in U$. Similarly, a C^1-function $g : U \to A_n$ is called right-monogenic if $g(\underline{x})D = \sum_{j=1}^{n} \frac{\partial g}{\partial x_j} e_j = 0$, for each $\underline{x} \in U$.

An example of a function which is both left- and right-monogenic is the function $G(\underline{x}) = \frac{1}{w_n} \frac{\underline{x}}{\|\underline{x}\|^n}$, where w_n is the surface area of the unit sphere in R^n. This function is the Clifford-Cauchy kernel, and plays the same role in Clifford analysis as the Cauchy kernel does in one-variable complex analysis.

Following simple geometric arguments given in [26], we have:

THEOREM 8.4.2
Suppose that $\begin{pmatrix} a & b \\ c & d \end{pmatrix} \in V(n)$, and $f, g : U \to A_n$ are, respectively, left- and right-monogenic functions. Suppose that S is a closed, bounded, and

oriented surface lying in U. Then

$$\int_S g\left(\underline{y}\right) n\left(\underline{y}\right) f\left(\underline{y}\right) d\sigma\left(\underline{y}\right)$$

$$= \int_{\phi^{-1}(S)} g\left(\phi\left(\underline{x}\right)\right) J(\widetilde{\phi,\underline{x}}) n\left(\underline{x}\right) J\left(\phi,\underline{x}\right) f\left(\phi\left(\underline{x}\right)\right) d\sigma\left(\underline{x}\right),$$

where $\underline{y} = \phi\left(\underline{x}\right) = \left(a\underline{x} + b\right)\left(c\underline{x} + d\right)^{-1}$, $n\left(\underline{y}\right)$ is the outward-pointing normal vector to S at $\phi\left(\underline{x}\right)$, $n\left(\underline{x}\right)$ is the outward-pointing normal vector to $\phi^{-1}(S)$ at $\underline{x}$, $J\left(\phi,\underline{x}\right) = \frac{\widetilde{c\underline{x}+d}}{\|c\underline{x}+d\|^n}$, and dS is the usual Lebesgue measure on S.

Suppose, now, that $S = S^{n-1}$, the unit sphere in R^n, and $h : S^{n-1} \to A_n$ is an L^2-integrable function, so that

$$Re\left(\int_{S^{n-1}} h\left(\underline{x}\right)\overline{h\left(\underline{x}\right)}dS^{n-1}\right) < +\infty,$$

where $Re(a)$ denotes the real part of, or identity component of, the element $a \in A_n$. Let us formally consider the singular Cauchy transform

$$P.V. \int_{S^{n-1}} G\left(\underline{x} - \underline{y}\right) n\left(\underline{x}\right) h\left(\underline{x}\right) dS^{n-1},$$

where $\underline{y} \in S^{n-1}$. This integral becomes

$$P.V. \int_{S^{n-1}} G\left(\phi\left(\underline{u}\right) - \phi\left(\underline{v}\right)\right) n\left(\phi\left(\underline{u}\right)\right) h\left(\phi\left(\underline{u}\right)\right) dS^{n-1}, \qquad (8.4.1)$$

where ϕ is the inverse Cayley transform. Now, as illustrated in [16], we have that

$$G\left(\eta\left(\underline{u}\right) - \eta\left(\underline{v}\right)\right) = J(\widetilde{\eta,\underline{u}})^{-1}\left(G\left(\underline{v} - \underline{u}\right)\right) J\left(\eta,\underline{v}\right)^{-1}, \qquad (8.4.2)$$

where $\eta\left(\underline{x}\right) = \left(a'\underline{x} + b'\right)\left(c'\underline{x} + d\right)^{-1}$ is a Möbius transformation. Consequently, the formal integral (1) can be reexpressed as

$$J\left(\phi,\underline{x}\right)^{-1} P.V. \int_{R^{n-1}} G\left(\underline{u} - \underline{v}\right) e_n J\left(\phi,\underline{u}\right) h\left(\phi\left(\underline{u}\right)\right) du^{n-1}. \qquad (8.4.3)$$

Now, a simple change of variable argument shows us that

$$\int_{S^{n-1}} \bar{h}\left(\underline{x}\right) h\left(\underline{x}\right) dS^{n-1} = \int_{R^{n-1}} \bar{h}\left(\phi\left(\underline{u}\right)\right) \overline{J\left(\phi,\underline{u}\right)} J\left(\phi,\underline{u}\right) h\left(\phi\left(\underline{u}\right)\right) du^{n-1}.$$

Consequently, $h\left(\underline{x}\right)$ is L^2-integrable over S^{n-1} if and only if $J\left(\phi,\underline{u}\right) h\left(\phi\left(\underline{u}\right)\right)$ is L^2-integrable over R^{n-1}.

Now, $G\left(\underline{u} - \underline{v}\right)|_{R^{n-1}} = \sum_{j=1}^{n-1} R_j\left(\underline{u} - \underline{v}\right) e_j$, where R_j is the jth Riesz transform, see [5]. Furthermore, [7], the Fourier transform of $\sum_{j=1}^{n-1} R_j e_j$ is

$ci\frac{\xi}{\|\underline{\xi}\|}$, where $\underline{\xi} \in R^{n-1}$, and $c \in R^+$. From Plancherel's theorem it follows that the convolution operator

$$P.V. \int_{R^{n-1}} G(\underline{u} - \underline{v}) e_n , du^{n-1}$$

is an L^2-bounded operator, see [5]. From the equality of expressions (8.4.1) and (8.4.3), and a simple change of variable argument, it follows that we have established:

PROPOSITION 8.4.3
The convolution operator $P.V. \int_{S^{n-1}} G(\underline{x} - \underline{y}) n(\underline{x}) , dS^{n-1}$ *is an* L^2-*bounded operator.*

The argument given here to establish Proposition 8.4.3 simplifies an argument given in [23] to deduce the same result.

It should be noted that compared to the general cases of considering the L^2-boundedness of the singular Cauchy transform over a Lipschitz surface (see, for instance, [7,14]), the cases where the surface is either R^{n-1} or S^{n-1} are quite simple. However, the arguments used to establish Proposition 8.4.3 readily extend, using the L^2-boundedness of the singular Cauchy transform over the graph of a Lipschitz function, from R^{n-1} to R, established in [7,14] and elsewhere, to give us the following special of a known result, see [7,14]:

THEOREM 8.4.4
Suppose that $\Sigma, \subseteq R^n$, *is the graph of a Lipschitz-continuous function* $\eta :$ $R^{n-1} \to R$, *and* $\begin{pmatrix} a & b \\ c & d \end{pmatrix} \in V(n)$. *Then the singular Cauchy transform* $P.V. \int_{\phi(\Sigma)} G(\underline{x} - \underline{y}) n(\underline{x}) d\phi(\Sigma)$ *is an* L^2-*bounded operator, where* $\phi(\underline{x}) = (a\underline{x} + b)(c\underline{x} + d)^{-1}$.

A particular case to consider would be the case where η has compact support, and ϕ is the Cayley transform. In this case, we obtain a Lipschitz perturbation of the sphere. As the identity component of the singular Cauchy transform is the double-layer potential operator, it follows that this operator is L^2-bounded over $\phi(\Sigma)$.

A consequence of Theorem 8.4.2 is that if $f\left((a\underline{x} + b)(c\underline{x} + d)^{-1}\right)$ is a left-monogenic function with respect to $\phi(\underline{x}) = (a\underline{x} + b)(c\underline{x} + d)^{-1}$, then $J(\phi, \underline{x}) f(\phi(\underline{x}))$ is left-monogenic with respect to $\underline{x}$. Also, if $h : \phi(\Sigma) \to A_n$ is L^2-integrable, then $\int_{\phi(\Sigma)} G(\underline{x} - y) n(\underline{x}) h(\underline{x}) d\phi(\Sigma)$ gives a left-monogenic function on $\phi(\Sigma^+) \cup \phi(\Sigma^-)$, where Σ^+ is the domain lying above Σ, and Σ^- is the domain lying below Σ, while $\phi(\Sigma^\pm)$ are the images under ϕ of $\Sigma^\pm$. It follows that $\int_\Sigma G(\underline{u} - \underline{v}) n(\underline{u}) J(\phi, \underline{u}) h(\underline{u}) d\Sigma$ is a left-monogenic function on $\Sigma^+ \cup \Sigma^-$. Moreover, [11], we have that

$$\lim_{\underline{v} \to \underline{u}'} \int_\Sigma G\left(\underline{u} - \underline{v}\right) n\left(\underline{u}\right) k\left(\underline{u}\right) d\Sigma = \tfrac{1}{2} k\left(\underline{u}'\right)$$
$$+ P.V. \int_\Sigma G\left(\underline{u} - \underline{u}'\right) n\left(\underline{u}\right) k\left(\underline{u}\right) d\Sigma$$

for almost all $\underline{u}' \in \Sigma$, for $\underline{v} \in \Sigma^+$, and all L^2-integrable functions k on Σ. Moreover, $\underline{v}$ approaches $\underline{u}'$ nontangentially. Consequently, we have via the Möbius transformation ϕ:

THEOREM 8.4.5
Suppose that $\phi\left(\underline{x}\right)$ is a Möbius transformation, and Σ is a Lipschitz graph. Then for each L^2-integrable function h on $\phi(\Sigma)$ we have that for almost all $\underline{y}' \in \phi\left(\Sigma^+\right)$,

$$\lim_{y \to y'} \int_{\phi(\Sigma)} G\left(\underline{x} - \underline{y}\right) n\left(\underline{x}\right) h\left(\underline{x}\right) d\phi(\Sigma)$$

$$= \frac{1}{2} h\left(\underline{y}'\right) + P.V. \int_{\phi(\Sigma)} G\left(\underline{x} - \underline{y}'\right) n\left(\underline{x}\right) h\left(\underline{x}\right) d\phi(\Sigma),$$

where $y \in \phi\left(\Sigma^+\right)$, and y approaches y' nontangentially.

A particular example to consider, again, is the case where $\phi(\Sigma)$ is a Lipschitz perturbation of the sphere, S^{n-1}. It should also be noted that for this case and all other cases, the Clifford-Haar wavelet/martingale arguments developed in [3,6,15] carry over to their images under Möbius transformation. This gives rise, under the Cayley transform, to a suitable definition of H^2-spaces over the Lipschitz-perturbed disk, bounded by the Lipschitz-perturbed sphere. In fact, following [11], one can always construct H^2-spaces using Clifford analysis over arbitrary Lipschitz domains.

Following [12,13], let us now introduce a real Banach algebra B, and consider $B \otimes_R A_n$. Also following [12,13], let us consider n mutually commuting, bounded operators $T_1, \ldots T_n \in B$. Following [12], we may consider the operator $\underline{T} = T_1 e_1 + \cdots + T_n e_n \in B \otimes R^n \subseteq B \otimes A_n$. The spectrum of $\underline{T}$ is defined [12] to be the set

$$\left\{\underline{\lambda} = \lambda_1 e_1 + \cdots + \lambda_n e_n \in R^n : \underline{T} - \underline{\lambda} I \text{ is not invertible}\right\},$$

and it is denoted by $Sp\left(\underline{T}\right)$.

Suppose now that $\begin{pmatrix} a & b \\ c & d \end{pmatrix} \in V(n)$. Let us consider the formal expression $(a\underline{T} + b)(c\underline{T} + d)^{-1}$. This expression is well defined, provided $d\tilde{c} \notin Sp\left(\underline{T}\right)$. Moreover, whenever $d\tilde{c} \notin Sp\left(\underline{T}\right)$, we have that $(a\underline{T} + b)(c\underline{T} + d)^{-1} \in B \otimes R^n$. So we can, in this case, place $(a\underline{T} + b)(c\underline{T} + d)^{-1} = P_1 e_1 + \cdots + P_n e_n$, where $P_1, \ldots, P_n \in B$, and it may be observed that the operators $P_1, \ldots, P_n$ are mutually commuting, bounded operators.

Again, following [12], we shall assume that n is even. In this case, $G(\underline{x}) = (-1)^{\frac{n}{2}} \underline{x}^{-n+1}$. In this case, we can introduce the operator $G(\underline{T} - \underline{y})$, see [12]. Also, for each left-monogenic function $f : U \to A_n$ and for each closed, bounded, oriented surface $S \subseteq U$, with $S \cap Sp(\underline{T}) = \emptyset$, we have, [12], that the integral

$$\int_S G(\underline{T} - \underline{x}) \, n(\underline{x}) \, f(\underline{x}) \, d\sigma(\underline{x})$$

is well defined. We may place $f(\underline{T}) = \int_S G(\underline{T} - \underline{x}) \, n(\underline{x}) \, f(\underline{x}) \, d\sigma(\underline{x})$.

The identity (2) generalizes to give us

$$G(\phi(\underline{T}) - \phi(\underline{x})) = \tilde{J}(\phi, \underline{T})^{-1} \, G(\underline{x} - \underline{T}) \, J(\phi, \underline{x})^{-1},$$

where $\phi(\underline{T}) = (a\underline{T} + b)(c\underline{T} + d)^{-1}$, with $\begin{pmatrix} a & b \\ c & d \end{pmatrix} \in V(n)$, and

$$J(\phi, \underline{T}) = \frac{(c\underline{T} + d)}{\left((c\underline{T} + d)(\widetilde{c\underline{T} + d})\right)^{\frac{n}{2}}}.$$

It follows that

$$\int_S G(\phi(\underline{T}) - \phi(\underline{x})) \, n(\phi(\underline{x})) \, f(\phi(\underline{x})) \, d\sigma(\phi, \underline{x})$$
$$= J(\phi, \underline{T})^{-1} \int_{\phi(S)} G(\underline{x} - \underline{T}) \, n(\underline{x}) \, J(\phi, \underline{x}) \, f(\phi(\underline{x})) \, d\sigma(\underline{x}).$$

In [29], X. Zhenyuan shows that if $f : U \to A_n(\mathbf{C})$ satisfies the equation $(D + \lambda)f = 0$, where $\lambda \in \mathbf{C}$, then for each closed bounded region $M, \subseteq U$, with piecewise C^1, or Lipschitz, boundary and each $\underline{u} \in \overset{\circ}{M}$, we have that

$$f(\underline{u}) = \int_{\partial M} E_{-\lambda}(\underline{u} - \underline{v}) \, n(\underline{v}) \, f(\underline{v}) \, d\sigma(\underline{u}), \qquad (8.4.4)$$

where $E_{-\lambda}(\underline{v}) = \sum_{k=0}^{\infty}(-1)^k \lambda^k G_{k+1}(\underline{v})$. Here, $G_1(\underline{v}) = G(\underline{v})$, $G_{2\ell}(\underline{v}) = \frac{A_{2\ell}}{\|\underline{v}\|^{n-2\ell}}$ and $G_{2\ell+1}(\underline{v}) = A_{2\ell+1}\frac{\underline{x}}{\|\underline{x}\|^{n-2\ell}}$ when n is odd, and $G_{2\ell}(\underline{v}) = \frac{A'_{2\ell}}{\|\underline{v}\|^{n-2\ell}}$ for $2 \leq 2\ell \leq n - 2$, $G(\underline{v}) = A'_{2\ell}(\underline{v}^{k-n} + B_{2\ell}\underline{v}^{k-n} \log \|\underline{v}\|)$ for $k \leq n$, and $G_{2\ell+1}(\underline{v}) = A'_{k+1}\frac{\underline{v}}{\|\underline{v}\|^{n-2\ell}}$ for $2 \leq 2\ell \leq n - 2$, when n is even. The coefficients A_k, A'_k, and B_k are real, and they are chosen so that $DG_{k+1}(\underline{v}) = G_k(\underline{v})$.

Suppose now that $\underline{u} = \phi(\underline{y})$ and $\underline{v} = \phi(\underline{x})$, where $\phi(\underline{x}) = (a\underline{x} + b)(c\underline{x} + d)^{-1}$, and $\begin{pmatrix} a & b \\ c & d \end{pmatrix} \in V(n)$. Then expression (8.4.4) becomes

$$f(\phi(\underline{y})) = \int_{\partial\phi^{-1}(M)} E_{-\lambda}(\phi(\underline{y}) - \phi(\underline{x})) \, \tilde{J}(\phi, \underline{x}) \, n(\underline{x}) \, J(\phi, \underline{x}) \, f(\underline{x}) \, d\sigma(\underline{x}).$$

Consequently, we have the Cauchy integral formula

$$J(\phi, \underline{y}) \, f(\phi(\underline{y})) \qquad (8.4.5)$$

$$= \int_{\partial \phi^{-1}(M)} J\left(\phi, \underline{y}\right) E_{-\lambda} \left(\phi\left(\underline{y}\right) - \phi\left(\underline{x}\right)\right) \tilde{J}\left(\phi, \underline{x}\right) n\left(\underline{x}\right) J\left(\phi, \underline{x}\right) f\left(\underline{x}\right) d\sigma\left(\underline{x}\right).$$

Moreover, the function $J\left(\phi, \underline{y}\right) f\left(\phi\left(\underline{y}\right)\right)$ is annihilated by the operator $D + \frac{\lambda}{\|c\underline{y}+d\|^2}$, see [23]. Also, we have that

$$\left(J\left(\phi, \underline{y}\right) E_{-\lambda}\left(\phi\left(\underline{y}\right) - \phi\left(\underline{x}\right)\right) \tilde{J}\left(\phi, \underline{x}\right)\right) \left(D - \frac{\lambda}{\|c\underline{y}+d\|^2}\right) = 0,$$

while

$$\left(D + \frac{\lambda}{\|c\underline{y}+d\|^2}\right) J\left(\phi, \underline{y}\right) E_{-\lambda}\left(\phi\left(\underline{y}\right) - \phi\left(\underline{x}\right)\right) \tilde{J}\left(\phi, \underline{x}\right) = 0.$$

In particular, when $\phi\left(\underline{x}\right) = \underline{x}^{-1}$, the formula (8.4.5) gives us a Cauchy integral formula for functions annihilated by the operator $D + \frac{\lambda}{\|\underline{x}\|^2}$. Unfortunately, the kernel appearing in (8.4.5) is not in closed form.

However, in [8] it is shown that one may associate Plemelj formulae to the operator $D + \lambda$. It now follows that one can use conformal covariance to associate Plemelj formulae to the operator $D + \frac{\lambda}{\|c\underline{x}+d\|^2}$. More specifically, on denoting the kernel $J\left(\phi, \underline{y}\right) E_{-\lambda}\left(\phi\left(\underline{y}\right) - \phi\left(\underline{x}\right)\right) \tilde{J}\left(\phi, \underline{x}\right)$ by $E_{(c\underline{x}+d)}\left(\underline{x}, \underline{y}\right)$, we have:

PROPOSITION 8.4.6
Suppose that M is a closed and bounded region in R^n, with piecewise C^1, or Lipschitz boundary. Suppose also that $g : \partial M \to A_n$ is a Hölder continuous function. Then

(a) *the integral P.V. $\int_{\partial M} E_{(c\underline{x}+d)}\left(\underline{x}, \underline{y}\right) n\left(\underline{x}\right) g\left(\underline{x}\right) d\partial M$ is well defined for almost all $\underline{y} \in \partial M$.*

(b) *$\int_{\partial M} E_{(c\underline{x}+d)}\left(\underline{x}, \underline{y}\right) n\left(\underline{x}\right) g\left(\underline{x}\right) d\partial M$ is annihilated by $D + \frac{\lambda}{\|c\underline{x}+d\|^2}$ on $R^n \backslash \partial M$.*

(c) *$\lim_{t\to 0} \int_{\partial M} E_{(c\underline{x}+d)}\left(\underline{x}, \eta(t)\right) n\left(\underline{x}\right) g\left(\underline{x}\right) d\sigma\left(\underline{x}\right) = \frac{1}{2} g(y) l + P.V. \int_{\partial M}$*
     $E_{(c\underline{x}+d)}\left(\underline{x} + \underline{y}\right) n\left(\underline{x}\right) g\left(\underline{x}\right) d\sigma\left(\underline{x}\right)$, where $\eta : (0,1) \to \overset{o}{M}$ is a C^1-map with $\lim_{t\to 0} \eta(t) = \underline{y}$, and $\lim_{t\to 0} \frac{d\eta}{dt}(t) \notin T\partial M\underline{y}$, for almost all $\underline{y} \in \partial M$.

(d) *$\lim_{t\to 0} \int_{\partial M} E_{(c\underline{x}+d)}\left(\underline{x}, \theta(t)\right) n\left(\underline{x}\right) g\left(\underline{x}\right) d\sigma\left(\underline{x}\right) = \frac{-1}{2} g\left(\underline{y}\right) l + P.V. \int_{\partial M}$*
     $E_{(c\underline{x}+d)}\left(\underline{x}, \underline{y}\right) n\left(\underline{x}\right) g\left(\underline{x}\right) d\sigma\left(\underline{x}\right)$, where $\theta : (0,1) \to R^n \backslash M$ is a C^1-map with $\lim_{t\to 0} \theta(t) = \underline{y}$, and $\lim_{t\to 0} \frac{d\theta}{dt}(t) \notin T\partial M\underline{y}$, for almost all $\underline{y} \in \partial M$.

Following [14,15], it may be observed that the assumption that the function g appearing in Proposition 8.4.6 is a Hölder continuous function can be weakened. To go further, we need the following lemma:

LEMMA 8.4.7

$$E_{c\underline{x}+d}\left(\underline{x},\underline{y}\right) = G_1\left(\underline{x}-\underline{y}\right)F_{c\underline{x}+d}\left(\underline{x},\underline{y}\right),$$

where $F_{c\underline{x}+d}\left(\underline{x},\underline{y}\right)$ is a C^1-function on $R^n \times R^n$, and $F_{c\underline{x}+d}\left(\underline{x},\underline{x}\right) = 1$.

PROOF Now

$$E_{c\underline{x}+d}\left(\underline{x},\underline{y}\right) = J\left(\phi,\underline{y}\right)E_{-\lambda}\left(\phi\left(\underline{y}\right)-\phi\left(\underline{x}\right)\right)\tilde{J}\left(\phi,\underline{x}\right),$$

where

$$E_{-\lambda}\left(\phi\left(\underline{y}\right)-\phi\left(\underline{x}\right)\right) = \sum_{k=0}^{\infty}(-1)^k\lambda^k G_{k+1}\left(\phi\left(\underline{y}\right)-\phi\left(\underline{x}\right)\right).$$

So

$$E_{c\underline{x}+d}\left(\underline{x},\underline{y}\right) = J\left(\phi,\underline{y}\right)\sum_{k=0}^{\infty}(-1)^k\lambda^k G_{k+1}\left(\phi\left(\underline{y}\right)-\phi\left(\underline{x}\right)\right)\tilde{J}\left(\phi,\underline{x}\right).$$

Now

$$J\left(\phi,\underline{y}\right)G_1\left(\phi\left(\underline{y}\right)-\phi\left(\underline{x}\right)\right)\tilde{J}\left(\phi,\underline{x}\right) = G_1\left(\underline{x}-\underline{y}\right).$$

Let us consider $J\left(\phi,\underline{y}\right)\dfrac{\left(\phi(\underline{x})-\phi(\underline{y})\right)}{\|\phi(\underline{x})-\phi(\underline{y})\|^{m-n}}\tilde{J}\left(\phi,\underline{x}\right)$, where $m \in \mathbf{Z}$ and $m < n$. We have that

$$J_1\left(\phi,\underline{y}\right)\left(\phi\left(\underline{x}\right)-\phi\left(\underline{y}\right)\right)\tilde{J}\left(\phi,\underline{x}\right)\|\phi\left(\underline{x}\right)-\phi\left(\underline{y}\right)\|^{n-m}$$

$$= G_1\left(\underline{x}-\underline{y}\right)\|\phi\left(\underline{x}\right)-\phi\left(\underline{y}\right)\|^{n-m}.$$

Let us also consider

$$\frac{J\left(\phi,\underline{y}\right)\tilde{J}\left(\phi,\underline{x}\right)}{\|\phi\left(\underline{x}\right)-\phi\left(\underline{y}\right)\|^m},$$

where $m \in \mathbf{Z}$ and $m < n-1$. Here,

$$\frac{J\left(\phi,\underline{y}\right)\tilde{J}\left(\phi,\underline{x}\right)}{\|\phi\left(\underline{x}\right)-\phi\left(\underline{y}\right)\|^m}$$

$$= \frac{\left(c\underline{y}+d\right)}{\|c\underline{y}+d\|^n}\frac{\left(\phi\left(\underline{x}\right)-\phi\left(\underline{y}\right)\right)^{-1}}{\|\phi\left(\underline{x}\right)-\phi\left(\underline{y}\right)\|^{n-2}}\frac{\left(\widetilde{c\underline{x}+d}\right)\left(\widetilde{c\underline{x}+d}\right)^{-1}}{\|c\underline{x}+d\|^n}$$

$$\left(\phi\left(\underline{x}\right) - \phi\left(\underline{y}\right)\right)\left(\widetilde{c\underline{x} + d}\right)\frac{1}{\left\|\phi\left(\underline{x}\right) - \phi\left(\underline{y}\right)\right\|^{m-2+2}}$$

$$= G_1\left(\underline{x} - \underline{y}\right)\left(\widetilde{c\underline{x} + d}\right)^{-1}\left(\phi\left(\underline{x}\right) - \phi\left(\underline{y}\right)\right)\left(\widetilde{c\underline{x} + d}\right)\cdot\frac{1}{\left\|\phi\left(\underline{x}\right) - \phi\left(\underline{y}\right)\right\|^{m-n+2}}$$

$$= G_1\left(\underline{x} - \underline{y}\right)\frac{\left(c\underline{x} + d\right)}{\left\|c\underline{x} + d\right\|}\left(\phi\left(\underline{x}\right) - \phi\left(\underline{y}\right)\right)\frac{\left(c\underline{x} + d\right)}{\left\|c\underline{x} + d\right\|}\cdot\frac{1}{\left\|\phi\left(\underline{x}\right) - \phi\left(\underline{y}\right)\right\|^{m-n+2}}.$$

The result now immediately follows for the case n odd. The result follows also for the case n even by similar arguments, and an application of l'Hopital's rule to the logarithmic part of G_k for $k > n$. ∎

From Lemma 8.4.7 we have:

THEOREM 8.4.8
Suppose that U is a bounded domain, and for $\begin{pmatrix} a & b \\ c & d \end{pmatrix} \in V(n)$ the function $\frac{1}{\|c\underline{x} + d\|^2}$ is well defined on $c\ell(U)$. Suppose also that $g : U \to A_n$ is a bounded C^1-function. Then $\left(D + \frac{1}{\|c\underline{y} + d\|^2}\right)\int_U E_{c\underline{x} + d}\left(\underline{x}, \underline{y}\right)g\left(\underline{x}\right)dx^n = -g\left(\underline{y}\right)$, for each $\underline{y} \in U$.

Using Theorem 8.4.8 and Proposition 8.4.6, we may now mimic arguments presented in [8] to obtain the following L^2-decomposition of square integrable, A_n-valued functions over a bounded domain.

THEOREM 8.4.9
Suppose that U is a bounded domain with piecewise C^1, or Lipschitz, boundary. Suppose also that $\begin{pmatrix} a & b \\ c & d \end{pmatrix} \in V(n)$, and $\frac{1}{\|c\underline{x} + d\|^2}$ is well defined on $c\ell(U)$. Then

$$L^2\left(U, A_n\right) = B^2_{c\underline{x} + d}\left(U, A_n\right) \oplus W_{c\underline{x} + d}\left(U, A_n\right),$$

where $L^2\left(U, A_n\right)$ is the A_n-module of A_n-valued square integrable functions defined on U, $B^2_{c\underline{x} + d}\left(U, A_n\right)$ is the Bergman module of A_n-valued square integrable functions annihilated on the left by the operator $D - \frac{1}{\|c\underline{x} + d\|^2}$, and $W_{c\underline{x} + d}\left(U, A_n\right)$ is the A_n-module $\left\{g : U \to A_n : \int \|g\left(\underline{x}\right)\|^2 dx^n < +\infty \text{ and for which there is a function } h_g : c\ell(U) \to A_n \text{ with } h_g|_{\partial c\ell U} = 0, \text{ and } \left(D + \frac{1}{\|c\underline{x} + d\|^2}\right)h_g = -g\right\}.$

Using the L^2-decomposition described in Theorem 8.4.9, we can now adapt arguments given in [8, Ch.4] to solve boundary value problems associated to the differential operator $\left(D + \frac{1}{\|c\underline{x}+d\|^2}\right)^2$. These boundary value problems are similar to those described in [8, Ch.4].

8.5 $V(\mathbf{C}^n)$ and Complex Clifford Analysis

In this section, we look at ways in which conformal covariance arise in complex Clifford analysis. In [21], we showed that, given a domain manifold, we could construct a differential operator D_M over its tangent bundle such that for each C^1-function $g : \overset{\circ}{M} \rightarrow A_n(\mathbf{C})$, we have that

$$D_M \frac{1}{w_n} \int_M G\left(\underline{z} - \underline{w}\right) g\left(\underline{z}\right) dz^n = g\left(\underline{w}\right), \qquad (8.5.1)$$

where w_n is the surface area of S^{n-1}, and $G\left(\underline{z} - \underline{w}\right)$ is the restriction of the holomorphic continuation of the Cauchy kernel to M. It should be noted that this function is well defined on M, in all dimensions.

When $M \subseteq R^n$, then $D_M = D$. As dz^n is a complex measure, and $(a\underline{z} + b)(c\underline{z} + d)^{-1}$ is a holomorphic function for each $\begin{pmatrix} a & b \\ c & d \end{pmatrix} \in V(\mathbf{C}^n)$, we have from Theorem 8.3.2 that arguments given in [16] and [22] extend automatically to give us:

PROPOSITION 8.5.1
Suppose that M is a domain manifold, and $\begin{pmatrix} a & b \\ c & d \end{pmatrix} \in V(\mathbf{C}^n)$ is such that $\phi\left(\underline{z}\right) = (a\underline{z} + b)(c\underline{z} + d)^{-1}$ is well defined for each $\underline{z} \in M$. Then, for each C^1-function $\psi : M \rightarrow A_n(\mathbf{C})$, with compact support, we have that

$$\psi\left(\phi\left(\underline{w}\right)\right) = \int_{\phi(M)} G\left(\phi\left(\underline{z}\right) - \phi\left(\underline{w}\right)\right) D_{\phi(M)} \psi\left(\phi\left(\underline{z}\right)\right) d\phi\left(\underline{z}\right)^n \qquad (8.5.2)$$

$$= J_1\left(\phi, \underline{w}\right)^{-1} \int_M G\left(\underline{z} - \underline{w}\right) J_{-1}\left(\phi, \underline{z}\right) D_{\phi(M)} \psi\left(\phi\left(\underline{z}\right)\right) dz^n,$$

where

$$J_1\left(\phi, \underline{w}\right) = \left(\widetilde{c\underline{w} + d}\right)\left((c\underline{w} + d)\left(\widetilde{c\underline{w} + d}\right)\right)^{\frac{-n}{2}},$$

and

$$J_{-1}\left(\phi, \underline{z}\right) = l\left(\widetilde{c\underline{z} + d}\right)\left((c\underline{z} + d)\left(\widetilde{c\underline{z} + d}\right)\right)^{\frac{-n}{2-1}}.$$

Using the identities (8.5.1) and (8.5.2), we now have

$$D_M J_1\left(\phi, \underline{w}\right) \psi\left(\phi\left(\underline{w}\right)\right) = J_{-1}\left(\phi, \underline{w}\right) D_{\phi(M)} \psi\left(\phi\left(\underline{w}\right)\right). \qquad (8.5.3)$$

Using a partition of unity argument, we now have:

THEOREM 8.5.2
Suppose that M is a domain manifold, and $\begin{pmatrix} a & b \\ c & d \end{pmatrix} \in V(\mathbf{C}^n)$ is such that $\phi(\underline{z}) = (a\underline{z} + b)(c\underline{z} + d)^{-1}$ is well defined on M. Then for each C^1-function $g : M \to A_n(\mathbf{C})$, we have that

$$D_M J_1(\phi, \underline{w}) g(\phi(\underline{w})) = J_{-1}(\phi, \underline{w}) D_{\phi(M)} g(\phi(\underline{w}))$$

It should be noted that the "functions" $J_1(\phi, \underline{w})$ and $J_{-1}(\phi, \underline{w})$ are not necessarily uniquely defined on domains in $\mathbf{C}^n$ when n is odd.

Theorem 8.5.2 shows us that the multiplication operators J_1 and J_{-1} are intertwining operators for the differential operators D_M and $D_{\phi(M)}$. This is in complete analogy to the Euclidean case described in [16,22]. It is now straightforward to adapt arguments given in [21], and combine them with arguments outlined in [16,22] to deduce that

$$D_M^k J_k(\phi, \underline{w}) g(\phi(\underline{w})) = J_{-k}(\phi, \underline{w}) D_{\phi(M)}^k g(\phi(\underline{w})), \qquad (8.5.4)$$

where $k \in \mathbf{N}$, g is now assumed to be a C^k-function, and

$$J_k(\phi, \underline{w}) = \left(\widetilde{c\underline{w} + d}\right)\left((c\underline{w} + d)\left(\widetilde{c\underline{w} + d}\right)\right)^{\frac{-n+k-1}{2}}$$

when k is odd,

$$J_k(\phi, \underline{w}) = \left((c\underline{w} + d)\left(\widetilde{c\underline{w} + d}\right)\right)^{\frac{-n+k}{2}}$$

when k is even;

$$J_{-k}(\phi, \underline{w}) = \left(\widetilde{c\underline{w} + d}\right)^{-1}\left((c\underline{w} + d)\left(\widetilde{c\underline{w} + d}\right)\right)^{\frac{-n-k+1}{2}}$$

when k is odd, and

$$J_{-k}(\phi, \underline{w}) = \left((c\underline{w} + d)\left(\widetilde{c\underline{w} + d}\right)\right)^{\frac{-n-k}{2}}$$

when k is even.

In [21], we show that if $g : M \to A_n(\mathbf{C})$ is the restriction to M of a holomorphic function $g^+ : \Omega \to A_n(\mathbf{C})$, where Ω is a domain of holomorphy containing M, then

$$D_M g = D_{\mathbf{C}} g^+|_M,$$

where $D_{\mathbf{C}} = \sum_{j=1}^{n} e_j \frac{\partial}{\partial z_j}$. It now follows from the homotopy deformation arguments described in [17,18,19] and the uniqueness of holomorphic continuation, that we have:

THEOREM 8.5.3

Suppose that $g^+ : \Omega \to A_n(\mathbf{C})$ *is a holomorphic function, and* $\begin{pmatrix} a & b \\ c & d \end{pmatrix} \in$ $V(\mathbf{C}^n)$, *then wherever* $\phi(\underline{z}) = (a\underline{z} + b)(c\underline{z} + d)^{-1}$ *is well defined on* Ω, *we have*

$$D_{\mathbf{C}}^k J_k(\phi, \underline{w}) g^+(\phi(\underline{z})) = J_{-k}(\phi, \underline{w}) D_{\mathbf{C}}^k g(\phi(\underline{z})). \qquad (8.5.5)$$

When n is even, expression (8.5.5) is well defined. However, when n is odd, the functions J_k and J_{-k} need not be uniquely defined, and so in these cases expression (8.5.5) should be interpreted as holding on some Riemann surface. A particular case to consider is the case where Ω is the cell of harmonicity for an annulus, or spherical shell, in R^n, with $n > 2$.

Theorem 8.5.3 shows us that $D_{\mathbf{C}}^k$ is a conformally covariant operator with intertwining operators J_k and J_{-k}. As a consequence of Theorems 8.3.3 and 8.5.3, and [17,21], we have:

PROPOSITION 8.5.4

Suppose that M *is a domain manifold, and* $\begin{pmatrix} a & b \\ c & d \end{pmatrix} \in V(\mathbf{C}^n)$ *is such that* $\phi(\underline{z}) = (a\underline{z} + b)(c\underline{z} + d)^{-1}$ *is well defined on* M. *Suppose also, that* $f : M^+ \to A_n(\mathbf{C})$ *is a holomorphic function satisfying* $D_{\mathbf{C}}^k f(\underline{z}) = 0$. *Then*

$$J_k(\phi, \underline{z}) f(\phi(\underline{z}))$$

is well defined on $\phi(M)^+$ *and is annihilated by* $D_{\mathbf{C}}^k$.

Note: When n is odd, and when n is even and $k > n$, it is often the case that the function $f(\underline{z})$ appearing in Proposition 8.5.1 is usually defined on some covering space of M^+. In these cases $J_k(c\underline{z} + d) f(\phi(\underline{z}))$ is defined on the corresponding covering space of $\phi(M)^+$. This follows as the fundamental solutions, G_k, to the operator D^k do not have unique holomorphic continuations to $\mathbf{C}^n \backslash N(\underline{0})$ in these cases.

Let us now consider the convolution

$$\int_{\phi(M)} G(\phi(\underline{z}) - \phi(\underline{w})) g(\phi(\underline{z})) d\phi(\underline{z})^n, \qquad (8.5.6)$$

where M is a domain manifold, $\begin{pmatrix} a & b \\ c & d \end{pmatrix} \in V(\mathbf{C}^n)$ satisfies the conditions given in Theorem 8.3.2, and $g(\phi(\underline{z}))$ is a well defined $A_n(\mathbf{C})$-valued function on M, and is such that the integral (8.5.6) is well defined, and $\underline{w} \in M$. Let us denote the function given by (8.5.6) by $Q_{\phi(M)} g(\phi(\underline{w}))$.

By a simple change of variables, we now have that expression (8.5.6) is equal to

$$J_1(\phi, \underline{w})^{-1} \int_M G(\underline{z} - \underline{w}) J_{-1}(\phi, \underline{z}) g(\phi(\underline{z})) d\underline{z}^n.$$

Consequently, we have that

$$Q_m J_{-1} \left(\phi, \underline{z} \right) g \left(\phi \left(\underline{z} \right) \right) = J_1 \left(\phi, \underline{z} \right) Q_{\phi(M)} g \left(\phi \left(\underline{z} \right) \right).$$

So, we have that J_{-1} and J_1 are intertwining operators for the convolution operators Q_M and $Q_{\phi(M)}$. Let us now denote the functions

$$\int_{\phi(M)} G_k \left(\phi \left(\underline{z} \right) - \phi \left(\underline{w} \right) \right) g \left(\phi \left(\underline{z} \right) \right) d\phi \left(\underline{z} \right)^n$$

by $Q_{k,\phi(M)} g \left(\phi \left(\underline{w} \right) \right)$, so that $Q_{1,\phi(M)} g = Q_{\phi(M)}$. Then, following [23], we have:

PROPOSITION 8.5.5

Suppose that M is a domain manifold, and $\begin{pmatrix} a & b \\ c & d \end{pmatrix} \in V \left(\mathbf{C}^n \right)$ is such that $\phi \left(\underline{z} \right) = \left(a\underline{z} + b \right) \left(c\underline{z} + d \right)^{-1}$ satisfies the condition given in Theorem 8.3.2. Then the identity

$$Q_{k,M} J_{-k} \left(\phi, \underline{z} \right) g \left(\phi \left(\underline{z} \right) \right) = J_k \left(\phi, \underline{z} \right) Q_{k,M} g \left(\phi \left(\underline{z} \right) \right) \qquad (8.5.7)$$

holds for all k when n is odd, and for $k < n$ when n is even.

The reason why the identity (8.5.7) breaks down when $k \geq n$, for n even, is that in these cases the kernel G_k contains a log term, which is not conformally covariant, see [16,22].

From Proposition 8.5.4, and identities (8.5.3) and (8.5.4), it follows that all the basic results and ideas presented in [17] over domains in R^n carry over for the domain manifolds presented here. In particular, in [16] we show that the methods used by Gürlebeck and Sprössig in [8], for solving the Dirichlet problem for the Laplacian over a bounded domain, can be extended using the conformal group and applied to unbounded domains. In [22], we use the operator D_M to show that the methods advocated by Gürlebeck and Sprössig in [8] extend to solving the Dirichlet problem for D_M^2 over bounded domain manifolds. As we can conformally transform bounded domain manifolds to unbounded domain manifolds, it follows that we can now set up and solve Dirichlet problems for D_M^2 over unbounded domain manifolds. Following arguments presented in the previous section, it is now relatively straightforward to use results from this section and [22] to establish and solve Dirichlet problems for the operator $D_M + \left(\left(c\underline{z} + d \right) \left(\widetilde{c\underline{z} + d} \right) \right)^{-1}$ for $\begin{pmatrix} a & b \\ c & d \end{pmatrix} \in V \left(\mathbf{C}^n \right)$ and M a bounded domain manifold.

8.6 The Bergman Kernel and Harmonic Measure

This conformal covariance enables us to set up a Bergman kernel over domain manifolds. First, we shall do this over bounded domain manifolds.

Suppose that M is a closed and bounded domain manifold. Then, following [24], we have for each $\underline{w} \in \overset{\circ}{M}$ a function $g(\underline{z}, \underline{w})$, with $\underline{z} \in M$, and

(i) $D_M^2 g(\underline{z}, \underline{w}) = 0$,

(ii) $g(\underline{z}, \underline{w}) = G_2(\underline{z} - \underline{w})$ for each $\underline{z} \in \partial M$.

The function $g(\underline{z}, \underline{w})$ is defined on $M \times \overset{\circ}{M}$ and is a C^1-function with respect to the variable $\underline{w}$.

Let us now consider the function

$$H(\underline{z}, \underline{w}) = G_2(\underline{z} - \underline{w}) - g(\underline{z}, \underline{w}).$$

Now, let us introduce the function $D_{M,\underline{z}} H(\underline{z}, \underline{w}) D_{M,\underline{w}}$, where $D_{M,\underline{z}}$ is the operator D_M acting with respect to the variable $\underline{z}$, while $D_{M,\underline{w}}$ is the operator D_M acting with respect to the variable $\underline{w}$. Here, the operator $D_{M,\underline{w}}$ acts on the right, while $D_{M,\underline{z}}$ acts on the left. We denote the function $D_{M,\underline{z}} H(\underline{z}, \underline{w}) D_{M,\underline{w}}$ by $B_M(\underline{z}, \underline{w})$. The construction of $B_M(\underline{z}, \underline{w})$ over domains in R^n has previously been described in [24].

By similar arguments to those given in [24], we have:

THEOREM 8.6.1
Suppose that M is a closed bounded domain manifold, and $f : M^+ \to A_n(\mathbf{C})$ satisfies $D_{\mathbf{C}} f = 0$. Moreover, the integral $\int_M \bar{f}(\underline{z}) f(\underline{z}) dz^n$ is bounded. Then,

$$\int_M \bar{f}(\underline{z}) B_M(\underline{z}, \underline{w}) dz^n = f(\underline{w})$$

for each $\underline{w} \in \overset{\circ}{M}$.

In [21], we show that if $g : M \to A_n(\mathbf{C})$ is a C^1-function, and $\int_M \bar{g}(\underline{z}) g(\underline{z}) dz^n$ is finite, then there is a function $f : M^+ \to A_n(\mathbf{C})$ satisfying $D_{\mathbf{C}} f = 0$, and there is a C^1-function $\psi : M \to A_n(\mathbf{C})$ with $\psi(\underline{z}) = 0$ on ∂M, such that $g(\underline{z}) = f(\underline{z}) + \psi(\underline{z})$, and

$$\int_M F(\underline{z}) \bar{\psi}(\underline{z}) dz^n = 0$$

for each $F : M^+ \to A_n(\mathbf{C})$ which satisfies

(i) $D_{\mathbf{C}} F = 0$

(ii) $\left\| \int_M \bar{F}(z) F(z) \, dz^n \right\| < +\infty.$

Consequently, we have that

$$\int_M \overline{g(z)} B_M(z, w) \, dz^n = f(w).$$

It follows that the kernel $B_M(z, w)$ is a projection operator. The Euclidean analogue of this result is established in [24].

Using an orthonormal basis $\{f_n\}_{n=1}^\infty$ for the inner product module

$$B^2 \ (M, A_n(\mathbf{C})) = \{f : M^+ \to A_n(\mathbf{C}) : D_\mathbf{C} f = 0 \text{ and } \left\| \int_M \bar{f}(z) f(z) \, dz^n \right\| < +\infty\},$$

and by uniform convergence arguments it may be observed that $D_{M,w} B_M^+$ $(z, w) = 0$ for each $z \in M$. It follows that for each $z \in M$, the function $B_M(z, w)$ extends to a holomorphic function $B_M^+(z, w)$ on M^+, and this function satisfies $B_M^+(z, w) D_\mathbf{C} = 0$. Moreover, as $D_{z,M}^2 g_M(z, w) = 0$, it follows that $g_M(z, w)$ is the restriction to M of a holomorphic function $g(z, w)$, which satisfies $\Delta_{n,\mathbf{C}} g(z, w) = 0$, where $\Delta_{n,\mathbf{C}} = \sum_{j=1}^n \frac{\partial^2}{\partial z_j^2}$. It follows that $D_\mathbf{C} B(z, w) = 0$, where $D_\mathbf{C}$ now acts on the variable z. It follows that $B_M^+(z, w)$ is the Bergman kernel for the inner-product module $B^2(M, A_n(\mathbf{C}))$.

Let us now consider the following homotopy.

DEFINITION 8.6.2 *Suppose that M is a closed and bounded domain manifold. Suppose also that*

$$h : M \times [0, 1] \to \mathbf{C}^n$$

is a homotopy satisfying the following conditions:

(i) $h(z, 0) = z$ *for each $z \in M$;*
(ii) $h(M, t)$ *is a domain manifold for each $t \in [0, 1]$;*
(iii) $h(z, t) = z$ *for each $z \in \partial M$ and $t \in [0, 1]$.*

Then the homotopy h is called a domain manifold homotopy.

We have introduced and used this type of homotopy in a number of our previous papers; see, for instance, [17,19]. In [19], we also illustrate that there is an abundance of these types of homotopies.

As ∂M remains fixed throughout this homotopy deformation, we have:

LEMMA 8.6.3
Suppose that M is a closed and bounded domain manifold, and h is a domain manifold homotopy. Furthermore, suppose that there is a neighborhood U_M, in M, of ∂M such that $h(z, t) = z$ for each $z \in U_M$ and

$t \in [0, 1]$. *Then*

$$B_{h(M,t)}\left(\underline{z}, \underline{w}\right) = B_M^+\left(\underline{z}, \underline{w}\right)$$

for each $\underline{z}, \underline{w} \in h(M, t)$ *and each* $t \in [0, 1]$.

By repeated application of Lemma 8.6.3 we obtain:

THEOREM 8.6.4
Suppose that M *is a closed and bounded domain manifold, and* h *is a domain manifold homotopy. Then*

$$B_{h(M,t)}\left(\underline{z}, \underline{w}\right) = B_M^+\left(\underline{z}, \underline{w}\right)$$

for each $\underline{z}, \underline{w} \in h(M, t)$ *and each* $t \in [0, 1]$.

As $M^+ = M'^+$ if M and M' are closed, bounded domain manifolds which can be deformed into each other via a domain manifold homotopy, then it follows that there is a unique Bergman kernel $B_M^+\left(\underline{z}, \underline{w}\right) \left(= B_{M'}^+\left(\underline{z}, \underline{w}\right)\right)$ on $M^+ \times M^+$. For this reason, it is simpler to write $B^+\left(\underline{z}, \underline{w}\right)$ for $B_M^+\left(\underline{z}, \underline{w}\right)$. Also, via Cauchy's theorem, we obtain

$$\int_M \bar{f}\left(\underline{z}\right) f\left(\underline{z}\right) dz^n = \int_{h(M,t)} \bar{f}\left(\underline{z}\right) f\left(\underline{z}\right) dz^n$$

for each domain manifold homotopy h.

Consequently, $B^2\left(M, A_n(\mathbf{C})\right) = B^2\left(h(M, t), A_n(\mathbf{C})\right)$ for each domain manifold homotopy h. It follows that the Bergman kernel $B\left(\underline{z}, \underline{w}\right)$ is, in fact, acting on a unique module of "square integrable" functions, which remains invariant under domain manifold homotopies. For this reason, we may denote the space by $B^2\left(M^+, A_n(\mathbf{C})\right)$, and we have that

$$B : B^2\left(M^+, A_n(\mathbf{C})\right) \to B^2\left(M^+, A_n(\mathbf{C})\right)$$

$$: B(f) = \int_M \bar{f}\left(\underline{z}\right) B\left(\underline{z}, \underline{w}\right) dz^n$$

is the identity map.

Using Theorem 8.3.2, we may conformally transform a bounded domain manifold to an unbounded domain manifold. Using the arguments developed in the previous section and in [23], it follows that we may adapt the arguments used to introduce a Bergman kernel over a bounded domain manifold to also introduce a Bergman kernel over an unbounded domain manifold. It is now straightforward to verify that this kernel has similar properties to the one we have introduced over bounded domain manifolds.

Let us now consider harmonic measure on the boundary of a domain manifold. We previously introduced such a measure in [21]. Suppose that

M is a closed bounded domain manifold, and $z_0 \in \overset{\circ}{M}$. Then, we may set up a complex-valued measure μ_{z_0} on ∂M such that $\int_{\partial M} h(z) \, d\mu_{z_0} = h(z_0)$ for each complex harmonic function $h(z)$, where $h(z)$ and $D_C \overline{h}(z)$ have continuous extensions to ∂M. The condition that $h(z)$ and $D_C h(z)$ have continuous extension to ∂M is a more stringent condition than one finds in the cases where $M \subseteq R^n$. This is because over general domain manifolds one does not, so far, have a suitable maximum principle.

Following [21], we now have a kernel $K_{\partial M}(z, w)$ such that

$$\int_{\partial M} h(z) K_{\partial M}(z, w) \, d\mu_{z_0} = h(w)$$

for each $w \in \overset{\circ}{M}$. Following the homotopy arguments that we have just introduced, it follows that the kernel $K_{\partial M}(z, w)$ is well defined over each domain manifold M' which is homotopic, via a domain manifold homotopy, to M. It follows that $K_{\partial M}(z, w)$ is well defined on $\partial M \times \Omega(M)$, where $\Omega(M)$ is a subdomain of M^+, comprising of the union of all interiors of domain manifolds which are homotopic, via domain manifold homotopies, to M. The domain $\Omega(M)$ is introduced in [17], and in [18] it is shown that $\Omega(M) = N(M^+)$, where $N(M^+) = \{z \in M^+ : \lambda z + (1 - \lambda)u \in M^+$ for each $\lambda \in [0, 1]$, and each $u \in N(z) \cap M\}$. The fact that $N(M^+)$ is not always identical to M^+ may be observed by considering the cell of harmonicity of a spherical shell.

From Theorem 8.3.4 we obtain:

PROPOSITION 8.6.5
Suppose that M is a domain manifold, and $\begin{pmatrix} a & b \\ c & d \end{pmatrix} \in V(\mathbf{C}^n)$ is such that $\phi(z) = (az + b)(cz + d)^{-1}$ is well defined on M. Then $\phi(N(M^+)) = N(\phi(M^+))$.

From Proposition 8.6.5 we obtain:

THEOREM 8.6.6
Suppose that M is a domain manifold, and $\begin{pmatrix} a & b \\ c & d \end{pmatrix} \in V(\mathbf{C}^n)$ is such that $\phi(z) = (az + b)(cz + d)^{-1}$ is well defined on M. Suppose also that $c\ell(\phi(M))$ is a bounded domain manifold. Then

$$\int_{\partial M} J_2(\phi, z) \, h(\phi(z)) \, J_2(\phi, z)^{-1} \, K_{\partial \phi(M)}(\phi(z), \phi(w)) \, J_2(\phi, w) \, d\hat{\mu}_{\phi(z_0)}(z)$$

$$= \int_{\partial \phi(M)} J_2(\phi, z) \, h(\phi(z)) \, K_{\partial M}(z, w) \, d\mu_{z_0}(z)$$

for each complex harmonic function $h(u)$ defined on $N(\phi(M^+))$, where

$h\left(\underline{u}\right)$ and $D_{\mathbf{C}}h\left(\underline{u}\right)$ have continuous extensions to $\partial\phi(M)$. Moreover, the measure $\hat{\mu}_{\phi\left(\underline{z_0}\right)}$ is defined to satisfy $\hat{\mu}_{\phi\left(\underline{z_0}\right)}(K) = \mu_{\phi\left(\underline{z_0}\right)}\left(\phi(K)\right)$ for each set $\phi(K)$ which is measurable with respect to $\mu_{\phi\left(\underline{z_0}\right)}$.

8.7   More on Conformal Covariance in Complex Clifford Analysis

In this concluding section, we return to point out some further results pertaining to conformal covariance in $\mathbf{C}^n$. Details will be discussed elsewhere.

In [17], we use the homotopies set up in the previous section to show that for each holomorphic function $g : \Omega \rightarrow A_n(\mathbf{C})$, we may construct a holomorphic function Qg defined on a subdomain of $\Omega \cap N(M^+)$, where M is a bounded domain manifold lying in Ω. Moreover, $D_{\mathbf{C}}Qg = -g$, and over each domain manifold M' which is homotopic to M via a domain manifold homotopy, we have that $Qg\left(\underline{w}\right) = Q_{M'}g\left(\underline{w}\right)$ for each $\underline{w} \in \overset{\circ}{M'}$. Using Proposition 8.5.5, and elementary uniqueness of holomorphic continuation arguments, we now have:

THEOREM 8.7.1

Suppose that M is a bounded domain manifold, and $\begin{pmatrix} a & b \\ c & d \end{pmatrix} \in V(\mathbf{C}^n)$ is such that $\phi\left(\underline{z}\right) = (a\underline{z} + b)(c\underline{z} + d)^{-1}$ is well defined on M. Then

$$QJ_{-1}\left(\phi, \underline{z}\right)g\left(\phi\left(\underline{z}\right)\right) = J_1\left(\phi, \underline{z}\right)Qg\left(\phi\left(\underline{z}\right)\right)$$

holds for any holomorphic function $g : \Omega \rightarrow A_n(\mathbf{C})$ defined in a neighborhood of M.

Theorem 8.7.1 shows us that J_{-1} and J_1 are intertwining operators for the operator Q. The previous theorem shows us that intertwining relations described in [22] in the Euclidean setting carry over to the several complex variable setting, via the homotopies described here, and in [19]. In [18], we also introduce a transform $Q_k g$ for each holomorphic function $g\left(\underline{z}\right)$, and we have that $D_{\mathbf{C}}^k Q_k g\left(\underline{z}\right) = (-1)^k g\left(\underline{z}\right)$.

In complete analogy to Theorem 8.7.1, we may use Proposition 8.5.5 to deduce that J_{-k} and J_k are intertwining operators for Q_k under conformal transformations satisfying the conditions laid out in Theorem 8.3.2, provided $k < n$ when n is even. The result holds for all k when n is odd. It should also be noted that the domain $\phi\left(\Omega \cap N(M^+)\right)$ need not be bounded. So, many results obtained in [18,19] over bounded domains can now be pushed through via Möbius transformations to the unbounded case. In particular, the Huygens' principle integral formulae described in

[18,19] over the domain manifolds described here may easily be seen to be conformally covariant with conformal weights $J_k(\phi, \underline{z})$, $k = 1, 2, \ldots, m, \ldots$. It follows that the continuous extension of solutions to the Dirac equation, or one of its iterates, defined over a cell of harmonicity, to an open dense subset of $\mathbf{C}^n$, described in [18,19], are conformally covariant with conformal weights $J_k(\phi, \underline{z})$, $k = 1, \ldots, m, \ldots$. It also follows that the integral formula given in [18, Thm. 23a] is conformally covariant for the case $k = 1$, with conformal weight $J_1(\phi, \underline{z})$, and that the Huygens' principle integral formula described in [25] is also conformally covariant.

Bibliography

[1] L. V. Ahlfors, *Clifford numbers and Möbius transformations in R^n*, Clifford Algebras and Their Applications in Mathematical Physics, (J. S. R. Chisholm and A. K. Common, Eds.) NATO ASI, Ser. C, Vol. **183**, D. Riedel, Dordrecht, 1986, 167–175.

[2] L. V. Ahlfors, *Möbius transformations in R^n expressed through 2×2 matrices of Clifford numbers*, Complex Variables **5**, (1986), 215–224.

[3] P. Auscher and Ph. Tchamitchian, *Bases d'ondelettes sur les courbes corde-arc, noyou de Cauchy et Espaces de Hardy associés*, Revista Matematica Iberoamericana **5**, (1989), 139–170.

[4] V. Avanissian, *Cellule D'Harmonicité et Prolongement Analytique Complexe*, Hermann, Paris, 1985.

[5] B. Bojarski, *Conformally covariant differential operators*, Proceedings, XXth Iranian Math. Congress, Tehran, 1989.

[6] G. Gaudry, R. Long, and T. Qian, A martingale proof of L^2-boundedness of Clifford-valued singular integrals, *Annali di Mathematica* **165**(1993) pp.369–394.

[7] J. Gilbert and M. A. M. Murray, *Clifford Algebras and Dirac Operators in Harmonic Analysis*, Cambridge Studies in Advanced Mathematics No.26, Cambridge, 1991.

[8] K. Gürlebeck and W. Sprössig, Quaternionic Analysis and Elliptic Boundary Value Problems, Birkhäuser Verlag, Basel, 1990.

[9] L. K. Hua, *Harmonic Analysis of Functions of Several Complex Variables in the Classical Domains*, Translations of Mathematical Monographs Vol.6, A.M.S., Providence, RI, 1963.

[10] M. Kolar, *Envelopes of holomorphy for solutions of the Laplace and Dirac equations*, Comment. Math. Univ. S. C., (1991), 479–494.

[11] C. Li, A. McIntosh, and S Semmes, *Convolution singular integrals on*

Lipschitz surfaces, J. of the A.M.S. **5**, (1992), 455–481.

[12] A. McIntosh and A. Pryde, *A functional calculus for several commuting operators*, Indiana Univ. Math. J. **36**, (1987), 421–439.

[13] A. McIntosh, A. Pryde, and W. Ricker, *Comparison of joint spectra for certain classes of commuting operators*, Studia Math. **88**, (1988), 23–36.

[14] A. McIntosh, *Clifford algebras and the higher-dimensional Cauchy integral*, Approximation Theory and Function Spaces, Banach Center Publications **22**, (1989), 253–267.

[15] M. Mitrea, *Clifford Wavelets, Singular Integrals, and Hardy Spaces*, Lecture Notes in Mathematics, No. 1575, Springer-Verlag, New York, 1994.

[16] J. Peetre and T. Qian, *Möbius covariance of iterated Dirac operators*, J. Australian Math. Soc., Ser. A, (to appear).

[17] J. Ryan, *Applications of complex Clifford analysis to the study of solutions to generalized Dirac and Klein-Gordon equations, with holomorphic potential*, J. Diff. Eq. **67**, (1987), 295–329.

[18] J. Ryan, *Dirac operators, Schrödinger-type operators in $\mathbf{C}^n$, and Huygens' principle*, J. Funct. Anal. **87**, (1989), 321–347.

[19] J. Ryan, *Cells of harmonicity and generalized Cauchy integral formulæ*, Proc. London Math. Soc. **60**, (1990), 295–318.

[20] J. Ryan, *Intertwining operators for iterated Dirac operators over Minkowski-type spaces*, J. Math. Anal. and its Appl. **177**, (1993), 1–23.

[21] J. Ryan, *Intrinsic Dirac operators in $\mathbf{C}^n$*, Adv. in Math., (to appear).

[22] J. Ryan, *Conformally covariant operators in Clifford analysis*, (to appear).

[23] M. V. Shapiro and N. L. Vasilevski, *Quaternionic ψ-hyperholomorphic functions, singular operators with quaternionic Cauchy kernel, and analogues of the Riemann boundary value problem I*, Complex Variables, (to appear).

[24] M. V. Shapiro and N. L. Vasilevski, *On the Bergman kernel function in Clifford analysis*, Clifford Algebras and their Appl. in Math. Physics, (F. Brackx, R. Delanghe, and H. Serras, Eds.), Kluwer Academic, Dordrecht, 1993, 183–192.

[25] V. Soucek, *Complex-quaternionic analysis applied to spin $\frac{1}{2}$ massless fields*, Complex Variables **1**, (1983), 327–346.

[26] A. Sudbery, *Quaternionic analysis*, Math. Proc. Cambridge Phil. Soc. **86**, (1979), 199–225.

[27] A. Unterberger and J. Unterberger, *A quantization of the Cartan domain BD1(q = 2) and operators on the light cone*, J. Funct. Anal. **72**, (1987), 279–319.

[28] K. Th. Vahlen, *Über Bewegungen und Complexe Zahlen*, Math. Ann. **55**, (1902), 585–593.

[29] X. Zhenyuan, *A function theory for the operator $D - \lambda$*, Complex Variables, **16**, (1991), 27–42.

Department of Mathematical Sciences,
University of Arkansas,
Fayetteville, Arkansas 72701
E-mail: jryan@comp.uark.edu

9

Singular Integrals with Monogenic Kernels on the m-Torus and their Lipschitz Perturbations

Tao Qian

ABSTRACT Explicit transformation formulas between the H^∞-Fourier multipliers on $L^p(\mathbf{T}^m)$ and the monogenic kernels of the associated singular integrals are given. It is indicated, as a generalization and variation of the Coifman-McIntosh-Meyer theorem on Lipschitz graphs, that these kernels give rise to bounded operators on Lipschitz perturbations of $\mathbf{T}^m$.

9.1 Introduction

It is well known that holomorphic bounded functions, defined in sectors of the complex plane containing the real axis, are $L^p(\mathbf{R})$-bounded Fourier multipliers, owing to Michlin's multiplier theorem. We shall call these functions H^∞-Fourier multipliers. The good thing about these functions is that, if we define an analogous notion of the standard Fourier transform on Lipschitz curves, then they still give rise to L^p-bounded Fourier multiplier operators on these curves (see [CM], [McQ1]). In [McQ1], [McQ2] and [McQ3] the authors characterize the corresponding kernels, by giving explicit transformation formulas between the kernels and the multipliers, and establish an even more general Fourier multiplier theory on Lipschitz curves.

In [LMcQ] the authors develop a higher-dimensional theory using Clifford monogenic functions on the kernel side, and holomorphic functions of several complex variables on the Fourier multiplier side. The generalization is not a trivial one, owing to the fact that Clifford numbers and Clifford monogenicity are not perfect analogues of the corresponding objects in complex analysis. For instance, Clifford multiplication is noncommutative, and the product of two Clifford monogenic functions is no longer a Clif-

0-8493-8481-8/96/$0.00+$.50
© 1996 by CRC Press

ford monogenic function in general. The use of the extended exponential functions $e(x, \eta)$ and the related projections $\chi_{\pm}$ are the key to solving the problem (see [LMcQ]).

At this point I wish to comment on extensions of the standard exponential functions $\exp(i < \mathbf{x}, \mathbf{y} >), \mathbf{x}, \mathbf{y} \in \mathbf{R}^m$. The extensions which are monogenic in both variables, each in either $\mathbf{R}^{m+1}$ or $\mathbf{C}^{m+1}$, together with some applications, were first published by F. Sommen in [S1], [S2]. It appears that A. McIntosh found these extensions independently. In 1986 McIntosh showed me his handwritten manuscript on the extensions together with some applications to Fourier transform, when we started working on the interrelationship between singular integrals with monogenic kernels and Fourier multipliers on curves and surfaces. We originally spent some time trying to use the extensions which are monogenic in both variables, but did not succeed. McIntosh subsequently realized that the right thing for our purpose is the extension which is monogenic in one variable in $\mathbf{R}^{m+1}$, and complex holomorphic in the other variable in $\mathbf{C}^m$. With the help of this extension we, together with Chun Li, built up a perfectly analogous theory to the one-dimensional case [LMcQ]. At the same time, this allowed us to develop the H^∞-functional calculus of the Dirac differential operator on Lipschitz surfaces.

In [Q1] the corresponding theory on the unit circle and its Lipschitz perturbations is studied. The basic method used in that paper is an adapted Poisson summation method based on the special properties of the kernel functions proved in [McQ1] and also in [McQ2]. The boundedness result is proved in [GQW]. The topic of singular integrals on star-shaped Lipschitz domains offers a rather direct theory to the potential solutions of the Dirichlet and the Neumann boundary value problems on these domains (see [FJR] and [V] for examples). We are able to restrict ourselves to the star-shaped Lipschitz domains because of the following observation: every simply connected Lipschitz domain of the complex plane is the image of a star-shaped Lipschitz domain under some conformal mapping and the fact that conformal mappings preserve harmonic functions.

The higher-dimensional theory is of interest of its own, and has direct applications to the corresponding boundary value problems as well. The present paper will deal with the m-torus and its Lipschitz perturbations and so formulate a higher-dimensional generalization of the theory developed in [Q1] and [GQW]. The method is based on Fourier transforms between functions defined on sectors developed in [LMcQ], and an adapted Poisson summation formula according to the special properties of the kernel functions proved in [LMcQ]. Our first three theorems together identify the class of bounded singular integral operators with right-monogenic periodic kernels, or equivalently, the class of bounded Fourier multiplier operators induced by the H^∞-functions, defined on periodic Lipschitz perturbations of $\mathbf{D}^m = [-\pi, \pi]^m$. The mapping : $x \to \exp(ix)$ then transfers the theory to Lipschitz perturbations of the m-torus $\mathbf{T}^m$.

The L^2-boundedness of these operators on $\mathbf{T}^m$ can be easily deduced from Plancherel's theorem. As in the one-dimensional case, however, this does not work for the boundedness of the above-mentioned operators on periodic Lipschitz perturbations of $\mathbf{D}^m$. Not surprisingly, ingredients of a proof of the boundedness are the same as those in the proofs of [CMM]'s theorem, its higher-dimensional extensions, and the $T(b)$ theorems as well. In [GQW] several proofs of the boundedness in the one-dimensional case are provided, and each of them can be adapted to the present higher-dimensional cases. In [Q1] there is another proof for the one-dimensional case. We therefore decided not to prove the boundedness assertion (ii) Th.3, and refer the interested reader to either [GQW], or [Q1]. [Q2] contains a transference proof of the boundedness, using the monogenic extensions of the Gauss-Weierstrass kernels, based on the boundedness result in [LMcS], or in [T]. We also include a corollary on monogenic extensions.

It is worthwhile mentioning, without going into any detail, that there is a H^∞-functional calculus theory, similar to the one on infinite Lipschitz graphs, of the Dirac differential operator $-i\mathbf{D} = -i(e_1\partial_1 + \cdots + e_m\partial_m)$ on periodic Lipschitz graphs. It can be deduced following the pattern in [LMcQ], by using the knowledge developed in this paper.

The author wishes to thank Alan McIntosh and John Ryan, for the helpful discussions on this topic during his visits to Washington University at St Louis and the University of Arkansas in April and May, 1993, respectively. Thanks are also due to John Ryan for pointing out to me that it is F. Sommen who had earlier published in [S2] the monogenic extensions of the exponential functions.

9.2 Preliminary

Denote the standard basis vectors of $\mathbf{R}^{m+1}$ by $e_1, ..., e_m, e_L$, and identify e_L with either e_0 or e_{m+1}, where $e_0^2 = 1, e_i^2 = -1, i = 1, ..., m+1, e_i e_j = -e_j e_i, 1 \le i < j \le m+1$. We shall adopt the way in [LMcQ] of embedding $\mathbf{R}^{m+1}$ into the real Clifford algebras, according to which we write an element $x \in \mathbf{R}^{m+1}$ as $x = \mathbf{x} + x_L e_L$, where $\mathbf{x} = x_1 e_1 + \cdots + x_m e_m \in \mathbf{R}^m$, and $x_L \in \mathbf{R}$. For simplicity we shall only deal with those cones which are rotationally symmetric about the x_L axis as studied in [LMcS]. The methods and results of this paper can be adapted to the nonrotationally symmetric case considered in [LMcQ].

Set $\mathbf{T} = \{\exp(i\theta) \mid -\pi \le \theta \le \pi\}$. $\mathbf{T}^m = \mathbf{T} \times \cdots \times \mathbf{T}$ (m *factors*), the m−torus. $\mathbf{D} = [-\pi, \pi], \mathbf{D}^m = \mathbf{D} \times \cdots \times \mathbf{D}$ (m *factors*). For $\mu \in (0, \frac{\pi}{2}], \mathbf{C}_{\mu,+} = \{0 \ne x = \mathbf{x} + x_L e_L \in \mathbf{R}^{m+1} \mid x_L > -|\mathbf{x}| \tan\mu\}, \mathbf{C}_{\mu,-} =$

$-\mathbf{C}_{\mu,+}$, and $\mathbf{S}_\mu = \mathbf{C}_{\mu,+} \cap \mathbf{C}_{\mu,-}$. For $\mathbf{Q} = \mathbf{S}_\mu, \mathbf{C}_{\mu,+}$ or $\mathbf{C}_{\mu,-}$, denote $\mathbf{Q}(\pi) = \mathbf{Q} \cap \{x \mid \mathbf{x} \in \mathbf{D}^m\}$, and $\mathbf{pQ} = \bigcup_{l \in \mathbf{Z}^m} (2\pi l + \mathbf{Q}(\pi))$, where we embed $\mathbf{Z}^m$ into $\mathbf{R}^m$ by setting $l = l_1 e_1 + \cdots + l_m e_m \in \mathbf{Z}^m, l_i \in \mathbf{Z}$, the set of the integers. $\mathbf{T}_\mu(\pi) = \{y = \mathbf{y} + y_L e_L \in \mathbf{R}^{m+1}_+ \mid y^\perp \subset \mathbf{S}_\mu, 0 < y_L \le \pi\}$. A function defined in $\mathbf{R}^{m+1}$ is said to be 2π-periodic, if it is 2π-periodic in every coordinate. Denote $\mathbf{S}_\mu(\mathbf{C}^m) = \{\zeta = \sum_{j=1}^m \zeta_j e_j = \sum_{j=1}^m (\xi_j + i\eta_j)e_j = \xi + i\eta \in \mathbf{C}^m \mid |\zeta|^2_\mathbf{C} \notin (-\infty, 0] \text{ and } |\eta| < Re(|\zeta|_\mathbf{C}) \tan\mu\}$, where $|\zeta|^2_\mathbf{C} = \sum_{j=1}^m \zeta_j^2$, and $Re(|\zeta|_\mathbf{C}) > 0$. The important functions in this theory are the projections $\chi_\pm(\zeta) = \frac{1}{2}(1 \pm i\zeta e_L |\zeta|^{-1}_\mathbf{C})$ and the exponential functions $e(x, \zeta) = e_+(x, \zeta) + e_-(x, \zeta)$, where $e_\pm(x, \zeta) = e^{i\langle \mathbf{x}, \zeta \rangle} e^{\mp x_L |\zeta|_\mathbf{C}} \chi_\pm(\zeta)$, which is formally from $e(x, \zeta) = e^{-ix \cdot \zeta} = e^{-i\mathbf{x}\zeta} e^{-ie_L x_L \zeta}$ and the corresponding rule between functions of one complex variable and functions of several complex variables induced by the projections $\chi_\pm$ indicated, for example, in [LMcQ].

Denote $D_l = \sum_{i=0}^m e_i \frac{\partial}{\partial x_i}$ and $D_r = \sum_{i=0}^m \frac{\partial}{\partial x_i} e_i$. For a Clifford-valued function f we define

$$D_l f = \sum_{i=0}^m e_i \frac{\partial f}{\partial x_i}, \quad D_r f = \sum_{i=0}^m \frac{\partial f}{\partial x_i} e_i.$$

A function f is said to be left-monogenic or right-monogenic, if $D_l f = 0$ or $D_r f = 0$, respectively. The good thing with the monogenicity is that if f is right-monogenic and g is left-monogenic in a neighborhood of a domain D with smooth boundary, then the Cauchy theorem holds:

$$\int_{\partial D} f(x)n(x)g(x)d\sigma(x) = 0,$$

where $n(x)$ is the outer normal on ∂D and $d\sigma(x)$ the area element on the boundary. The Cauchy formulas also hold (see [LMcQ]).

Let $n = \mathbf{n} + n_L e_L$ be a unit vector in $\mathbf{R}^{m+1}$. We shall use the open half tubes in $\mathbf{R}^{m+1}$: $C_n^\pm = \{x \in \mathbf{R}^{m+1} \mid \mathbf{x} \in \mathbf{D}^m, \pm\langle x, n \rangle > 0\}$, and the real m-dimension surface $n(\mathbf{C}^m)$ in $\mathbf{C}^m$, defined by

$$n(\mathbf{C}^m) = \{\zeta = \xi + i\eta \in \mathbf{C}^m \mid |\zeta|^2_\mathbf{C} \notin (-\infty, 0] \text{ and } n_L \eta = Re(|\zeta|_\mathbf{C})\mathbf{n}\}.$$

See [LMcQ] for some equivalent characterizations of $n(\mathbf{C}^m)$ and the relation that $n \in \mathbf{T}_\mu(\pi)$ if and only if $n(\mathbf{C}^m) \subset \mathbf{S}_\mu(\mathbf{C}^m)$, where

$$\mathbf{S}_\mu(\mathbf{C}^m) = \{\zeta = \xi + i\eta \in \mathbf{C}^m \mid |\zeta|^2_\mathbf{C} \notin (-\infty, 0] \text{ and } |\eta| < Re(|\zeta|_\mathbf{C}) \tan\mu\}.$$

Functions $e_\pm(x, \zeta)$ satisfy the following relations:

$$|e_\pm(x, \zeta)| = e^{-\langle \mathbf{x}, \eta \rangle \mp x_L Re|\zeta|_\mathbf{C}} |\chi_\pm(\zeta)|$$
$$\le \frac{\sec(\mu)}{\sqrt{2}} e^{\mp \langle \mathbf{x}, \mathbf{n} \rangle Re|\zeta|_\mathbf{C}/n_L}, \quad \zeta \in n(\mathbf{C}^m) \subset \mathbf{S}_\mu(\mathbf{C}^m).$$

We shall denote by K_ω the class $K(S_{N_\mu})$ in page 3 of [LMcQ], where $\omega = \mu + \mu_N$.

9.3   Fourier Transforms

THEOREM 9.3.1
Let $\omega \in (0, \frac{\pi}{2}], b \in H^\infty(\mathbf{S}_\omega(\mathbf{C}^m))$, and $(\phi, \underline{\phi})$ a pair of functions determined by b according to Th.4.3 in [LMcQ]. Then there exists a pair of functions $(\Phi, \underline{\Phi})$ defined in $\mathbf{pS}_\omega$ and $\mathbf{T}_\omega(\pi)$, respectively, such that for every $\mu \in (0, \omega)$

(i)   *Φ is unique, modulo constants, 2π-periodic and right-monogenic in $\mathbf{pS}_\omega$ and*

$$|\Phi(x)| \le \frac{C_{\omega,\mu}}{|x|^m}, \quad x \in \mathbf{S}_\mu(\pi).$$

Moreover,

$$\Phi(x) = \phi(x) + \phi_0(x) + c, \quad x \in \mathbf{S}_\omega(\pi),$$

where ϕ_0 is right-monogenic and bounded in any $\mathbf{S}_\mu(\pi)$.

(ii)   *$\underline{\Phi}$ is determined, modulo constants, by Φ, continuous in $\mathbf{T}_\omega(\pi)$, $\|\underline{\Phi}\|_{L^\infty(\mathbf{T}_\mu(\pi))} \le C_{\omega,\mu}\|b\|_\infty$, and*

$$\underline{\Phi}(y) - \underline{\Phi}(z) = \int_{A(y,z)} \Phi(x)n(x)dS_x, \quad y, z \in \mathbf{T}_\omega(\pi),$$

where $A(y, z)$ is a smoothly oriented m-manifold in $\mathbf{S}_\omega(\pi)$ joining the $(m-1)$-sphere $S_y = \{x \in \mathbf{R}^{m+1}|\ \langle x, y \rangle = 0, |x| = |y|\}$ and the similarly defined $(m-1)$-sphere S_z.

(iii)   *Parseval's identity holds*

$$(2\pi)^m \quad \sum_{l \in \mathbf{Z}^m} b(l)\hat{F}(-l)$$
$$= \lim_{\epsilon \to 0} \ \left\{ \int_{x \in \mathbf{D}^m, |x| \ge \epsilon} \Phi(x)e_L F(x)dx + \underline{\Phi}(\epsilon e_L)F(0) \right\}$$

for any 2π-periodic and smooth function F defined on $\mathbf{R}^m$, where $\hat{F}$ stands for the nth standard Fourier coefficient of F, and $b(0) = (\frac{1}{2\pi})^m \underline{\Phi}(\pi)$.

PROOF   Let

$$\Phi(x) = (2\pi)^m \sum_{l \in \mathbf{Z}^m} \phi(x + 2\pi l), \quad x \in \mathbf{pS}_\omega, \qquad (9.3.1)$$

where $l = l_1 e_1 + \cdots + l_m e_m, l_i \in \mathbf{Z}$, and the summation is in the following sense: there exists a subsequence $(n_k) \subset (n)$ such that the partial sums

$$s_{n_k}(x) = (2\pi)^m \sum_{n=0}^{n_k} \sum_{\max\{|l_i|:1 \le i \le m\}=n} \phi(x + 2\pi l), \quad x \in \mathbf{pS}_\omega$$

converge, as $k \to \infty$, uniformly in any smaller set $\mathbf{pS}_\mu$, where $\mu \in (0, \omega)$.

∎

To verify the summability and that the sum satisfies the required properties we employ the decomposition $\phi = \phi^+ + \phi^-$ in Th.3.1 of [LMcQ]. We shall prove, for a suitably chosen (n_k), that both the partial sums

$$s_{n_k}^\pm(x) = (2\pi)^m \sum_{n=0}^{n_k} \sum_{\max\{|l_i|:1\leq i \leq m\}=n} \phi^\pm(x + 2\pi l), \quad x \in \mathbf{pC}_{\mu,\pm}, \quad (9.3.2)$$

converge in the indicated regions, respectively.

This, however, is justified by the summability of the following two series:

$$\sum_1 = \lim_{k\to\infty} \sum_{n=1}^{n_k} \sum_{\max\{|l_i|:1\leq i \leq m\}=n} (\phi^\pm(x + 2\pi l) - \phi^\pm(2\pi l)),$$

and

$$\sum_2 = \lim_{k\to\infty} \sum_{n=1}^{n_k} \sum_{\max\{|l_i|:1\leq i \leq m\}=n} \phi^\pm(2\pi l).$$

Since $\phi^\pm$ are right-monogenic, using Cauchy's formula we obtain

$$|\phi^\pm(x + 2\pi l) - \phi^\pm(2\pi l)| \leq \frac{C_\mu}{|l|^{m+1}} \leq \frac{C_\mu}{n^{m+1}}, \quad x \in \mathbf{pC}_{\mu,\pm},$$

where $\max\{|l_i| : 1 \leq i \leq m\} = n$. This, together with the fact that for any fixed n there are at most $2mn^{m-1}$ entries in the sum $\sum_{\max\{|l_i|:1\leq i \leq m\}=n}$, shows that the sum $\sum_1$ is dominated, entry by entry, by the series $C_\mu \sum_n \frac{1}{n^2}$ $< \infty$. This proves the summability of $\sum_1$.

To prove the summability of $\sum_2$ we need the following lemma.

LEMMA 9.3.2
There are bounded and differentiable Clifford-valued functions $\underline{\psi}^\pm : (0, \infty)$
$\to \mathbf{R}^{(m+1)}$ such that

$$(\underline{\psi}^\pm)'(r) = \pm \int_{\partial(r\mathbf{D}^m)} \phi^\pm(\mathbf{x}) e_L d\mathbf{x},$$

where $r\mathbf{D}^m = \{r\mathbf{x} \mid \mathbf{x} \in \mathbf{D}^m\}$, and $\partial(r\mathbf{D}^m)$ its boundary.

PROOF [**Proof of the Lemma**] Define

$$\underline{\psi}^\pm(r) = \int_{\substack{\{x=\mathbf{X}+x_L e_L \ \mid \ \mathbf{X}\in\partial(r\mathbf{D}^m),0<\pm x_L<\pi r, \\ \text{or,} \\ \mathbf{X}\in(r\mathbf{D}^m)^o,\pm x_L=\pi r\}}} \phi^\pm(x) n(x) dS(x),$$

where $n(x)$ denotes the outer normal, and $dS(x)$ the area measure on the surface. The boundedness of $\psi^\pm$ can be verified by using the Cauchy theorem for Clifford monogenic functions and the size condition of $\phi^\pm$.

Now we continue the proof of Th. 9.3.1. The partial sum of $\sum_2$ can be decomposed into

$$
\sum_{n=1}^{n_k} \sum_{\max\{|l_i|:1\leq i\leq m\}=n} \phi^\pm(2\pi l)
$$

$$
= \int_{2\pi}^{2(n_k+1)\pi} (\psi^\pm)'(r) + \sum_{n=1}^{n_k}\Big(\sum_{\max\{|l_i|\}=n} \phi^\pm(2\pi l) - \psi^\pm(r_n)\Big)
$$

$$
= \psi^\pm(2(n_k+1)\pi) - \psi^\pm(2\pi) + \sum_{n=1}^{n_k}\Big(\sum_{\max\{|l_i|\}=n} \phi^\pm(2\pi l) - \psi^\pm(r_n)\Big),
$$

where $2n\pi < r_n < 2(n+1)\pi$. Using the kernel size condition again, the series part is dominated by $C_\mu \sum_{n=1} \frac{1}{n^2}$. The boundedness of $\psi^\pm$ then implies the existence of a sequence (n_k) for which the limit $\lim_{k\to\infty} \psi^\pm(2(n_k+1)\pi)$ exists. We therefore conclude the summability of the series defining $\Phi^\pm$, and so that of $\Phi = \Phi^+ + \Phi^-$ as well. A more careful examination of $\sum_1$ gives the decompositions of $\Phi^\pm$, and therefore that of Φ indicated in the assertion (i). The periodicity is obvious.

Now we prove (ii) and (iii). Set $b^{\pm,\delta}(\zeta) = b^\pm \exp(\mp\delta|\zeta|_\mathbf{C})$. Let $\phi^\pm$ and $\phi^{\pm,\delta}$ be associated with $b^\pm$ and $b^{\pm,\delta}$, respectively, in the pattern of Th.4.1 of [LMcQ]. It is easy to see that $\phi^{\pm,\delta}(\cdot) = \phi^\pm(\cdot \pm \delta e_L)$ and the latter are the inverse Fourier transforms of $b^{\pm,\delta}$. We now formulate, according to the above procedure, the corresponding right-monogenic and periodic functions $\Phi^\pm$ and $\Phi^{\pm,\delta}$ defined on $\mathbf{pC}_{\omega,\pm}$, respectively, with the estimate in (i) in $\mathbf{S}_\mu(\pi)$. Furthermore, we have the decomposition

$$
\Phi^{\pm,\delta}(x) = \phi^{\pm,\delta}(x) + (\phi^{\pm,\delta})_0(x) + c_0^{\pm,\delta},
$$

and

$$
\Phi^\pm(x) = \phi^\pm(x) + (\phi^\pm)_0(x) + c_0^\pm, \quad x \in \mathbf{C}_{\omega,\pm}(\pi),
$$

where $(\phi^{\pm,\delta})_0$ and $(\phi^\pm)_0$ are bounded and right-monogenic. Note that we can employ a same sequence (n_k) to determine all the constants $c_0^{\pm,\delta}$ and $c_0^\pm$. Since the convergence of the series defining $\phi^{\pm,\delta}$ and $c^{\pm,\delta}$ is uniform in $\delta \to 0$ as $n_k \to \infty$, we can change order of the limit procedures $n_k \to \infty$ and $\delta \to 0$. So we conclude, $c_0^{\pm,\delta} \to c_0^\pm$, and $(\phi^{\pm,\delta})_0 \to (\phi^\pm)_0$ uniformly in $\mathbf{C}_{\mu,\pm}$, respectively. We then have $\Phi^{\pm,\delta}(x) \to \Phi^\pm(x)$ locally and uniformly in $\mathbf{C}_{\mu,\pm}(\pi)$, respectively, as $\delta \to 0$. Since $\Phi^{\pm,\delta} \in L^\infty(\mathbf{D}^m)$, and the series that define $\Phi^{\pm,\delta}$ are uniformly convergent with respect to the sequence

(n_k), using the formula (9.3.1), we have, for all $0 \neq \xi \in \mathbf{R}^m$,

$$\int_{\mathbf{D}^m} \Phi^{\pm,\delta}(\mathbf{x}) e_L e(-\mathbf{x}, \xi) d\mathbf{x} = (2\pi)^m \int_{\mathbf{R}^m} \phi^{\pm,\delta}(\mathbf{x}) e_L e(-\mathbf{x}, \xi) d\mathbf{x}$$

$$= (2\pi)^m b^{\pm,\delta}(\xi),$$

according to the relations established in [LMcQ]. It concludes that

$$\{b^{\pm,\delta}(l)\}_{l \neq 0}$$

are the standard Fourier coefficients of $\Phi^{\pm,\delta}$. We now first establish the assertions (ii) and (iii) for $\Phi^{\pm}$. This is done by taking the limit $\delta \to 0$ in the standard Parseval formulas for the L^∞ functions $\Phi^{\pm,\delta}$. Here in writing out the result formulas, we need to employ functions $\underline{\Phi}^{\pm}$, defined by

$$\underline{\Phi}^{\pm}(y) = \int_{A_\pm(y)} \Phi(x) n(x) dS(x), \qquad y \in \mathbf{T}_\omega(\pi),$$

where $A_\pm(y)$ are some smooth and oriented m-manifolds in $\mathbf{C}_{\omega,\pm}$, respectively, joining the $(m-1)$-sphere $S_y = \{x \in \mathbf{R}^{m+1} | < x, y >= 0, |x| = |y|\}$ (see [LMcQ], [Q1]).

Having proved the Parseval formula for $\Phi^{\pm}$, on letting $\underline{\Phi} = \underline{\Phi}^{+} + \underline{\Phi}^{-}$ and using the relation $\Phi = \Phi^{+} + \Phi^{-}$, we conclude the Parseval formula for Φ. Note that adding constants $c^{\pm}$ to $\Phi^{\pm}$ induces the same additive constants $c^{\pm}$ to $\underline{\Phi}^{\pm}$, respectively. Accordingly, the values $b^{\pm}(0)$ and $b(0)$ should be redefined as $b^{\pm}(0) + c^{\pm}$ and $b(0) + c^{+} + c^{-}$, respectively, to make the corresponding Parseval formulas still hold. The proof is complete. ∎

REMARK The series (9.3.1) may not converge if the summability is otherwise defined. In fact, the series is not absolutely convergent.

The function pair $(\Phi, \underline{\Phi})$ concluded in Th. 9.3.1 can be regarded as the inverse Fourier transform of the function b. The following theorem deals with the Fourier transform, in the sense consistent to Th. 9.3.1, of a given such function pair.

THEOREM 9.3.3
Let $\omega \in (0, \frac{\pi}{2}]$ and $(\Phi, \underline{\Phi})$ be a pair of functions defined on $\mathbf{pS}_\omega$ and $\mathbf{T}_\omega(\pi)$, respectively, satisfying

(i)   *Φ is 2π-periodic and right-monogenic in $\mathbf{pS}_\omega$ and there is a constant c_0 such that*

$$|\Phi(x)| \leq \frac{c_0}{|x|^m}, \qquad x \in \mathbf{S}_\omega(\pi).$$

(ii) Φ *is uniformly continuous in* $\mathbf{T}_\omega(\pi)$, *and there is a constant* c_1 *such that* $\|\Phi\|_{L^\infty(\mathbf{T}_\omega(\pi))} \le c_1$, *and*

$$\Phi(y) - \Phi(z) = \int_{A(y,z)} \Phi(x) n(x) dS_x, \quad y, z \in \mathbf{T}_\omega,$$

where $A(y,z)$ *is defined in Th. 9.3.1. Then for every* $\mu \in (0, \omega)$ *there exists a function* $b^\mu \in H^\infty(\mathbf{S}_\mu(\mathbf{C}^m))$ *such that*

$$\|b^\mu\|_{H^\infty(\mathbf{S}_\mu(\mathbf{C}^m))} \le C_{\omega,\mu}(c_0 + c_1),$$

and the function pair determined by the function b^μ *according to Th. 9.3.1 is identical, modulo constants, to* $(\Phi, \underline{\Phi})$. *Moreover,* $b^\mu = b^{\mu,+} + b^{\mu,-}$, *where*

$$b^{\mu,pm}(\zeta) = \lim_{\epsilon \to 0} \tfrac{1}{(2\pi)^m}$$
$$\left\{ \int_{r^\pm(\epsilon, |\zeta|^{-1}) \cup s^\pm(|\zeta|^{-1}, \theta) \cup \sigma^\pm(|\zeta|^{-1}, \theta)} \Phi(x) e_L e(-x, \zeta) \; dx + \underline{\Phi}(\epsilon e_L) \right\},$$
$$\zeta \in \mathbf{S}_{\mu,\pm},$$

respectively, where $\theta = \frac{\mu + \omega}{2}$, *and for* $\rho \le \pi$,

$$r^\pm(\epsilon, \rho) = \{x \mid \epsilon \le |x| \le \rho\},$$

$$s^\pm(\rho, \theta) = \{x \mid |x| = \rho, -|\mathbf{x}| \tan \theta \le \pm x_L \le 0\},$$

$$\sigma^\pm(\rho, \theta) = \{x \mid \pm x_L = -|\mathbf{x}| \tan \theta, |x| > \rho, \mathbf{x} \in \mathbf{D}^m\},$$

and for $\rho > \pi$,

$$r^\pm(\epsilon, \rho) = r^\pm(\epsilon, \pi), s^\pm(\rho, \theta) = s^\pm(\pi, \theta),$$

$$\sigma^\pm(\rho, \theta) = \sigma^\pm(\pi, \theta).$$

PROOF Fix a $\mu \in (0, \omega)$, and write b^μ as b in the rest of the proof. Set

$$(b^\pm)_\epsilon(\zeta) = \frac{1}{(2\pi)^m} \int_{A^\pm(\epsilon, \theta, |\zeta|^{-1})} \Phi^\pm(x) e_L e(-x, \zeta) dx + \underline{\Phi}^\pm(\epsilon e_L), \quad (9.3.3)$$

where

$$A^\pm(\epsilon, \theta, \rho) = r^\pm(\epsilon, \rho) \cup s^\pm(\rho, \theta) \cup \sigma^\pm(\rho, \theta),$$

$$\Phi^\pm(x) = \pm \lim_{\epsilon \to 0}$$
$$\left\{ \int_{H^\pm(\epsilon, \omega), |y| > \epsilon} \Phi(y) n(y) K(x - y) dS_y + \underline{\Phi}(\epsilon e_L) K(x) \right\},$$
$$x \in \mathbf{C}_{\omega,\pm},$$

where $H^\pm(\epsilon, \omega)$ are the n-surfaces defined by

$$H^\pm(\epsilon, \omega) = \{x \pm \epsilon \tan \omega \mid x = \mathbf{x} + x_L e_L, x_L = \mp |\mathbf{x}| \tan \omega, \mathbf{x} \in \mathbf{D}^m\},$$

and K is the Poisson sum of $k(x) = \frac{1}{\sigma_m} \frac{\bar{x}}{|x|^{m+1}}$ in the sense indicated in the proof of Th. 9.3.1, and σ_m is the volume of the unit m-sphere in $\mathbf{R}^{m+1}$, and $\underline{\Phi}^{\pm}$ are correspondingly formulated as in the proof of Th. 9.3.1. It can be verified using a similar method as in Th.6.1 in [LMcS] that $\Phi^{\pm}$ satisfy the estimate in the assumption (i) in the regions $\mathbf{C}_{\theta,\pm}$, respectively. Owing to the fact that when $0 < h \to \infty$,

$$\int_{A^{\pm}(\epsilon,\theta,\rho)\pm he_L} \Phi^{\pm}(x)e_L e_{\mp}(-x,\zeta)dx \to 0,$$

and (9.3.3) reduces to

$$(b^{\pm})_{\epsilon}(\zeta) = \int_{A^{\pm}(\epsilon,\theta,\rho)} \Phi^{\pm}(\mathbf{x})e_L e_{\pm}(-\mathbf{x},\zeta)d\mathbf{x} + \underline{\Phi}^{\pm}(\epsilon e_L). \qquad (9.3.4)$$

We shall now only deal with the case $b = b^{+}$, as the case $b = b^{-}$ is similar. Below we shall proceed as in [McQ1] and [LMcQ] with some minor changes. We need the following estimates. Let $\zeta \in n(\mathbf{C}^m) \subset \mathbf{S}_{\mu}(\mathbf{C}^m)$.

(a)   Whenever $0 < \epsilon < s \leq \pi$ and $s \approx |\zeta|^{-1}$,

$$(b)_{\epsilon,s}(\zeta) = \int_{\epsilon \leq |\mathbf{x}| \leq s} \Phi^{+}(\mathbf{x})e_L e_{+}(-\mathbf{x},\zeta)d\mathbf{x} + \underline{\Phi}^{+}(\epsilon e_L)$$

is uniformly bounded by $C(c_0 + c_1)$. Moreover, the uniform limit $\lim_{\epsilon \to 0}$ exists for such s and $\zeta \in \mathbf{S}_{\omega}(\mathbf{C}^m)$. In fact,

$$\int_{\epsilon \leq |\mathbf{x}| \leq s} \Phi^{+}(\mathbf{x})e_L e_{+}(-\mathbf{x},\zeta)d\mathbf{x} =$$

$$\int_{\epsilon \leq |\mathbf{x}| \leq s} \Phi^{+}(\mathbf{x})e_L(e_{+}(-\mathbf{x},\zeta) - 1)d\mathbf{x} + \int_{\epsilon \leq |\mathbf{x}| \leq s} \Phi^{+}(\mathbf{x})e_L d\mathbf{x}.$$

The first integral is dominated by Cc_0 and uniformly convergent as $\epsilon \to 0+$, and the second integral is equal to $\underline{\Phi}^{+}(se_L) - \underline{\Phi}^{+}(\epsilon e_L)$. We then conclude the desired bound for $(b)_{\epsilon,s}$

(b)   Let $\rho = |\zeta|^{-1} \leq \pi$. Denote by $S^m(s)$ the m-dimensional s-sphere centered at the origin in $\mathbf{R}^{m+1}$. Then the integral

$$\int_{S^m(\rho) \cap \mathbf{C}_{\theta,+}} |\Phi^{+}(x)n(x)e_{+}(-x,\zeta)|dS_x$$

is bounded by $C_{\theta}c_0$. The proof of this fact is similar to the proof of Th.4.1 in [LMcQ].

(c)   Let $\rho = |\zeta|^{-1} \leq \pi$. Then the integral $\int_{\sigma^{+}(\rho,\theta))} \Phi^{+}(x)n(x)e_{+}(-x,\zeta)dS_x$ is uniformly bounded by $\frac{C_{\mu}c_0}{\theta - \mu}$ as showed in the proof of Th.4.1 in [LMcQ].

For a fixed ζ, if $|\zeta|^{-1} \leq \pi$, applying the assertion (a) to (9.3.4) we conclude the convergence of $b_{\epsilon}(\zeta)$ as $\epsilon \to 0$. The integral (9.3.4) is equal to a sum of $(b)_{\epsilon,|\zeta|^{-1}}(\zeta)$ and some integrals on some closed subsets of $S^{m-1}(|\zeta|^{-1}) \cap \mathbf{C}_{\theta,+}$, and on a surface of the type $\sigma^+(|\zeta|^{-1},\theta)$. Invoking $(a),(b)$ and (c), we conclude that $(b)_\epsilon$ is uniformly dominated by the desired bounds. If $|\zeta|^{-1} > \pi$, the argument and estimate in (a) can still be used to the integral over the set $r^+(\epsilon,\pi)$ to conclude that the integral is uniformly bounded and has a limit as $\epsilon \to 0$. To prove that the integrals over $s^+(|\zeta|^{-1},\theta)$ and $\sigma^+(|\zeta|^{-1},\theta)$ are bounded we use Cauchy's theorem to change contour of the integration, and so to integrate on the set on $\{x \mid \mathbf{x} \in \partial\mathbf{D}^m\}$ with the boundary $\{x \mid \mathbf{x} \in \partial\mathbf{D}^m\} \cap \sigma^+(\pi,\theta)$. It is easy to show then the integral on the last-mentioned set is bounded. To prove the inverse Fourier transform of b being the given pair $(\Phi,\underline{\Phi})$, modulo constants, we proceed as in the case $m = 1$ in Th.2 in [Q1]. ∎

9.4 Singular Integral Operators and Fourier Multiplier Operators on Periodic Lipschitz Surfaces

By a periodic Lipschitz perturbation of $\mathbf{D}^m$ we mean a surface $\Gamma = \{\mathbf{x} + e_L G(\mathbf{x}) \mid \mathbf{x} \in \mathbf{D}^m, G : \mathbf{R}^m \to \mathbf{R}$ is $2\pi -$ periodic and Lipschitz $\}$. Denote $N = \|\nabla G\|_\infty$, $m_0 = \min\{G(\mathbf{x}) : \mathbf{x} \in \mathbf{D}^m\}$ and $M_0 = \max\{G(\mathbf{x}) : \mathbf{x} \in \mathbf{D}^m\}$. Without loss of generality, we may assume $m_0 < 0 < M_0$. Let $\mathcal{A}(\Gamma) = \{F \mid F$ is $2\pi -$ periodic and left $-$ monogenic in $m_0 - \delta < x_L < M_0 + \delta$ for some $\delta > 0\}$. $\mathcal{A}(\Gamma)$ is dense in $L^2(\Gamma, dS)$ that can be proved using a similar method as in [CM] or [GQW].

Let $F \in \mathcal{A}(\Gamma)$. It is known, from the classical Fourier series theory, that

$$F(\mathbf{x}) = \sum_{l \in \mathbf{Z}^m} e(\mathbf{x},l)\hat{F}(l), \qquad \mathbf{x} \in \mathbf{D}^m,$$

where the convergence is absolute and uniform in $\mathbf{D}^m$. Now we shall show that the left-monogenic extension of $F(\mathbf{x})$ is equal to the sum of all the left-monogenic extensions of the entries on the right-hand side for all x in the rectangle on which F is defined, and the convergence is locally uniform and absolute. For the case $m = 1$, the assertion follows from Laurant series theory of complex analysis. For the higher-dimensional cases we shall make use of the decomposition of the extended exponential function. In fact, using the functions $e_\pm(x,l)$ one can formally write the series

$$F^+(x) = \sum_{l \in \mathbf{Z}^m} e_+(x,l)\hat{F}(l), \qquad x_L > m - \delta,$$

and

$$F^-(x) = \sum_{l \in \mathbf{Z}^m} e_-(x,l)\hat{F}(l), \quad x_L < M + \delta,$$

where

$$e_\pm(x,\zeta) = e^{i\langle \mathbf{x},\zeta \rangle} e^{\mp x_L|\zeta|} \mathbf{C}\chi_\pm(\zeta).$$

Now we prove that the series defining $F^\pm$ locally absolutely and uniformly converge to some left-monogenic functions in the indicated regions, respectively. Therefore, the series $F(x) = F^+(x) + F^-(x), x_L \in (m_0 - \delta, M_0 + \delta)$ is of the same convergent property. Denote $F_h(\mathbf{x}) = F(\mathbf{x} + he_L)$. For $h \in (m_0 - \delta, M_0 + \delta)$, where δ is associated with F in the definition of $\mathcal{A}(\Gamma)$, Cauchy's theorem implies

$$\hat{F}(l) = \frac{1}{(2\pi)^m} \int_{\mathbf{D}^m} e(\mathbf{x} \mp he_L, l) F(\mathbf{x} \mp he_L) d\mathbf{x}$$
$$= (e^{\pm h|l|}\mathbf{C}\chi_+(l) + e^{\mp h|l|}\mathbf{C}\chi_-(l))\hat{F}_{\mp h}(l).$$

Then we have

$$F^\pm(x)$$
$$= \sum_{l \in \mathbf{Z}^m} e(\mathbf{x},l) e^{\mp x_L|l|} \mathbf{C}\chi_\pm(l)(e^{\pm h|l|}\mathbf{C}\chi_+(l) + e^{\mp h|l|}\mathbf{C}\chi_-(l))\hat{F}_{\mp h}(l)$$
$$= \sum_{l \in \mathbf{Z}^m} e(\mathbf{x},l) e^{\pm(h-x_L)|l|} \mathbf{C}\chi_\pm(l)\hat{F}_{\mp h}(l).$$

To see the convergence of the series defining $F^+(x)$ for $x_L > m_0 - \delta$ we choose $h \in (m_0 - \delta, x_L)$; and to see the convergence of $F^-(x)$ for $x_L < M_0 + \delta$ we choose $h \in (x_L, M_0 + \delta)$. The argument then implies the desired assertion.

Because of the type of convergence, for $b \in H^\infty(\mathbf{S}_\omega(\mathbf{C}^m))$ the series

$$\sum_{0 \neq l \in \mathbf{Z}} b(l)e(x,l)\hat{F}(l)$$

locally absolutely and uniformly converges. Define $M_b : \mathcal{A}(\Gamma) \to \mathcal{A}(\Gamma)$, by

$$M_b F(x) = (2\pi)^m \sum_{0 \neq l \in \mathbf{Z}} b(l)e(x,l)\hat{F}(l).$$

We postpone to include a corollary of Th. 9.3.1, Th. 9.3.3 on monogenic extension (see [Q1] for the version in complex analysis). By virtue of the above argument, we have

COROLLARY 9.4.1
Let $b \in H^\infty(\mathbf{S}_\omega(\mathbf{C}^m))$. Then the functions

$$\Phi^\pm(x) = \sum_{l \in \mathbf{Z}^m} b(l) e_\pm(x, l), \quad \pm x_L > 0$$

can be monogenically extended to $\mathbf{pC}_{\omega,\pm}$, respectively, and the function Φ defined by $\Phi = \Phi^+ + \Phi^- = \sum_{l \in \mathbf{Z}^m} b(l) e(x, l)$ satisfies the three conditions in Th. 9.3.1. Conversely, if a function pair $(\Phi, \underline{\Phi})$ satisfies the assumptions in Th. 9.3.3 with respect to $\omega \in [0, \frac{\pi}{2})$, then for every $\mu \in (0, \omega)$ there exists a function $b^\mu \in H^\infty(\mathbf{S}_\mu)$ such that $\Phi(x) = \sum_{l \in \mathbf{Z}^m} b^\mu(l) e(x, l)$ in the above-mentioned sense.

For a pair of functions $(\Phi, \underline{\Phi})$ determined by the function b according to Th. 9.3.1 one can formulate

$$T_{(\Phi, \underline{\Phi})} F(x) = \lim_{\epsilon \to 0} \left\{ \int_{|x-y| > \epsilon, y \in \Gamma} \Phi(x - y) n(y) F(y) dS_y + \underline{\Phi}(\epsilon t(x)) F(x) \right\},$$

where $t(x)$ is the outer normal of Γ at x.
We have the following result.

THEOREM 9.4.2
Let $\omega \in (\arctan N, \frac{\pi}{2}], b \in H^\infty(\mathbf{S}_\omega(\mathbf{C}^m))$, and $(\Phi, \underline{\Phi})$ be the function pair associated with b in the pattern of Th. 9.3.1. Then the following hold

1. $T_{(\Phi, \underline{\Phi})}$ *is well defined from $\mathcal{A}(\Gamma) \to \mathcal{A}(\Gamma)$, and, modulo a constant multiple of the identity operator, $T_{(\Phi, \underline{\Phi})} = M_b$.*

2. M_b *is extensible to a bounded operator on $L^2(\Gamma)$ with operator norm dominated by $C\|b\|_\infty$.*

PROOF To prove (i) we use the functions $b_x^{\pm,\delta}(\zeta) = e(x, \zeta) b^{\pm,\delta}(\zeta)$, where $b^{\pm,\delta}(\zeta) = b^\pm \exp(\mp \delta |\zeta|_{\mathbf{C}})$ as used in the proof of Th. 9.3.1. By denoting $(\Phi^{\pm,\delta}, 0)$ the function pairs associated with $b^{\pm,\delta}$ in the pattern of Th. 9.3.1, $(\Phi^{\pm,\delta}(x + \cdot), 0)$ are associated with $b_x^{\pm,\delta}$ in the same pattern. Applying Parseval's identity to $b_x^{\pm,\delta}$ and $(\Phi^{\pm,\delta}(x + \cdot), 0)$ and then taking the limit $\delta \to 0$ in the identity, we conclude the desired relation (also see the proof of (i) Th.5 in [Q1]). ∎

For a proof of (ii) the reader is referred to either [GQW], or [Q1], or [Q2].
We state, without a proof, the following converse result of (ii) Th. 9.4.2. For a proof the reader is referred to [McQ2].

THEOREM 9.4.3

Let $\omega \in (\arctan N, \frac{\pi}{2}]$, Φ be a right-monogenic function on $\mathbf{pS}_\omega(\pi)$ satisfying the estimate (i) of Th. 9.3.3 in $\mathbf{S}_\omega(\pi)$. If T is a bounded operator on $L^2(\Gamma)$, and

$$TF(x) = \int_\Gamma \Phi(x - y)n(y)F(y)dS(y), \qquad x \notin \text{supp } (F),$$

for all $F \in C_c^o(\Gamma)$, the class of continuous functions on Γ with compact support, then there exists a unique differentiable function $\underline{\Phi}$, bounded in every $\mathbf{T}_\mu(\pi), \mu \in (0, \omega)$, satisfying

$$\frac{d}{dr}\underline{\Phi}(rn) = \int_{S^m(r) \cap \{x | <x,n> = 0\}} \Phi(x)n(x)dS(x),$$

where n is a unit vector in $\mathbf{T}_\mu(\pi), S^m(r)$ is the m-dimensional r-sphere, such that

$$TF = T_{(\Phi,\underline{\Phi})}.$$

Bibliography

[CM]   R. Coifman and Y. Meyer, *Fourier analysis of multilinear convolutions, Calderón's theorem, and analysis on Lipschitz curves*, Lecture Notes in Mathematics, **779**, Springer-Verlag, New York, 1980, 104–122.

[FJR]   E. Fabes, M. Jodeit, Jr and N. Riviere, *Potential techniques for boundary problems on C^1 domains*, Acta Math. **141** (1978), 165-186.

[GLQ]   G. Gaudry, R.L. Long and T. Qian, *A martingale proof of L^2-boundedness of Clifford-valued singular integrals*, Annali di Mathematica Pura Ed Applicata, **165** (1993), pp 369-394.

[GQW]   G. Gaudry, T. Qian and S-L. Wang, *Boundedness of singular integral operators with holomorphic kernels on star-shaped Lipschitz curves*, preprint.

[LMcS] C. Li, A. McIntosh and S. Semmes, *Convolution singular integral operators on Lipschitz surfaces*, Journal of the America Mathematical Societ **5** (1992), 455-481.

[LMcQ] C. Li, A. McIntosh and T. Qian, *Clifford algebras, Fourier transforms and singular convolution operators on Lipschitz surfaces*, to appear in Revista Mathematica Iberoamericana, 10, (1994), 665–721.

[McQ1] A. McIntosh and T. Qian, *Convolution singular integral opera-tors on Lipschitz curves*, Lecture Notes in Mathematics, Springer-Verlag, New York, 1991, 142–162.

[McQ2] A. McIntosh and T. Qian, *A note on singular integrals with holo-morphic kernels*, Approximation Theory and its Applications, 6(4), 1990, 40–57.

[McQ3] A. McIntosh and T. Qian, *Lp Fourier multipliers along Lipschitz curves*, Trans. of A.M.S., 333(1), 1992, 157–176.

[Q1] T. Qian, *Singular integrals with holomorphic kernels and H^{∞}-Fourier multipliers on star-shaped Lipschitz curves*, preprint.

[Q2] T. Qian, *Transference from Lipschitz graphs to periodic Lipschitz*, preprint.

[S1] F. Sommen *Plane waves, biregular functions and hypercomplx Fourier analysis*, Proceedings of the 13th Winter School on ab-stract analysis, SRNI, 5-12 January, 1985. Supplemento ai Ren-diconti del Circolo Matematico di Palermo, Serie II - numero **9** (1985), 205-219.

[S2] F. Sommen, *Microfunctions with values in a Clifford algebra II*, Scientific Papers of the College of Arts and Sciences, The Univer-sity of Tokyo, 36(1), 1986, 15–37.

[T] T. Tao, *Convolution operators generated by right-monogenic and harmonic kernels*, preprint.

[V] G. Verchota, *Layer Potentials and Regularity for the Dirichlet Problem for Laplace's Equation in Lipschitz Domains*, J. of Funct. Anal. **59** (1984), 572-611.

Department of Mathematics,
The University of New England,
Armidale, NSW 2351,
AUSTRALIA
E-mail: tao@neumann.une.edu.au

10

Scattering Theory for Orthogonal Wavelets

Palle E.T. Jorgensen*

ABSTRACT We apply the Lax-Phillips wave equation scattering theory to multiresolutions associated with wavelets. For wavelet scattering, the translation symmetry, the scaling operator, and the scaling function are identified in the scattering theoretic spectral transform; the scaling function is shown to be analytic; and an analytic spectral function is identified as an invariant for multiresolutions, normalized so that the Haar wavelet corresponds to the constant function. For the study of the functional equation, we introduce almost periodic spaces and establish a general convergence for the infinite product formula with the limit in the L^2-space of the corresponding Bohr group.

10.1 Introduction

The theory of wavelets has received much recent attention (see, e.g., [Chu 1–2], [Dau] and [Mey], and the references therein), both because of its immediate practical use, and also because of new theoretical possibilities.

In this paper, we want to follow up on a certain operator theoretic method which turns out to apply both to the solutions to the wave equations, and to the analysis of wavelets. The wave operator,

$$L = -\left(\frac{\partial}{\partial x_0}\right)^2 + \left(\frac{\partial}{\partial x_1}\right)^2 + \cdots + \left(\frac{\partial}{\partial x_n}\right)^2 \qquad (10.1.1)$$

has the familiar factorization $D^2 = L$ where D is the associated Dirac operator, treated in detail by several of the authors of the present volume,

---

*Work supported in part by the NSF, NATO, and a University of Iowa Faculty Scholar Award.

1991 *Mathematics Subject Classification.* 42C15, 47D45, 47A40, 58F08

acting on Clifford vector-valued functions (see [GM]). We recall that the abstract approach (see [LP]) to the wave equation from (10.2.1) with obstacle perturbation starts with an identification of associated *incoming* and *outgoing subspaces*. These in turn are scaled by the unitary one-parameter group which solves the corresponding wave equation. The obstacle perturbation is then identified by the scattering operator acting between the two spaces, and it is spectral analyzed by the introduction of operator-valued Hardy spaces. Here we show that this classical Hardy space-approach to the Lax-Phillips theory may be adapted to the multiresolutions which are now used for the construction of wavelets (see [Dau]), with the dilation groups now doing the scaling of the respective spaces. But our analysis also indicates that the Hardy spaces based instead on $\bar{D}$ (see [GM, Chapters 2.5 and 5.3]) rather than $\partial/\partial\bar{z}$ may further be adapted to this general philosophy. However, presently we shall do the analysis only for the latter case, and we hope that this book will help inspire more work on the former.

For the classical wave equations, the Lax-Phillips scattering theory (see [LP]) provides a tool for comparing waves with obstacle scattering with the solutions to the free wave equations corresponding to *incoming* and *outgoing* states. We show in this paper, that the idea of incoming vs. outgoing states is useful at two levels for *wavelets* which are defined from standard *multiresolutions*. It turns out that there are separate scattering operators associated both with the *scaling*, and with the *integral translations*. We then combine the two approaches to obtain a *spectral scattering function* which turns out to be computable. For the line case, it compares the spectral theory of left vs. right integral translates for L^2-wave functions; and for the multidimensional case in the Hilbert space $L^2(\mathbb{R}^N)$, $N > 1$, there is a multiresolution associated with lattices $\Gamma \subset \mathbb{R}^N$, and it is pointed out that each *reflection symmetry* in Γ corresponds to a separate spectral *scattering function*. These functions turn out to be trivial for the translation invariant wavelets; or when $N = 1$, for the classical Haar wavelet. (For other work on the $\mathbb{R}^N$-lattice case see, e.g., [Mad] and [JP2].)

10.2 Preliminaries

Let M be a complex Hilbert space, and let $\mathbb{T}$ denote the circle group, i.e.,

$$\mathbb{T} = \{z \in \mathbb{C} : |z| = 1\}. \tag{10.2.1}$$

Writing $z = e^{i\theta}$, $0 \leq \theta < 2\pi$, the normalized Haar measure on $\mathbb{T}$ becomes $(2\pi)^{-1}\, d\theta$ where $d\theta$ is the usual Lebesgue measure. We shall assume M separable and consider L^2-functions on $\mathbb{T}$ with values in M; $f : \mathbb{T} \to M$, norm

$$\|f\| = \left(\int_{\mathbb{T}} \|f(z)\|^2 \, dz \right)^{1/2} \tag{10.2.2}$$

where the norm *under* the integral sign is the original norm from M. This L^2- space will be denoted $\mathcal{L}^2(M)$, and the corresponding Hardy space $\mathcal{H}^2(M)$. (For more details see, e.g., [He] and [SNF].) Using vector-valued Fourier series, elements in $\mathcal{L}^2(M)$ may be expanded into orthogonal series,

$$f(z) = \sum_n z^n w_n \tag{10.2.3}$$

where the summation is over $n \in \mathbb{Z}$, and the vector coefficients $w_n \in M$. Then

$$\|f\|^2 = \sum_n \|w_n\|^2 \tag{10.2.4}$$

by the Parseval theorem see, e.g., [Ru1, He]. The corresponding expansion

$$h(z) = \sum_{n=0}^{\infty} z^n w_n, \qquad (w_n \in M) \tag{10.2.5}$$

represents the general element in $\mathcal{H}^2(M)$, and, by the Paley-Wiener theorem (see, e.g., [Ru1] and [JM]), the elements h in $\mathcal{H}^2(M)$ may be characterized alternatively by the analytic continuation property; i.e., characterized by analytic continuation $H(z)$ defined, for z in $B = \{z \in \mathbb{C} : |z| < 1\}$, with the function $H : B \to M$ analytic, and

$$\sup_{0 < r < 1} \int \|H(re^{i\theta})\|^2 \, d\theta < \infty.$$

There is a boundary limit-function,

$$\lim_{r \to 1} H(re^{i\theta}),$$

defined a.e. in $\theta \in [0, 2\pi)$ by virtue of Fatou's theorem. The corresponding (a.e. defined) function $\tilde{h}$ then satisfies

$$\int_0^{2\pi} \|\tilde{h}(\theta)\|^2 \, d\theta = \sup_{0 < r < 1} \int \|H(re^{i\theta})\|^2 \, d\theta$$

If conversely $h \in \mathcal{H}^2(M)$ is given, then $h(e^{i\theta}) = \tilde{h}(\theta)$ a.e., and it follows that the three functions $h(e^{i\theta})$, $\tilde{h}(\theta)$, and $H(z)$, (the latter defined for $z \in B$,) may be isometrically identified. In particular, it follows that

$$\int_0^{2\pi} e^{-ik\theta} \tilde{h}(\theta) \, d\theta = 0 \tag{10.2.6}$$

for all $k = -1, -2, \ldots$; and the representation (10.2.5) becomes

$$\tilde{h}(\theta) = \sum_{n=0}^{\infty} e^{in\theta} w_n \qquad (10.2.7)$$

where

$$w_n = \frac{1}{2\pi} \int_0^{2\pi} e^{-in\theta} \tilde{h}(\theta) \, d\theta \qquad (10.2.8)$$

for $n = 0, 1, 2, \ldots$, with the integral convergent in the Hilbert space M (which was given at the outset). In the sequel, we shall abuse notation and denote any one of the three functions h, H, or $\tilde{h}$, by the single letter h; and we will then add comments (as appropriate, on which interpretation is to be adopted).

Linear operators T in $\mathcal{L}^2(M)$ which commute with multiplication,

$$f(z) \mapsto zf(z), \qquad (\forall f \in \mathcal{L}^2(M)) \qquad (10.2.9)$$

are called *decomposable*. It is known that every such T may be represented as a measurable operator field, $z \mapsto \tilde{T}(z)$, defined on $\mathbb{T}$ and taking values in $\mathcal{B}(M) = $ (all bounded operators on M), with

$$(Tf)(z) = \tilde{T}(z)(f(z)), \qquad \forall f \in \mathcal{L}^2(M), \quad z \in \mathbb{T}. \qquad (10.2.10)$$

If $\tilde{T}(z) : M \to M$ is constant a.e. z in $\mathbb{T}$, then we say that the operator T is *spatial*, or *spatial on* M.

Let $\mathcal{K}$ be a given separable complex Hilbert space, and let V be a closed subspace in $\mathcal{K}$. Let $U : \mathcal{K} \to \mathcal{K}$ be a unitary operator. We will say that the subspace V is *saturated* if

$$\bigwedge_k U^k(V) = 0, \qquad \text{and} \qquad (10.2.11)$$

$$\bigvee_k U^k(V) = \mathcal{K} \qquad (10.2.12)$$

Here the symbols $\bigwedge$ and $\bigvee$ are from lattice theory, where $\bigwedge_k$ referring to the intersection $\bigcap_{k=-\infty}^{\infty}$, $k \in \mathbb{Z}$, for the closed subspaces, while $\bigvee_k$ refers to the closed span of the spaces $U^k(V)$, again with k ranging over $\mathbb{Z}$. Finally, we shall say that V is *invariant* if $U(V) \subset V$. We shall say that then

$$W := V \ominus U(V) = \{f \in V : \langle f, Ug \rangle = 0, \quad \forall g \in V\} \qquad (10.2.13)$$

is the *defect space*; also called the *multiplicity space*, see below.

10.3 Scattering Theory

We will study *wavelets* which arise from *multiresolutions*. The scaling will be dyadic for the most part, although our theory will also apply to the general case, and to higher dimensions. For the general case, the Hilbert space is $\mathcal{K} = L^2(\mathbb{R}^N)$, and a pair Γ, A is given where $\Gamma \subset \mathbb{R}^N$ is a lattice, and A is an invertible N by N matrix satisfying the integrality condition

$$A(\Gamma) \subset \Gamma \qquad (10.3.1)$$

with proper inclusion. Then

$$U f(x) = (\det A)^{-1/2} f(A^{-1}x) \qquad (10.3.2)$$

defines a unitary operator in $\mathcal{K}$. We shall study closed subspaces V in $\mathcal{K}$ such that

$$U(V) \subset V, \qquad (10.3.3)$$

and this also will be a proper inclusion; the subspace V will correspond to the lattice Γ, and properness in (10.3.1) correspond to that of (10.3.3). In application of our scattering theory, we will restrict primarily to $\mathcal{K} = L^2(\mathbb{R})$, $\Gamma = \mathbb{Z}$, and

$$U f(x) = 2^{-1/2} f\left(\frac{x}{2}\right) \qquad (10.3.4)$$

as a special case of (10.3.2), but we mention how results for (10.3.4) generalize to (10.3.2).

LEMMA 10.3.1
Let $\mathcal{K}$ be a Hilbert space, $U : \mathcal{K} \to \mathcal{K}$ a unitary operator, and $V \subset \mathcal{K}$ a closed subspace which is invariant under U and saturated. Let

$$W = V \ominus UV \qquad (10.3.5)$$

be the defect space from (10.2.13). Then there is a unitary isomorphism,

$$R : \mathcal{K} \to \mathcal{L}^2(W) \qquad (10.3.6)$$

mapping $\mathcal{K}$ onto $\mathcal{L}^2(W)$, and mapping the subspace V onto $\mathcal{H}^2(W)$, i.e.,

$$R(V) = \mathcal{H}^2(W). \qquad (10.3.7)$$

Moreover R is uniquely determined up to a spatial unitary of W by this, and the intertwining property

$$RU = zR \qquad (10.3.8)$$

where z denotes the multiplication operator on $\mathcal{L}^2(W)$, i.e., the operator $f(z) \mapsto zf(z)$, defined for $f \in \mathcal{L}^2(W)$.

REMARK   Note that the assumptions apply equally well to the cases when $\mathcal{K}$ is $L^2(\mathbb{R})$ and $L^2(\mathbb{R}^N)$, $N > 1$; in each case subspaces V and W are considered, but their properties will depend on the pair (Γ, A), see (10.3.1), and on the spectral properties of the associated unitary operator U defined in (10.3.2) above.

PROOF   [**Proof of Lemma 10.3.1**] The operator R is called the scattering theory *spectral representation*, and the lemma is a variant of *the Lax-Phillips result* on that representation (see [LP], [CS] and [JM]), but we need some more details in the construction, and we also need a representation theoretic approach, so details will be sketched below. Traditionally, the uniqueness part comes from the inner/outer factorization for operator valued Hardy class functions, see, e.g., [SNF, LP] and [FS], and also use of the Beurling-Lax-Halmos invariant subspace theorem [He]. The present approach is based instead directly on a certain *Schrödinger representation* which we proceed to describe. For $k \in \mathbb{Z}$, define the spaces $U^k(V)$, and note that $U^{k+1}(V) \subset U^k(V)$. If E_k denotes the orthogonal projection of $\mathcal{K}$ onto $U^k V \ominus U^{k+1}(V)$, then we have the *orthogonal* resolution

$$\sum_{k \in \mathbb{Z}} E_k = I \qquad (10.3.9)$$

by virtue of (10.2.11) and (10.2.12), where I denotes the identity operator on $\mathcal{K}$. We further have

$$U E_k = E_{k+1} U, \qquad \forall k \in \mathbb{Z}, \qquad (10.3.10)$$

so define a representation $J(\theta)$ of $\mathbb{T} = \{e^{i\theta} : \theta \in [0, 2\pi)\}$ by

$$J(\theta) = \sum_{k \in \mathbb{Z}} e^{-ik\theta} E_k \qquad (10.3.11)$$

and note that

$$U J(\theta) U^{-1} = e^{i\theta} J(\theta) \qquad (10.3.12)$$

The two formulas (10.3.10) and (10.3.12) are equivalent, and they state that we have a unitary representation of the canonical commutation relation, in discrete form, see [LP] and [Jo]. We then apply the *Stone-von Neumann uniqueness theorem*, and get this system intertwined with a canonical system built on $W := V \ominus U(V)$ from (10.3.5). The corresponding intertwining operator R will be the one asserted in the lemma. Let functions in $\mathcal{L}^2(W)$ be identified with 2π-periodic vector functions taking values in W, and consider the operators $\tilde{U}$ and $\tilde{J}(\theta)$ acting on $\mathcal{L}^2(W)$, and defined as follows:

$$(\tilde{U} f)(\xi) = e^{i\xi} f(\xi) \qquad (10.3.13)$$

$$\tilde{J}(\theta) f(\xi) = f(\xi - \theta) \qquad (10.3.14)$$

given for $f : \mathbb{R} \to W$, 2π-periodic, $\xi, \theta \in \mathbb{R}$, and (10.3.14) representing 2π-periodic translations. Clearly then

$$\tilde{U}\tilde{J}(\theta)\tilde{U}^{-1} = e^{i\theta}\tilde{J}(\theta) \qquad (10.3.15)$$

as an operator identity on $\mathcal{L}^2(W)$; and $\tilde{U}$ is the operator, denoted z above, and is also multiplication by $z = e^{i\theta}$ when complex notation is used for functions on $\mathbb{T}$. The Stone-von Neumann theorem asserts existence of a *unitary isomorphism*, $R : \mathcal{K} \to \mathcal{L}^2(W)$, such that

$$RU = \tilde{U}R \qquad \text{and} \qquad (10.3.16)$$

$$RJ(\theta) = \tilde{J}(\theta)R \qquad (10.3.17)$$

as operator-identities on $\mathcal{L}^2(W)$. Since V is a spectral subspace for $J(\theta)$ on $\mathcal{K}$, and $\mathcal{H}^2(W)$ is one for $\tilde{J}(\theta)$ on $\mathcal{L}^2(W)$, it follows from (10.3.10) that

$$R(V) = \mathcal{H}^2(W) \simeq W \oplus zW \oplus z^2W \oplus \cdots . \qquad (10.3.18)$$

Moreover (10.3.16) is the other asserted formula (10.3.8) in the lemma.

Both conclusions regarding R in the lemma are in fact equivalent to the intertwining property summarized in (10.3.13)–(10.3.15). So if we have two intertwiners $R_i : \mathcal{K} \to \mathcal{L}^2(W)$, $i = 1, 2$; then $R_1 R_2^{-1}$ commutes with the representation $\{\tilde{U}, \tilde{T}(\theta)\}$; and it is known [Jo] that this *commutant* is just the *spatial* operators, i.e., there is an $A \in \mathcal{B}(W)$ such that

$$R_1 R_2^{-1} f(z) = A(f(z)), \qquad (10.3.19)$$

$\forall f \in \mathcal{L}^2(W)$, $\forall z \in \mathbb{T}$. Since $R_1 R_2^{-1}$ is unitary on $\mathcal{L}^2(W)$, it follows that A must then be unitary as an operator on W; and (10.3.19) is the asserted uniqueness up to a *spatial* operator;—the spatial operators are like the constants for the scalar-valued theory, see [Ru, p. 376]. The conclusion may also be stated as the assertion that the $\{U, J(\theta)\}$ representation on $\mathcal{K}$ is unitarily equivalent to a direct sum of identical copies of the Schrödinger representation (see [Jo]); the dimension of W then counts the multiplicity of this single representation. $\blacksquare$

REMARK As it stands, the lemma says nothing about the multiplicity space, other than a counting of its dimension. When the multiresolution is introduced, we will see that the space W will contain the associated wavelet, and also reveal the spectral theoretic properties of the given multiresolution under consideration.

COROLLARY 10.3.2
Let U be a unitary operator in a Hilbert space $\mathcal{K}$, and let V_j, $j = 1, 2$, be closed subspaces which are both invariant and saturated. Then the cor-

responding two multiplicity spaces $W_j := V_j \ominus UV_j$ *are isometrically isomorphic, and will be identified* $W_j \simeq M$. *If* R_j, $j = 1, 2$, *are the spectral representations, then the operator* $R_1 R_2^{-1}$ *is decomposable; i.e., there is a measurable field of unitaries,* $z \mapsto A(z) \in \mathcal{B}(M)$, $A(z) : M \to M$ *unitary for a.e.* $z \in \mathbb{T}$, *and*

$$R_1 R_2^{-1} f(z) = A(z)(f(z)), \qquad \forall f \in \mathcal{L}^2(M), \quad z \in \mathbb{T}.$$

COROLLARY 10.3.3

Let U *be unitary in a Hilbert space* $\mathcal{K}$, *and let* $V_\pm$ *be orthogonal closed subspaces which are both saturated. Assume*

$$UV_+ \subset V_+, \qquad and \qquad U^{-1}V_- \subset V_- \qquad (10.3.20)$$

Then again the multiplicity spaces are the same (also denoted M *) up to isometric isomorphism. We pick spectral representations* $R_\pm$ *such that*

$$R_+(V_+) = \mathcal{H}^2(M), \qquad and \qquad (10.3.21)$$

$$R_-(V_-) = \mathcal{H}^2_-(M) := \cdots \oplus z^{-2}M \oplus z^{-1}M, \qquad (10.3.22)$$

i.e., $\mathcal{H}^2_-(M) := \sum^{\oplus}_{k \leq -1} z^k M$. *Moreover the operator,* $R_- R_+^{-1} := S$ *is unitary, decomposable, and analytic, i.e., is in the operator valued* $\mathcal{L}^\infty$-*analytic Hardy space. We have a unitary field,* $z \mapsto S(z)$, *with* $S(z)$ *unitary in* M, *(a.e.* $z \in \mathbb{T}$), *and*

$$(Sf)(z) = S(z)(f(z)) \qquad \forall f \in \mathcal{L}^2(M), \quad z \in \mathbb{T}; \qquad (10.3.23)$$

and S *leaving* $\mathcal{H}^2(M)$ *invariant; equivalently*

$$\int_{\mathbb{T}} z^k S(z)\, dz = 0, \qquad \forall k \in \mathbb{N}, \mathbb{T}$$

see also (10.2.6) above.

The proof of the two corollaries may be taken almost verbatim from Lax-Phillips [LP, pp. 41–45]; but they are not stated in this form there. The last point in Corollary 10.3.3 on the analyticity of the scattering spectral operator S in (10.3.23) is based on the Fourès-Segal theorem [FS].

REMARK If the two spaces $V_\pm$ are not orthogonal, but still satisfy the other conditions in the corollary, then the operator, $S = R_- R_+^{-1}$ still exists, and it is unitary in $\mathcal{L}^2(M)$ and decomposable. In fact, the *orthogonality* property $V_+ \perp V_-$ holds if S is also analytic. The analytic property in turn is equivalent (by [FS]) to S leaving each $z^n \mathcal{H}^2(M)$ invariant.

We will use S for comparison of multiresolutions, and the typical cases will lead to nonanalytic S.

The spaces $V_\pm$ are called incoming resp., outgoing in the terminology of scattering theory. We shall study wavelets in $L^2(\mathbb{R})$ which are derived from multiresolutions. We show how incoming/outgoing spaces may be associated to systems of multiresolutions, but if the multiresolutions are "substantially" different, typically the spaces V_+ and V_- will *not* be orthogonal. But in Section 10.6 we show that there is a different approach to an analytic scattering operator which uses all the structures of multiresolutions and effectively "separate" those that are "different".

10.4   Wavelets

An orthogonal dyadic *wavelet* on $\mathbb{R}$ is a function $\psi \in L^2(\mathbb{R})$ such that the double-indexed sequence

$$\{\psi(2^j x + k)\} \qquad j, k \in \mathbb{Z} \tag{10.4.1}$$

is an orthogonal basis of $L^2(\mathbb{R})$. We refer to the books [Chu1, Dau, Mey] for more detail, and recall *the Haar-wavelet*, where ψ is then given by

$$\psi(x) = \begin{array}{ll} 1, & \text{for } 0 \le x \le 1/2 \\ -1, & \text{for } 1/2 < x \le 1 \\ 0, & \text{otherwise .} \end{array}$$

Let $U : L^2(\mathbb{R}) \to L^2(\mathbb{R})$ be the unitary scaling operator

$$(Uf)(x) = 2^{-1/2} f(2^{-1} x) \tag{10.4.2}$$

The *scaling* condition on the given wavelet ψ will be defined in terms of a φ-function, see [Chu1, p. 119], i.e., some $\varphi \in L^2(\mathbb{R})$ such that $\{\varphi(x - k)\}_{k \in \mathbb{Z}}$ is a *Riesz-basis* for a subspace V_0 in $L^2(\mathbb{R})$. (See also the estimates (10.5.3) listed below.) Moreover, we require the following two conditions:

(i) The normalized φ_0 given by

$$\left(\sum_k |\hat{\varphi}(\omega + k2\pi)|^2 \right)^{1/2} \hat{\varphi}_0(\omega) = \hat{\varphi}(\omega), \qquad \text{a.e.,} \tag{10.4.3}$$

satisfies

$$\overline{\hat{\varphi}_0(\omega + \pi)} \hat{\psi}(2\omega) = e^{i\omega} \overline{\hat{\varphi}_0(2\omega + 2\pi)} \hat{\varphi}_0(\omega) \qquad \text{a.e.} \tag{10.4.4}$$

and

(ii) $UV_0 \subset V_0$.

We now take $\mathcal{K} = L^2(\mathbb{R})$, and let W_0 be the closed subspace spanned by $\{\psi(x + k)\}_{k \in \mathbb{Z}}$ where ψ is some orthogonal wavelet.

THEOREM 10.4.1
Let ψ be a scaled orthogonal wavelet, and let W_0 be the corresponding generating subspace. Let U be the scaling operator (10.3.4). Then there is a saturated invariant subspace V_0 such that

$$W_0 = U^{-1}(V_0) \ominus V_0 \tag{10.4.5}$$

and a spectral representation R such that

$$R(V_0) = z\mathcal{H}^2(W_0) \tag{10.4.6}$$

PROOF   Let

$$a_k = \frac{1}{2\pi} \int_0^{2\pi} e^{-ik\omega} p(\omega)\, d\omega,$$

for $k \in \mathbb{Z}$; then (ii) translates into the assertion

$$\hat{\varphi}(2\omega) = p(\omega)\varphi(\omega), \qquad \text{a.e.,} \tag{10.4.7}$$

or equivalently

$$\varphi(x) = \sum_k a_k \varphi(2x - k)\sqrt{2} \tag{10.4.8}$$

where $\sum_k |a_k|^2 < \infty$. We are getting a transform (for later use), $\Phi(z) = R(\varphi)(z)$ such that (10.4.9) turns into,

$$z\Phi(z) = \sum_{k \in \mathbb{Z}} a_k (RT_{k\varphi})(z),$$

and the operator U in (10.4.3) becomes multiplication by z, the complex transform variable. If V_0 denotes the closed subspace in $L_2(\mathbb{R})$ spanned by $\{\varphi(x - k)\}_{k \in \mathbb{Z}}$; then, by (ii), $U(V_0) \subset V_0$ where U is the scaling operator (10.3.4). Defining

$$V_j := U^{-j}(V_0), \qquad j \in \mathbb{Z}, \tag{10.4.9}$$

then $V_j \subset V_{j+1}$, $\forall j \in \mathbb{Z}$; and we claim that

$$V_1 \ominus V_0 = W_0 \tag{10.4.10}$$

where W_0 is the closed subspace in $L^2(\mathbb{R})$ spanned by $\{\psi(x - k)\}_{k \in \mathbb{Z}}$.

Since φ is a Riesz basis for V_0, it follows from [Chu1, p. 75] that there is a well-defined $\varphi_0 \in L^2(\mathbb{R})$ such that

$$\left(\sum_{k \in \mathbb{Z}} |\hat{\varphi}(\omega + k2\pi)|^2 \right)^{1/2} \hat{\varphi}_0(\omega) = \hat{\varphi}(\omega), \qquad \text{a.e.,}$$

and the sequence $\{\varphi_0(x - k)\}_{k \in \mathbb{Z}}$ is an *orthogonal* basis for the same space V_0. Moreover (10.4.11) translates into

$$\hat{\psi}(2\omega) = e^{i\omega}\overline{m_0(\omega + \pi)}\hat{\varphi}_0(\omega), \qquad \text{a.e.} \qquad (10.4.11)$$

where m_0 is 2π-periodic and determined by $\hat{\varphi}_0(2\omega) = m_0(\omega)\hat{\varphi}_0(\omega)$, a.e.

It follows that $\psi \in V_1$. So we must show that ψ is orthogonal to V_0. For the inner products, we have

$$\langle \psi, \varphi_0(\cdot - k) \rangle = \int_{\mathbb{R}} e^{-ik\omega}\hat{\psi}(\omega)\hat{\varphi}_0(\omega)\, d\omega$$

$$= \frac{1}{2\pi} \int_0^{2\pi} e^{-ik\omega} \sum_{l \in \mathbb{Z}} \hat{\psi}(\omega + l2\pi)\overline{\hat{\varphi}_0(\omega + l2\pi)}\, d\omega$$

We abbreviate the $l \in \mathbb{Z}$ summation, $\text{Per}\{\hat{\psi}(\omega)\overline{\hat{\varphi}_0(\omega)}\}$, so orthogonality amounts to the identity

$$\text{Per}\{\hat{\psi}\overline{\hat{\varphi}_0}\} = 0 \qquad \text{a.e.} \qquad (10.4.12)$$

Now make a substitution of (10.4.12) and

$$\hat{\varphi}_0(\omega) = m_0\left(\frac{\omega}{2}\right)\hat{\varphi}_0\left(\frac{\omega}{2}\right) \qquad \text{a.e.} \qquad (10.4.13)$$

into (10.4.13). Since the functions $\{\varphi_0(x - k)\}_{k \in \mathbb{Z}}$ are orthogonal, we get

$$|m_0(\omega)|^2 + |m_0(\omega + \pi)|^2 = 1 \qquad \text{a.e.} \qquad (10.4.14)$$

Using the obvious identity:

$$e^{i\omega}\overline{m_0(\omega + \pi)}\,\overline{m_0(\omega)} + e^{i(\omega+\pi)}\overline{m_0(\omega)}\,\overline{m_0(\omega + \pi)} = 0 \qquad \text{a.e.,}$$

we finally arrive at

$$\text{Per}\left(\hat{\psi}\overline{\hat{\varphi}_0}\right) = \sum_{k \in \mathbb{Z}} \hat{\psi}(\omega + 2\pi k)\overline{\hat{\varphi}_0(\omega + 2\pi k)} = 0 \qquad \text{a.e.}$$

∎

This concludes the proof of (10.4.13); and, defining,

$$W_j := U^{-j}(W_0), \qquad \forall j \in \mathbb{Z} \qquad (10.4.15)$$

and using (10.4.11), we also get

$$V_{j+1} \ominus V_j = W_j, \qquad (j \in \mathbb{Z}),$$

or equivalently

$$V_{j+1} = V_j \oplus W_j. \qquad (10.4.16)$$

But from (10.4.12) and (10.4.17), we then get that each W_j is spanned by $\{\psi(2^j x - k)\}_{k \in \mathbb{Z}}$. When *both indices* $j, k \in \mathbb{Z}$ *vary*, these functions form an orthonormal *basis* for $L^2(\mathbb{R})$; and, in particular, we have

$$\sum_{j \in \mathbb{Z}}^{\oplus} W_j = L^2(\mathbb{R}). \tag{10.4.17}$$

This means that V_1 is saturated. By the same argument, V_0 is saturated as well, see (10.4.10); and we have

$$\bigvee_{j \in \mathbb{Z}} V_j = L^2(\mathbb{R}) \qquad \text{and} \qquad \bigwedge_{j \in \mathbb{Z}} V_j = \{0\}. \tag{10.4.18}$$

Now Lemma 10.3.1 applies directly to the pair of spaces V_1, and $W_0 = V_1 \ominus V_0$. The spectral representation R, satisfying $R(V_1) = \mathcal{H}^2(W_0)$, will also satisfy the desired conclusions (10.4.7) of Theorem 10.4.1.

REMARK Let ψ denote the function which is given by (10.4.12) and which generates the Haar wavelet. Then the subspace W_0 is generated by the translates $\{\psi(x+k)\}_{k \in \mathbb{Z}}$; and we may take V_0 to be similarly generated by $\varphi = \chi_I$ where $I = [0, 1)$ is the unit-interval. The representation R from Theorem 10.4.1 is then normalized by $R\psi = \psi$, where $\psi \in W_0$ is viewed as the element in $\mathcal{L}^2(W_0) = \sum_{j \in \mathbb{Z}}^{\oplus} z^j W_0$ which is *the constant function*; and

$$(R\varphi)(z) = \sum_{1}^{\infty} \left(\frac{z}{2}\right)^n \psi = \frac{z}{2 - z} \psi.$$

10.5   Multiresolutions

Let M be a Hilbert space; and let $\mathcal{L}^2(M)$ and $\mathcal{H}^2(M)$ be the spaces from Section 10.2 of vector M-valued functions, defined on $\mathbb{T}$. Recall the functions in the Hardy space $\mathcal{H}^2(M)$ are characterized by the familiar analytic continuation property, for $|z| < 1$. We say that a bounded linear operator A in $\mathcal{L}^2(M)$ is *analytic* if it maps $z^n \mathcal{H}^2(M)$ into itself for all $n = 0, 1, \ldots$, i.e., $A(z^n \mathcal{H}^2(M)) \subset z^n \mathcal{H}^2(M)$.

We now recall the definition of *multiresolutions* for $L^2(\mathbb{R})$, for more details see, e.g., [Chu1, Dau, Mey]. On the scalar-valued Hilbert space $L^2(\mathbb{R})$, we have the usual operators:

$$Uf(x) = 2^{-1/2} f(2^{-1} x) \tag{10.5.1}$$

and

$$T_k f(x) = f(x - k), \tag{10.5.2}$$

defined for $f \in L^2(\mathbb{R})$, $x \in \mathbb{R}$, and $k \in \mathbb{Z}$. (We also have nonintegral translations, but we shall restrict, for the moment, to $k \in \mathbb{Z}$ in formula (10.5.2).)

A *multiresolution* is given by a closed subspace $V_0 \subset L^2(\mathbb{R})$, such that V_0 is *invariant* for U, and also *saturated*. Moreover, it is assumed that V_0 is further invariant under the translation group $\{T_k\}_{k \in \mathbb{Z}}$, and that V_0 contains a Riesz generator φ. The defining condition for φ amounts to the following *a priori* estimate (two positive constants a_1, a_2 given):

$$a_1 \sum_k |c_k|^2 \leq \left\| \sum_k c_k T_k \varphi \right\|^2 \leq a_2 \sum_k |c_k|^2 \tag{10.5.3}$$

where the summation is over all $k \in \mathbb{Z}$; and the condition is for *all* $(c_k) \in l^2$. The function φ is called a *scaling function*.

Using Corollary 10.3.2 above, for the abstract scattering operator, we shall now show that, *for any multiresolution*, the spectral representation of *any scaling function* is analytic, and, as such, generated by a corresponding wavelet.

We now calculate the spectral transform R, (from Corollary 10.3.2 and Theorem 10.4.1) for the *general* multiresolution and point out that the formula for $(R\varphi)(z)$ which we established for the Haar wavelet is in fact a special case of a general one.

Let $\mathcal{H}$ be the *scalar-valued* Hardy space: elements A in $\mathcal{H}$ will be written

$$A(z) = \sum_{n=0}^{\infty} a_n z^n, \qquad a_n \in \mathbb{C}; \tag{10.5.4}$$

and, for the norm, we have

$$\|A\|^2 = \sum_{n=0}^{\infty} |a_n|^2. \tag{10.5.5}$$

A function $f \in L^2(\mathbb{R})$ is said to be *normalized* if $\int_{\mathbb{R}} |f(x)|^2 \, dx = 1$; and it is said to be *pure* if the translates $\{f(x - k)\}_{k \in \mathbb{Z}}$ are mutually orthogonal. Recall that the orthogonality is equivalent to

$$\sum_{k \in \mathbb{Z}} |\hat{f}(\omega + 2\pi k)|^2 = 1 \qquad \text{a.e.} \tag{10.5.6}$$

THEOREM 10.5.1
Let (V_j) be a multiresolution for $L^2(\mathbb{R})$, and let $W_0 := V_1 \ominus V_0$. Pick generators $\varphi \in V_0$ and $\psi \in W_0$ which are both normalized and pure, and

let R be the corresponding spectral transform

$$R : L^2(\mathbb{R}) \to \mathcal{L}^2(W_0) \tag{10.5.7}$$

normalized by $R\psi = \psi \in W_0$.
   Then there is a unique

$$A(z) = a_1 z + a_2 z^2 + \ldots \in \mathcal{H} \tag{10.5.8}$$

such that

$$(R\varphi)(z) = A(z)\psi \tag{10.5.9}$$

for all $z \in \mathbb{C}$ with $|z| \le 1$.

PROOF   From Lemma 10.3.1, we have

$$R(V_0) = z\mathcal{H}^2(W_0),$$

and ψ ($\in W_0$) is a wavelet. So there are functions $(A_k(z))_{k\in\mathbb{Z}}$ in $\mathcal{H}$ as in (10.5.4) such that

$$(R\varphi)(z) = \sum_{k\in\mathbb{Z}} A_k(z) T_k \psi \tag{10.5.10}$$

where T_k denotes the translation group (10.5.2), ψ is the wavelet (so $T_k \psi \in W_0$), and

$$R(2^{-j/2}\psi(2^{-j}x - k)) = z^j(T_k\psi), \tag{10.5.11}$$

for $j = 1, 2, \ldots$, and $k \in \mathbb{Z}$. Since R is unitary, we also have (see (10.5.8)):

$$1 = \sum_{k\in\mathbb{Z}} \|A_k\|^2 \tag{10.5.12}$$

$$= \sum_k \sum_{j=1}^{\infty} |a_{kj}|^2.$$

We must show that, in (10.5.10), all the terms corresponding to $k \ne 0$ vanish identically. For this, we need the spectrum of the translation group. The commutation relation is

$$UT_t = T_{2t}U, \qquad t \in \mathbb{R} \tag{10.5.13}$$

which can be easily checked; we shall need it, at the moment, only for integral values of t. Suppose A is given by (10.5.8) and $\lambda \in \mathbb{R}$; then we adopt the notation

$$A(2)\lambda = \sum_j a_j 2^j \lambda \tag{10.5.14}$$

From (10.5.13), we get that $(T_k)_{k \in \mathbb{Z}}$ has Lebesgue spectrum with uniform multiplicity: Let χ_λ, $(\lambda \in \mathbb{R}/2\pi\mathbb{Z})$ be a field of orthogonal generalized eigen-functions; and

$$A_k(z)T_k\chi_\lambda = e^{ik\lambda}A_k(z)\chi_\lambda \qquad (10.5.15)$$

$$= e^{ik\lambda}\chi_{A_k(2)\lambda}$$

$$= e^{ik\lambda}\sum_j a_j\chi_{2^j\lambda}$$

The spectral representation for $(T_k)_{k \in \mathbb{Z}}$ has the form

$$T_k = \frac{1}{2\pi}\int_0^{2\pi} e^{i\lambda k}\,dF(\lambda) \qquad (10.5.16)$$

where $F(\cdot)$ denotes the orthogonal spectral resolution; (notation $dF(\lambda)\chi = \chi_\lambda$) and $\frac{1}{2\pi}\int_0^{2\pi} dF(\lambda)$ represents the identity operator. Now, apply this to both sides of formula (10.5.15), and substitute (10.5.16) on the right-hand side. It follows from the orthogonality of the spectral resolution that the terms in (10.5.15), corresponding to $k \neq 0$, indeed vanish. For more details on the spectral type of the representation (10.5.16), see [BJ]. The desired conclusion (10.5.9) now follows.

In Section 10.7 below, we shall consider a wavelet system based on $\{T_k, U\}$ but with the operators realized in a different Hilbert space, which, it turns out, allows a completely different spectral type. For this case, the spectral resolution (10.5.16) is particularly simple.

A typical example where the scattering operator from Corollary 10.3.2 can be shown to exist arises from an arbitrarily given *multiresolution*, and its Fourier transform, as follows. Let $(V_j)_{j \in \mathbb{Z}}$ be a given multiresolution in $L^2(\mathbb{R})$, and let $\hat{}$ denote the Fourier transform, i.e.,

$$\hat{f}(\omega) = \int_{\mathbb{R}} e^{-i\omega x}f(x)\,dx, \qquad \omega \in \mathbb{R}. \qquad (10.5.17)$$

The scaling operator U is as in (4.3). The two spaces $V_\pm$ from Corollary 10.3.3 may be defined as:

$$V_+ := V_1 \qquad \text{and} \qquad V_- := \hat{V}_0 \qquad (10.5.18)$$

But the two spaces are typically *not* orthogonal.

We also note that, in general, the spaces $(\hat{V}_j)_{j \in \mathbb{Z}}$ will *not* form a multiresolution. ∎

10.6   The Integral Translation Group

We now study the group of *integral* translations $\{T_k\}_{k\in\mathbb{Z}}$, see (10.5.2); and identify how the *scaling functions* for given multiresolutions transform under integral translates to the right resp., to the left. A scattering theory is developed for this pairing; and we show that there is a certain scalar-valued *inner function* which serves as a *spectral invariant*. It is computable, it separates distinct examples, and it is trivial precisely for the Haar wavelet.

COROLLARY 10.6.1

Let $\{V_j\}_{j\in\mathbb{Z}}$ be a (dyadic) multiresolution for $L^2(\mathbb{R})$ with scaling and translation operators given by (10.5.1)–(10.5.2). Let $W_0 := V_1 \ominus V_0$, and let R be the spectral representation

$$R : L^2(\mathbb{R}) \to \mathcal{L}^2(W_0)$$

from Corollary 10.3.2 normalized by $R(V_1) = \mathcal{H}^2(W_0)$. Then the operator group

$$\{RT_kR^{-1}\}_{k\in\mathbb{Z}} \qquad on \qquad \mathcal{L}^2(W_0) \qquad\qquad (10.6.1)$$

consists of operators which are meromorphic in the following sense. There is a sequence $(B_k(z))_{k\in\mathbb{Z}} \subset \mathcal{H}$ of scalar-valued Hardy-functions such that

$$\tilde{T}_k(A(z)\psi) = B_k(z)\psi, \qquad k \in \mathbb{Z}_+, \quad |z| \leq 1, \qquad (10.6.2)$$

where

$$\tilde{T}_k := RT_kR^{-1}, \qquad\qquad (10.6.3)$$

$A(z) \in \mathcal{H}$ is the function from (10.5.9) representing the scaling function φ, and $\psi \in W_0$ is the wavelet. Formula (10.6.2) is an identity for elements in $\mathcal{H}^2(W_0)$.

The case of (10.6.2) corresponding to the negative range $\mathbb{Z}_- = -\mathbb{Z}_+$ of the translation index (i.e., translations to the left by integral steps) is covered by

$$(R\varphi_{-1})(z) = A_*(z)\psi_{-1} \qquad\qquad (10.6.4)$$

for some scalar-valued Hardy function $A_* \in z\mathcal{H}$. Moreover, for $k \in \mathbb{Z}_+$, we have

$$\tilde{T}_{-k}(A(z)\psi) = B_k^*(z)\psi_{-1} \qquad\qquad (10.6.5)$$

for some sequence $B_1^*, B_2^*, \ldots \in z\mathcal{H}$ where

$$\varphi_{-1} = T_{-1}\varphi = \varphi(x+1) \in V_0$$

and

$$\psi_{-1} = T_{-1}\psi \in W_0$$

representing unit-translations to the left on the basic generating functions, i.e., the scaling function and the wavelet.

REMARK For the Haar wavelet, see 10.4.2 and (10.5.11), and Remark 10.4.2 above, we have the formulas

$$A(z) = \frac{z}{2-z}, \qquad B_1(z) = \frac{z(z-1)}{2-z}$$

and

$$\frac{B_1(z)}{A(z)} = z - 1$$

for the functions in $\mathcal{H}$ which are described in Theorem 10.5.1 above. If $k \geq 1$, i.e., $k = 1, 2, \ldots$, then

$$\frac{B_k(z)}{A(z)} = z^{k-1}(z-1). \tag{10.6.6}$$

In this example, the function A_* from (10.6.4) is given by the formula

$$A_*(z) = -A(z), \qquad |z| \leq 1; \tag{10.6.7}$$

but, in general, the formula for A_* may be more complicated.

PROOF Corollary 10.6.1 is a direct consequence of Theorem 10.5.1 above. In fact, formula (10.6.2) is the assertion

$$(R\varphi_k)(z) = B_k(z)\psi \tag{10.6.8}$$

where $\varphi_k := T_k\varphi$, $k \in \mathbb{Z}_+$. But, with k fixed, notice that the two vectors φ_k and ψ satisfy the conditions in Theorem 10.5.1. Clearly $\varphi_k \in V_0$ is a scaling function, and the orthogonality follows from:

$$\langle \varphi_k, \psi \rangle = \langle \varphi, T_{-k}\psi \rangle = 0$$

since $T_{-k}\psi \in W_0$.

Formula (10.6.4) is the assertion

$$(R\varphi_{-1})(z) = A_*(z)\psi_{-1} \tag{10.6.9}$$

where $\varphi_{-1} := T_{-1}\varphi$. But again Theorem 10.5.1 applies since the two vectors $\varphi_{-1} \in V_0$ and $\psi_{-1} \in W_0$ satisfy the conditions from the theorem. The theorem applies to a pair of elements in $L^2(\mathbb{R})$ subject to some assumptions vis-a-vis a given multiresolution; and the pair φ_{-1}, ψ_{-1} satisfies the conditions.

Finally (10.6.5) in the corollary follows from the above two arguments combined. ∎

THEOREM 10.6.2

Let (V_j) be a multiresolution for $L^2(\mathbb{R})$, $W_0 := V_1 \ominus V_0$; and let $\varphi \in V_0$ and $\psi \in W_0$ be the generating functions, chosen normalized and pure, i.e., φ the scaling function, and ψ the wavelet. Let $(B_k(z))$ and $(B_k^*(z))$, $k \in \mathbb{Z}_+$, be the sequences in the Hardy space $\mathcal{H}$ which are determined by (10.6.2) and (10.6.5) in Corollary 10.6.1. Then there is an inner function $C(z)$ such that

$$zC(z)B_k(z) = B_k^*(z) \qquad (10.6.10)$$

holds in $\{z \in \mathbb{C} : |z| \le 1\}$, and for all $k \in \mathbb{Z}_+$.

PROOF   The functions $\varphi \in V_0$ and $\psi \in W_0$ are chosen as specified above, normalized and pure. (Recall if the orthogonality is not satisfied for φ at the outset, then it may be replaced by one that is orthogonal by the usual construction, see (10.4.4) above.) Consider the two subspaces $\mathcal{K}_\pm$ in $L^2(\mathbb{R})$ defined by:

$$\mathcal{K}_+ := \overline{\operatorname{span}_{j \in \mathbb{Z}}} \{U^j T_k \varphi : k = 0, 1, 2, \ldots\}$$

and

$$\mathcal{K}_- := \overline{\operatorname{span}_{j \in \mathbb{Z}}} \{U^j T_{-k} \varphi : k \in \mathbb{Z}_+\},$$

and a *reflection* operator

$$\Omega : \mathcal{K}_+ \longrightarrow \mathcal{K}_- \qquad (10.6.11)$$

defined by

$$\Omega(U^j T_k \varphi) := \omega_k U^j T_{-k} \varphi, \qquad \text{for } \forall j \in \mathbb{Z} \quad \text{and} \quad k \in \mathbb{Z}_+, \qquad (10.6.12)$$

with constants ω_k put in to respect the relations.

Recall the generating spaces

$$V_j := \overline{\operatorname{span}} \{U^j T_k \varphi : k \in \mathbb{Z}\}.$$

They are nested by $V_j \subset V_{j+1}$; we have

$$V_j^\pm := \mathcal{K}_\pm \cap V_j$$

satisfying

$$V_j = V_j^+ \oplus V_j^-, \qquad \forall j \in \mathbb{Z}.$$

Since the vectors on each side of (10.6.12) are mutually orthogonal as k varies, but with j fixed, we conclude that Ω is well defined and isometric as a linear operator; we may set $\Omega \varphi = \hat{\varphi}$; (but that is not important.)

Let R be the spectral representation from Lemma 10.3.1, normalized by $R\psi = \psi$, and $R(V_1) = \mathcal{H}^2(W_0)$. Define two operators $R_\pm$ by restriction to the respective spaces $\mathcal{K}_\pm$ as follows

$$R_\pm := R|_{\mathcal{K}_\pm} \tag{10.6.13}$$

and a third operator S by

$$S := R_- \Omega R_+^{-1} \tag{10.6.14}$$

We have

$$R(\mathcal{K}_+) = \mathcal{H}\psi, \qquad \text{and} \qquad R(\mathcal{K}_-) = (z\mathcal{H})\psi_- \tag{10.6.15}$$

where $\mathcal{H}$ is the scalar-valued Hardy space, and $\psi_- = T_-\psi = \psi(x + 1)$. Combining (10.6.14) and (10.6.15), we note that S may then be viewed as an operator in $\mathcal{H}$, mapping $\mathcal{H}$ into itself. ∎

LEMMA 10.6.3
When S in (10.6.14) is viewed as an operator in $\mathcal{H}$, then it commutes with multiplication by z, i.e.,

$$Sz = zS \qquad \text{on } \mathcal{H} \tag{10.6.16}$$

as an operator identity, and with z representing the unilateral shift operator as it acts on $\mathcal{H}$.

PROOF   We will check the commutation (10.6.16) on a basis. Let B_k and B_k^* be the sequences in $\mathcal{H}$, defined for $k \in \mathbb{Z}_+$ from (10.6.2) and (10.6.5) in Corollary 10.6.1. Then

$$
\begin{aligned}
(Sz)(B_k(z)\psi) &= R\Omega R^{-1}(zB_k(z)\psi) \\
&= R\Omega U R^{-1}(B_k(z)\psi) \\
&= R\Omega U T_k \varphi \\
&= \omega_k R U T_{-k} \varphi \\
&= \omega_k z B_k^*(z)\psi_{-1} \\
&= \omega_k z R(T_{-k}\varphi)(z) \\
&= z R\Omega T_k \varphi \\
&= z(R\Omega R^{-1})B_k(z)\psi \\
&= (zS)(B_k(z)\psi).
\end{aligned}
$$

It follows from the properties of the scaling function and the wavelet that this indeed checks out formula (10.6.16) on a basis in $\mathcal{H}$, so (10.6.16) holds as an operator identity on $\mathcal{H}$ as claimed. ∎

To finish the proof of the theorem, note that the Beurling-Lax-Halmos theorem applies to $S(\mathcal{H})$ and $z\mathcal{H}$; and we get an inner function $C(z)$, i.e. analytic and $|C(z)| = 1$ a.e. $z \in \mathbb{T}$, such that

$$S(f(z)\psi) = zC(z)f(z)\psi_- \tag{10.6.17}$$

holds for all $f(z) \in \mathcal{H}$. We have just verified that

$$S(B_k(z)\psi) = B_k^*(z)\psi_{-1} \tag{10.6.18}$$

and we may conclude therefore that

$$zC(z)B_k(z) = B_k^*(z), \qquad \forall k \in \mathbb{Z}_+, \quad |z| \leq 1, \tag{10.6.19}$$

which is the conclusion of the theorem.

DEFINITION 10.6.4   *The function $C(z)$ in (10.6.17) is called the spectral function, and the operator S in (10.6.14) the wavelet scattering operator; we now show that it is a measure for how "different" the given wavelet ψ is from the Haar example (10.4.2) above.*

REMARK   It can be checked for the Haar wavelet in (10.4.2) and Remarks 10.6.2 and 10.4.2 that the identity

$$B_k(z) = B_{k+1}^*, \qquad \forall k \in \mathbb{Z}_+. \tag{10.6.20}$$

So by (10.6.19) we conclude that $C(z) \equiv 1$ a.e. $z \in \mathbb{T}$, for this example. Nontriviality of the spectral function is thus an invariant "measure" for scattering away from the Haar example, and applies to any wavelet ψ which is defined from a multiresolution.

10.7   The Besicovich Spaces

In [BJ] and [JP1–2], we showed that some spectral sets, defined relative to multiresolutions, may be understood using the theory of almost periodic functions. In this section, we show that the Besicovich-completions yield a new and different spectral theory for the theory, and that the *functional equation*, defining the multiresolution scaling function admits effective algorithms which converge in the (almost periodic) completions, as opposed to the traditional Hilbert space $L^2(\mathbb{R})$ which has been used almost exclusively in wavelet theory up to now.

Let Λ_2 be the additive group consisting of finite sums $\sum_j 2^j m_j$ where $j \in \mathbb{Z}$, and $m_j \in \mathbb{Z}$; it is the ring of 2-adic integers. The corresponding compact dual group will be denoted G_2. Since Λ_2 is torsion free, it follows that G_2 is connected. The inclusion $\Lambda_2 \subset \mathbb{R}$ dualizes to get us a continuous homomorphism $\Phi : \mathbb{R} \to G_2$ with dense image. Hence G_2 is also a Bohr compactification $b_{\Lambda_2}(\mathbb{R})$ of $\mathbb{R}$ subject to frequences in Λ_2. The Pontryagin pairing between G_2 and Λ_2 will be denoted $\langle \chi, \lambda \rangle = \chi(\lambda)$ when points χ in G_2 are realized as characters on Λ_2, i.e., $\chi : \Lambda_2 \to \mathbb{T}$ such that $\chi(\lambda + \lambda') = \chi(\lambda)\chi(\lambda')$, $\lambda, \lambda' \in \Lambda_2$; or $\langle \chi, \lambda + \lambda' \rangle = \langle \chi, \lambda \rangle \langle \chi, \lambda' \rangle$. (For more details, see [Bes] and [Ru2].) The homomorphism Φ is specified by

$$\langle \Phi(r), \lambda \rangle = e^{ir\lambda}, \qquad r \in \mathbb{R}, \quad \lambda \in \Lambda_2. \tag{10.7.1}$$

In the representation $\lambda = \sum_{j \in \mathbb{Z}} 2^j m_j$ (finite sum), we may pick $m_j \in \{0, 1\}$. The dyadic digits for 2λ may be worked out by recursion, and we get 2χ well defined by the duality

$$\langle 2\chi, \lambda \rangle := \langle \chi, 2\lambda \rangle = \chi(\lambda)^2, \qquad \forall \chi \in G_2, \quad \forall \lambda \in \Lambda_2. \tag{10.7.2}$$

This automorphism $\chi \mapsto 2\chi$ is *ergodic* on G_2 by [BJ, Proposition 1], and this feature of the almost periodic approach affects both the *spectral theory* and the *existence* results for multiresolution scaling functions as we proceed to show.

Recall first that the *"traditional"* approach to multiresolutions is based on a unitary representation of the matrix group

$$M_2(\mathbb{Z}) := \left\{ \begin{pmatrix} 2^{-j} & k2^{-j} \\ 0 & 1 \end{pmatrix} : j, k \in \mathbb{Z} \right\} \tag{10.7.3}$$

acting on $L^2(\mathbb{R})$. The *present approach* will be based instead on a unitary representation of

$$M_2(\Lambda) := \left\{ \begin{pmatrix} 2^{-j} & \lambda \\ 0 & 1 \end{pmatrix} : \lambda \in \Lambda_2, j \in \mathbb{Z} \right\}$$

acting now on $L^2(G_2)$ where this L^2-space is defined from the normalized Haar measure on G_2, and denoted $d\chi$. But $L^2(G_2)$ *is canonically isomorphic to the Besicovich space $B_2(\Lambda_2)$* of functions f on $\mathbb{R}$ such that

$$\lim_{T \to \infty} \frac{1}{2T} \int_{-T}^{T} |f(x)|^2 \, dx := \|f\|_{B_2}^2$$

is well defined and finite; and moreover that the generalized Fourier transform

$$c_\lambda := \lim_{T \to \infty} \frac{1}{2T} \int_{-T}^{T} f(x)e^{-i\lambda x} \, dx \tag{10.7.4}$$

is supported on Λ_2, and

$$\|f\|_{B_2}^2 = \sum_{\lambda \in \Lambda_2} |c_\lambda|^2. \tag{10.7.5}$$

(Under the isomorphism, the limit in (10.7.5) and above, will then become the Haar measure on G_2.)

Besicovich showed, conversely, that every $c \in l^2(\Lambda_2)$ corresponds to a unique function f on $\mathbb{R}$ which is almost periodic in the mean-square sense, reproduces c by (10.7.5), and satisfies (10.7.6).

Our present representation of $M_2(\Lambda_2)$ acting on $L^2(G_2) \simeq B_2(\Lambda_2)$ will be given by the pair $(U, T_\lambda)_{\lambda \in \Lambda_2}$ where

$$Uf(\chi) = f(2\chi), \qquad \chi \in G_2 \tag{10.7.6}$$

$$T_\lambda f(\chi) = \langle \lambda, \chi \rangle f(\chi), \qquad \lambda \in \Lambda_2 \tag{10.7.7}$$

These formulas correspond to the Fourier transformed version of our familiar two formulas (10.5.1) and (10.5.2) from above. The present commutation relation

$$UT_\lambda U^{-1} = T_{2\lambda}, \qquad \lambda \in \Lambda_2$$

corresponds to (10.5.13) above, and states that we have a unitary representation of the group $M_2(\Lambda_2)$ on $L^2(G_2)$; and it can be checked that it is in fact *irreducible*, a feature which is specific to this setting.

As in Section 10.5 above, we will study solutions to the equation

$$f(2\chi) = \sum_{k \in \mathbb{Z}} c_k (T_k f)(\chi) \qquad \text{a.e. } \chi \in G_2 \tag{10.7.8}$$

when some $(c_k) \in l^2(\mathbb{Z})$ is given. Defining

$$M(\chi) = \sum_{k \in \mathbb{Z}} c_k \langle \chi, k \rangle, \qquad \forall \chi \in G_2 \tag{10.7.9}$$

notice that (10.7.9) becomes the functional equation

$$f(2\chi) = M(\chi) f(\chi) \qquad \text{a.e. } \chi \in G_2 \tag{10.7.10}$$

Moreover (10.7.10) is the unique extension to G_2 of the 2π-periodic function

$$M(x) = \sum_{k \in \mathbb{Z}} c_k e^{ikx}, \qquad \forall x \in \mathbb{R}. \tag{10.7.11}$$

These formulas (10.7.10)–(10.7.12) are analogues of (10.4.9) and (10.4.14)–(10.4.15) in Section 10.4 for the "classical" theory.

THEOREM 10.7.1
(i) *Pick* $(c_k) \in l^2(\mathbb{Z})$ *and assume that the associated function M in (10.7.10) takes values in* $\mathbb{T}$, *i.e., that* $|M(\chi)| = 1$ *a.e.; then*

$$f(\chi) := \lim_{J \to \infty} \frac{1}{J} \sum_{j=1}^{J} \prod_{i=1}^{j} M(2^{-i}\chi) \tag{10.7.12}$$

is well defined as an element in $L^2(G_2)$, and it satisfies the function equation (10.7.11).

(ii) For $f \in L^2(G_2)$ the Riesz-estimate, with constants $0 < b \leq B < \infty$,

$$b \sum_k |a_k|^2 \leq \left\| \sum_k a_k T_k f \right\|_{L^2(G_2)}^2 \leq B \sum_k |a_k|^2 \tag{10.7.13}$$

for all $(c_k) \in l^2(\mathbb{Z})$, summation over $\mathbb{Z}$, is equivalent to

$$b \leq \int_H |f(\xi + \chi)|^2 \, d\xi \leq B \qquad a.e. \quad \chi \in G_2 \tag{10.7.14}$$

where

$$H := \{\xi \in G_2 : \langle \xi, k \rangle = 1, \qquad \forall k \in \mathbb{Z}\}. \tag{10.7.15}$$

PROOF   Let $(c_k) \in l^2(\mathbb{Z})$ be given, and let M satisfy the condition in (i). On the product space $\mathbb{T} \times G_2$, define

$$Q(z, \chi) := (M(\chi)z, 2\chi).$$

The expression in (10.7.13) comes from Q via the von Neumann-Ulam ergodic theorem [Hal, p. 93]. Defining $h(z, \chi) = M(\chi)z$, so that $Q(z, \chi) = (h(z, \chi), 2\chi)$, we get arithmetic average

$$\frac{1}{J} \sum_{j=1}^J h(Q^j(z, \chi))$$

as the expression on the right-hand side. On the product space $\mathbb{T} \times G_2$, we take the respective Haar measures on the factors. The resulting product measure is $Q-$invariant; so the *ergodic theorem* applies (see [Hal]). It follows that the limit (10.7.13) is well defined, a.e. in G_2; and that the limit function f is in $L^2(G_2)$. When χ is replaced by 2χ on the right-hand side in (10.7.13), the factor $M(\chi)$ comes out, and the functional equation (10.7.11) results.

The proof of (ii) is standard Pontryagin duality: note $(G_2/H)\hat{} \simeq \mathbb{Z}$, and the integral $\int_{G_2} d\chi$ may be factored as: $\int_{G_2/H} \cdots \int_H$, i.e., as the invariant measure on the quotient, and the Haar measure on the subgroup H. The middle term in (10.7.14) is

$$\int_{G_2} |F(\chi)f(\chi)|^2 \, d\chi = \int_{G_2/H} |F(\chi)|^2 \int_H |f(\xi + \chi)|^2 \, d\xi \, d\chi$$

where $F(\chi) = \sum_{k \in \mathbb{Z}} a_k \langle k, \chi \rangle$. Since $G_2/H \simeq \mathbb{T} \simeq \hat{\mathbb{Z}}$, we get

$$\int_{G_2/H} |F(\chi)|^2 \, d\chi = \sum_{k \in \mathbb{Z}} |a_k|^2,$$

and the assertion in (ii) follows.   ∎

THEOREM 10.7.2
Let $f \in L^2(G_2)$ be given, and suppose it satisfies (10.7.11) and (10.7.14); i.e., the functional equation for some 2π-periodic multiplier, and the Riesz-basis property on the closed subspace V_0 in $L^2(G_2)$ spanned by $\{T_k f : k \in \mathbb{Z}\}$. Then we have the following multiresolution properties:

$$U(V_0) \quad \subset V_0, \tag{10.7.16}$$

$$\bigwedge_{j \in \mathbb{Z}} U^j(V_0) = \{ \text{ the constant functions } \}, \qquad and \tag{10.7.17}$$

$$\bigvee_{j \in \mathbb{Z}} U^j(V_0) = L^2(G_2) \tag{10.7.18}$$

where (T_k, U) is given by (10.7.7)–(10.7.8).

PROOF   Property (10.7.17) follows from (10.7.11) and (10.7.9). Each finite sum on the right in (10.7.9) is clearly in V_0; but when (10.7.14) is also assumed, it follows that the $l^2(\mathbb{Z})$-summation is still in V_0; so $Uf \in V_0$. The same argument applies to $UT_k f$, when $k \in \mathbb{Z}$, proving (10.7.17).

It is clear that the constant function is in the intersection (10.7.18). They must be equal, by taking orthogonal complements, if we check (10.7.19). Define the spectrum of subspaces in $L^2(G_2)$ as the support of the transform (10.7.5). From (10.7.14) it follows that the spectrum of V_0 must be $\mathbb{Z}$; and the spectrum of $V_j := U^{-j}(V_0)$ is $2^{-j}\mathbb{Z}$. Hence the spectrum of $\bigvee_{j \in \mathbb{Z}} V_j$ must be Λ_2, which is to say that this closed subspace is in fact all of $L^2(G_2)$. ∎

Condition (10.7.18) is a little different from (10.2.11) and (10.4.19) above. But the scattering theory from Section 10.3, and the multiresolutions from Section 10.5, still apply to this modified setting, see, e.g., [CS] and [SNF]. The main point of the completion G_2 is that $L^2(G_2)$ allows more multiresolutions.

Bibliography

[Bes]   A. S. Besicovich, *Almost periodic functions*, Dover, New York, 1954.

[BJ]   B. Brenken and P. E. T. Jorgensen, *A family of dilation crossed product algebras*, J. Operator Theory **25**, (1991), 299–308.

[Chu1] C. K. Chui, *An introduction to wavelets*, Academic Press, Boston, 1992.

[Chu2] C. K. Chui (Ed.), *Wavelets, A tutorial in theory and applications*, Academic Press, Boston, 1992.

[CS] M. Cotlar and C. Sadosky, *Transference of metrics induced by unitary couplings ...*, J. Functional Anal. **111**, (1993), 473–488.

[Dau] I. Daubechies, *Ten lectures on wavelets*, CMBS 61, SIAM, Philadelphia, 1992.

[FS] Y. Fourès and I. E. Segal, *Causality and analyticity*, Trans. Amer. Math. Soc. **78**, (1955), 385–504.

[GM] J. E. Gilbert and M.A.M. Murray, *Clifford algebras and Dirac operators in harmonic analysis*, Cambridge University Press, Cambridge, 1991.

[Hal] P. R. Halmos, *Ergodic theory*, Chelsea, New York, 1956.

[He] H. Helson, *Lectures on invariant subspaces*, Academic Press, New York, 1964.

[JM] P. E. T. Jorgensen and P. S. Muhly, *Selfadjoint extensions satisfying the Weyl operator commutation relations*, J. d'Analyse Math. **37**, (1980), 46–99.

[Jo] P. E. T. Jorgensen, *Operators and representation theory*, North Holland-Elsevier, Amsterdam, 1988.

[JP1] P. E. T. Jorgensen and S. Pedersen, *Spectral theory for Borel sets in $\mathbb{R}^n$ of finite measure*, J. Funct. Anal. **107**, (1992), 72–104.

[JP2] P. E. T. Jorgensen and S. Pedersen, *Harmonic analysis of fractal measures induced by representations of a certain C^*-algebra*, J. Funct. Anal. **125**, (1994), 90–110.

[LP] P. D. Lax and R. S. Phillips, *Scattering theory*, 2nd ed., Academic Press, New York, 1989.

[Mad] W. R. Madych, *Some elementary properties of multiresolution analyses of $L^2(\mathbb{R}^n)$* in Wavelets, A tutorial in theory and applications, C. K. Chui (Ed.), Academic Press, Boston, 1992, pp. 259–294.

[Mey] Y. Meyer, *Ondelettes*, Hermann, Paris, 1990.

[Ru1] W. Rudin, *Real and complex analysis*, 2nd ed., McGraw-Hill, New York, 1974.

[Ru2] W. Rudin, *Fourier analysis on groups*, 2nd ed., Wiley Interscience, New York, 1990.

[Str] R. S. Strichartz, *Besicovitch meets Wiener: Fourier expansions and fractal measures*, Bull. Amer. Math. Soc. **20**, (1989), 55–59.

[SNF]   B. Sz.-Nagy and C. Foias, *Harmonic analysis of operators on Hilbert space*, Translation and revision of French edition 1964, Akad. Kiadó, Budapest, North Holland-Elsevier, Amsterdam, 1970.

Department of Mathematics,
University of Iowa,
Iowa City, Iowa 52242-1466
E-mail: jorgen@math.uiowa.edu

Acoustic Scattering, Galerkin Estimates and Clifford Algebras

Björn Jawerth* and Marius Mitrea[†]

11.1 Introduction

We study boundary value problems for the quaternionic Helmholtz operator $\Delta - a$, $a \in \mathbb{H}$, in a bounded Lipschitz domain Ω of $\mathbb{R}^3$. More specifically, we consider the (interior/exterior) Dirichlet problems

$$(D_\pm) \begin{cases} (\Delta - a)u = 0 \text{ in } \Omega_\pm, \\ u^* \in L^2(\partial\Omega), \\ u|_{\partial\Omega} = f \in L^2(\partial\Omega), \end{cases}$$

where $\Omega_+ := \Omega$, $\Omega_- := \mathbb{R}^3 \setminus \bar{\Omega}$, Δ is the usual Laplace operator in $\mathbb{R}^3$ and $a \in \mathbb{H}$ is such that $\operatorname{Re} a \neq 0$. Here u^* is the usual nontangential maximal function of u and the boundary trace is taken in the nontangential convergence sense. For the exterior Dirichlet problem, u is also assumed to satisfy the so-called Sommerfeld radiation condition (see, e.g., [CK])

$$\langle \frac{X}{|X|}, \nabla u(X) \rangle - b\,u(X) = o(\frac{1}{|X|}), \tag{11.1.1}$$

as $|X| \longrightarrow \infty$, uniformly in all directions $X/|X|$ or $\mathbb{R}^3$, where $b \in \mathbb{H}$ is such that $\operatorname{Re} b \leq 0$ and $b^2 = a$ (note that such a b always exists and is unique). We also consider the (interior/exterior) Neumann problems

$$(N_\pm) \begin{cases} (\Delta - a)u = 0 \text{ in } \Omega_\pm, \\ (\nabla u)^* \in L^2(\partial\Omega), \\ \frac{\partial u}{\partial n}\big|_{\partial\Omega} = f \in L^2(\partial\Omega), \end{cases}$$

where n denotes the outward unit normal to $\partial\Omega$.

*Supported in part by ONR Grant N001-90-J-1343.

[†]Supported in part by ONR Grant N001-90-J-1343 and DARPA Grant 749620-93-1-0083.

1991 *Mathematics Subject Classification.* Primary 42C10, 42B20; Secondary 30G35.

Our first result concerns the solvability of $(D_\pm)$, $(N_\pm)$ in terms of the double-layer acoustic potential operator D (see §11.2 for precise definitions).

THEOREM 11.1.1
The operators $\pm\frac{1}{2}I + D$ are invertible on the Sobolev spaces $H^s(\partial\Omega)$ for any $0 \le s \le 1$. In particular, for any $f \in L^2(\partial\Omega)$, the unique solutions of the Dirichlet problems $(D_\pm)$ are given by

$$u_\pm(X) := -\frac{1}{4\pi} \int_{\partial\Omega} \frac{\partial}{\partial n(Y)} \left(\frac{e^{b|X-Y|}}{|X-Y|} \right)$$

$$[(\pm\frac{1}{2}I + D)^{-1}f](Y)\, dS(Y), \quad X \in \Omega_\pm,$$

whereas for any $f \in L^2(\partial\Omega)$, the unique solutions of the Neumann problems $(N_\pm)$ are given by

$$u_\pm(X) = -\frac{1}{4\pi} \int_{\partial\Omega} \frac{e^{b|X-Y|}}{|X-Y|} [(\mp\frac{1}{2}I + D^*)^{-1}f](Y)\, dS(Y), \quad X \in \Omega_\pm,$$

(11.1.2)

where D^* is the formal transpose of D.

Actually these results are easily seen to be valid for $f \in L^p(\partial\Omega)$ with $2 - \epsilon < p < \infty$ and $1 < p < 2 + \epsilon$, respectively, for some positive ϵ depending solely on the Lipschitz character of Ω, but this is something we shall not pursue in this paper. See [DK], [Ve]. For smoother domains, and smoother boundary data, related problems have been considered in [GS].

Note that, in this form, the only parts not concretely computable in the integral representations (11.1.2)-(11.1.3) appear to be those involving $(\pm\frac{1}{2}I + D)^{-1}$ and $(\pm\frac{1}{2}I + D^*)^{-1}$.

To explain our next result, i.e., a Galerkin approximation scheme (with variable mesh size) for the solutions of the boundary integral equations of the acoustic scattering, we still need some more notation. Let $\{\Omega_h\}_h$ be a family of smooth domains approximating Ω in a special sense which will be made precise later (see §11.2). Among other things, the boundary of Ω will be assumed to be in a homeomorphic correspondence F_h with the boundary of Ω_h, and we shall often identify functions defined on $\partial\Omega_h$ with their pull-back on $\partial\Omega$ via F_h.

The functions f_h approximating the solution f of the equation $(\frac{1}{2}I + D)f = g$, for $g \in L^2(\partial\Omega)$, are going to be selected from a distinguished subspace X_h of $L^2(\partial\Omega_h)$ called *finite element space localized on the scale* $\varrho(h)$. The most common example of such a space can be constructed as follows. Let $\{E_h^j\}_{j \in \Lambda_h}$ be a finite partition of $\partial\Omega_h$ satisfying $\mathrm{diam}\,(E_h^j) \le \varrho(h)$ for all $j \in \Lambda_h$. Then take X_h to be the space of functions defined on $\partial\Omega_h$ and which are constant on each set of this partition.

For $P \in \partial\Omega_h$ and $r > 0$, we introduce a measure of the maximal curvature of $\partial\Omega_h$ near P by setting

$$\kappa(P, r) := \sup_{|P-Q|<r} |\nabla^2 \varphi(Q)|,$$

where φ is the smooth function whose graph defines $\partial\Omega_h$ in a neighborhood of $P \in \partial\Omega_h$.

Let Π denote the projection operator from $L^2(\partial\Omega_h)$ onto X_h and let D_h stand for the singular double-layer potential operator on $\partial\Omega_h$. Also, fix some points $P_h^j \in E_h^j$.

THEOREM 11.1.2
Suppose that $\lim_{h\to 0} \varrho(h) = 0$. Then there there exist $h_0 > 0$ and $\gamma > 1$ such that if $0 < h < h_0$, and if

$$\lim_{h \to 0} \max_j \kappa(P_h^j, \gamma\varrho(h))\varrho(h) = 0, \tag{11.1.3}$$

then

1. *The operators $\Pi_h(\pm\frac{1}{2}I + D_h)$ are invertible on X_h.*
2. *If $g \in L^2(\partial\Omega)$ and if $f_h^\pm \in X_h$ are defined by $\Pi_h(\pm\frac{1}{2}I + D_h)f_h^\pm = \Pi_h(g \circ F_h^{-1})$, $0 < h < h_0$, then $\|f_h^\pm\|_{L^2} \leq C\|g\|_{L^2}$.*
3. *If $f^\pm \in L^2(\partial\Omega)$ denote the unique solutions of $(\pm\frac{1}{2}I + D)f^\pm = g$, then*

    $$\|f_h \circ F_h - f\|_{L^2} \longrightarrow 0 \quad as \quad h \longrightarrow 0.$$

4. *Assume that the datum g actually belongs to the Sobolev space $H^s(\partial\Omega)$ for some $0 < s \leq 1$. Also, suppose that $\varrho(h) = \mathcal{O}(h)$ as h tends to zero. Then we have the error estimate*

    $$\|f_h^\pm \circ F_h - f^\pm\|_{L^2} \leq C h^s \|g\|_{H^s}. \tag{11.1.4}$$

We point out that, as will become apparent from its proof, Theorem 11.1.2 is actually valid in L^p, with $2 - \epsilon < p < \infty$ for the points (1), (2), (3), and $2 - \epsilon < p < 2 + \epsilon$ for the point (4). Here $\epsilon = \epsilon(\Omega) > 0$ depends exclusively on the Lipschitz character of Ω.

Similar results hold in the somewhat simpler case when Ω is the unbounded domain in $\mathbb{R}^3$ lying above the graph of a Lipschitz function.

Finally, we stress that, with natural modifications, these results continue to hold in arbitrary dimensions and with $\mathbb{H}$ replaced by the Clifford algebra canonically associated with $\mathbb{R}^m$ endowed with the usual Euclidean metric (see, e.g., [BDS], [Mi]). Our restriction to the case treated here is only to lighten the exposition.

11.2 Definitions and Preliminary Results

In this section we recall basic notations and results used throughout the paper. For the sake of brevity, we shall try to keep some of the definitions discussed in the introduction.

The orthodox notation for quaternions is

$$\mathbb{H} := \{a = a_0 + ia_1 + a_2 j + a_3 k \,;\, a_0, a_1, a_2, a_3 \in \mathbb{R}\},$$

where the imaginary units i, j, k satisfy $i^2 = j^2 = -1$ and $ij = -ji = k$. *The conjugate* $\bar{a}$ *of* $a = a_0 + ia_1 + a_2 j + a_3 k \in \mathbb{H}$ *is defined by* $\bar{a} := a_0 - ia_1 - a_2 j - a_3 k$. Obviously, $\overline{xy} = \bar{y}\,\bar{x}$. Also, *the real part of* $a \in \mathbb{H}$ *is given by* Re $a := a_0$.

Note that Re $a = $ Re $\bar{a} = \frac{1}{2}(a + \bar{a})$, and that Re $(xy) = $ Re (yx). Finally, $\mathbb{H}$ becomes a normed algebra when endowed with the Euclidean norm $|a|^2 := a_0^2 + a_1^2 + a_2^2 + a_3^2$, and we have $|a|^2 = $ Re $(a\bar{a})$. We shall also need the exponential mapping, defined for $a \in \mathbb{H}$ by

$$e^a := \sum_{n=0}^{\infty} \frac{a^n}{n!}.$$

A simple calculation shows that $|e^a| = e^{\text{Re } a}$. The following simple observation will also be of importance for us.

LEMMA 11.2.1
If $a, u \in \mathbb{H}$, *then* $|ua\bar{u}| \geq |\text{Re } a||u|^2$.

PROOF

$$
\begin{aligned}
|ua\bar{u}| \geq |\text{Re } (ua\bar{u})|[.07in] &= \tfrac{1}{2}|\text{Re } (ua\bar{u} + \overline{ua\bar{u}})| \\
&= \tfrac{1}{2}|\text{Re } (u(a + \bar{a})\bar{u})| = |\text{Re } a||\text{Re } (u\bar{u})| \\
&= |\text{Re } a||u|^2.
\end{aligned}
$$

∎

LEMMA 11.2.2
Let $b \in \mathbb{H}$ *with* Re $b \leq 0$. *Then the quaternionic-valued function* $\Phi \in L^1_{loc}(\mathbb{R}^3)$ *given by*

$$\Phi(X) := -\frac{1}{4\pi} \frac{e^{b|X|}}{|X|}, \quad X \in \mathbb{R}^3 \setminus \{0\},$$

is an elementary solution for the quaternionic Helmholtz operator $\triangle - b^2$ *in* $\mathbb{R}^3$.

The proof is identical with the one for the scalar case, hence omitted.

In the second part of this section we shall summarize some basic facts about Lipschitz domains and acoustic layer potentials on Lipschitz boundaries. Let us first recall the following definition.

DEFINITION 11.2.3 *Consider Ω a bounded open set of $\mathbb{R}^3$ with connected boundary. Call Ω Lipschitz domain, with Lipschitz constant $\leq M$ (respectively, C^∞ domain), if for any $P \in \partial\Omega$ there exists $r, h > 0$ and a coordinate system in $\mathbb{R}^3$ (isometric to the usual one) $\{y_0, y_1, y_2\}$ with origin at P and a Lipschitz continuous function with Lipschitz constant $\leq M$, $\varphi : \mathbb{R}^2 \longrightarrow \mathbb{R}$ (respectively, C^∞ function), such that, if $C(r, h)$ denotes the cylinder $\{(y_1, y_2); |y_j| < r, j = 1, 2\} \times (0, h) \subset \mathbb{R}^{n+1}$, then*

$$\Omega \cap C(r, h) = \{Y = (y_0, y_1, y_2) \,;\, |y_j| < r \text{ all } 1 \leq j \leq n \text{ and } y_0 > \varphi(y_1, y_2)\},$$

$$\partial\Omega \cap C(r, h) = \{Y = (y_0, y_1, y_2); |y_j| < r \text{ all } 1 \leq j \leq n \text{ and } y_0 = \varphi(y_1, y_2)\}.$$

We shall call $\partial\Omega \cap C(r, h)$ (or more accurately, $(\partial\Omega \cap C(r, h, \varphi)))$ a patch for $\partial\Omega$.

Since $\partial\Omega$ is compact, it is always possible to cover $\partial\Omega$ with finitely many patches $(U_\nu, \varphi_\nu)_\nu$. Call such a collection an atlas for $\partial\Omega$. Also, let dS stand for the canonical surface area on $\partial\Omega$.

Let Ω be a Lipschitz domain in $\mathbb{R}^3$, and let $(U_\nu, \varphi_\nu)_\nu$ be an atlas for $\partial\Omega$. Also, fix $(\theta_\nu)_\nu$ a partition of unity subordinated to the covering $(U_\nu)_\nu$. A quaternionic-valued measure u on $\partial\Omega$ is said to belong to the Sobolev space $H^s(\partial\Omega)$ with $0 \leq s \leq 1$, if

$$(1 + |\xi|^2)^{s/2}[(\theta_\nu u) \circ z_\nu]^\wedge(\xi) \in L^2(\mathbb{R}^2)$$

for all ν, where $z_\nu(x) := (\varphi_\nu(x), x)$ and $\wedge$ is the usual Fourier transform in $\mathbb{R}^2$. Note that $H^0(\partial\Omega)$ is isomorphic to $L^2(\partial\Omega)$, the Hilbert space of square integrable, quaternionic-valued functions on $\partial\Omega$, while $H^1(\partial\Omega)$ is the space of functions $f \in L^2(\partial\Omega)$ for which $\nabla_x(\theta_\nu f)(\varphi(x), x)$ belongs to $L^2(\mathbb{R}^2)$. We endow this space with the natural norm $\|f\|_{H^1} := \|f\|_{L^2} + \sum_\nu \|\nabla_x(\theta_\nu u)(\varphi_\nu(x), x)\|_{L^2}$. Also, for $0 < s < 1$, we endow $H^s(\partial\Omega)$ with the norm

$$\|u\|_{H^s} := \left(\int_{\partial\Omega} |u|^2 dS + \int_{\partial\Omega} \int_{\partial\Omega} \frac{|u(P) - u(Q)|^2}{|P - Q|^{2+2s}} \, dS(P) dS(Q) \right)^{1/2}.$$

For a quaternionic-valued function u defined in $\Omega_\pm$, the *nontangential maximal function* u^* is given by $u^*(X) := \sup_{Y \in \Gamma_\pm(X)} |u(Y)|$. Here $\Gamma_\pm(X)$ denote the interiors of the two components (in Ω_+ and in Ω_-) of a regular family of circular, doubly truncated cones $\{\Gamma(X); X \in \partial\Omega\}$, with vertex at X, as defined in, e.g., [Ve].

In the sequel, we shall find it very useful to approximate, in a suitable sense, a given Lipschitz domain with a sequence of C^∞ domains. More specifically, we note the following Nečas-Verchota type result.

LEMMA 11.2.4

Let Ω be a Lipschitz domain with Lipschitz constant less than or equal to M. Then there exists a sequence $\{\Omega_h\}_{0<h<1}$ of subdomains of Ω such that:

1. $\overline{\Omega_h}$ is a compact subset of Ω for all h.

2. Ω_h is a C^∞ domain for any h.

3. There exists a constant $C > 0$ such that, for any h, Ω_h is a Lipschitz domain with Lipschitz constant $\leq CM$.

4. $\kappa(h) \leq C/h$, where $\kappa(h)$ denotes the least upper bound of the absolute values of the curvature of $\partial\Omega_h$.

5. There exists a Lipschitz diffeomorphism $F_h : \partial\Omega \longrightarrow \partial\Omega_h$ such that the Lipschitz constants of F_h and F_h^{-1} are uniformly bounded in h.

6. $|F_h(P) - P| \leq Ch$ for all $P \in \partial\Omega$.

7. The Jacobian of F_h is bounded from above and from below by finite strictly positive constants, and converges to 1 almost everywhere in $\partial\Omega$.

8. There exists a finite open covering $(U_\nu)_\nu$ of $\partial\Omega$ with patches, such that, for any h, $F_h(U_\nu)$ is a patch for $\partial\Omega_h$.

The proof follows from a careful inspection of the arguments of [Ve], [DV]. We note that it is also possible to approximate Ω with a family of smooth domains from the exterior part, i.e., (1) from above should be replaced by $\overline{\Omega} \subset\subset \Omega_h$ for all h.

The double-layer acoustic potential operator $\mathcal{D}$ is defined for $f \in L^2(\partial\Omega)$ by

$$\mathcal{D}f(X) := -\frac{1}{4\pi} \int_{\partial\Omega} \frac{\partial}{\partial n(Y)} \left(\frac{e^{b|X-Y|}}{|X - Y|} \right) f(Y)\, dS(Y),\ X \in \mathbb{R}^3 \setminus \partial\Omega.$$

(11.2.1)

The next two theorems follow by standard techniques from the profound results of [CMM].

THEOREM 11.2.5

(i) $(\mathcal{D}f)^* \in L^2(\partial\Omega)$ for any f in $L^2(\partial\Omega)$, and $\|(\mathcal{D}f)^*\|_{L^2} \lesssim \|f\|_{L^2}$. Also, for any function f in $H^1(\partial\Omega)$, $\|(\nabla\mathcal{D}f)^*\|_{L^2} \lesssim \|f\|_{H^1}$.

(ii)   The singular double-layer acoustic potential operator,

$$\mathcal{D}f(X) := \lim_{\substack{\epsilon \to 0 \\ \substack{|X-Y|>\epsilon \\ Y \in \partial\Omega}}} \int \frac{\partial}{\partial n(Y)} \left(\frac{e^{b|X-Y|}}{|X-Y|} \right) f(Y)\, dS(Y), \quad X \in \partial\Omega,$$

(11.2.2)

is a bounded mapping of $L^2(\partial\Omega)$ and of $H^1(\partial\Omega)$.

(iii)   For any $f \in L^2(\partial\Omega)$, one has the jump-relations

$$\lim_{\substack{Y \to X \\ Y \in \Gamma_\pm(X)}} \mathcal{D}f(Y) = \pm\frac{1}{2}f(X) + \mathcal{D}f(X), \quad \text{for a.e. } X \in \partial\Omega.$$

(iv)   If $\mathcal{D}_0$ is the usual singular double-layer potential operator for the Laplacian, then the operator $\mathcal{D} - \mathcal{D}_0$ is compact in $L^2(\partial\Omega)$. Furthermore, if $\{\Omega_h\}_h$ is a smooth approximating sequence of subdomains of Ω as in Lemma 11.2.4, then $(\mathcal{D}_h - \mathcal{D}_{0h}) \circ F_h$ converges to $\mathcal{D} - \mathcal{D}_0$ in the strong operator norm in $L^2(\partial\Omega)$; here the index h indicates that the double layer potential operator is the one corresponding to $\partial\Omega_h$.

(v)   For any $f \in L^2(\partial\Omega)$, the function $\mathcal{D}f$ satisfies the Sommerfeld radiation condition in $\mathbb{R}^3 \setminus \overline{\Omega}$.

The single layer acoustic potential operator $\mathcal{S}$ is defined for $f \in L^2(\partial\Omega)$ by

$$\mathcal{S}f(X) := -\frac{1}{4\pi} \int_{\partial\Omega} \frac{e^{b|X-Y|}}{|X-Y|} f(Y)\, dS(Y), \quad X \in \mathbb{R}^3.$$

(11.2.3)

THEOREM 11.2.6

(i)   For any $f \in L^2(\partial\Omega)$, $(\mathcal{S}f)^*$ and $(\nabla\mathcal{S}f)^*$ belong to $L^2(\partial\Omega)$ and

$$\|(\mathcal{S}f)^*\|_{L^2} + \|(\nabla\mathcal{S}f)^*\|_{L^2} \lesssim \|f\|_{L^2}.$$

Also, $\mathcal{S} : L^2(\partial\Omega) \longrightarrow H^1(\partial\Omega)$ is well defined and bounded.

(ii)   One has the trace formulae

$$\lim_{\substack{Y \to X \\ Y \in \Gamma_\pm(X)}} \mathcal{S}f(Y) = \mathcal{S}f(X), \quad \text{for a.e. } X \in \partial\Omega.$$

(iii)   Concerning the normal and tangential derivatives of $\mathcal{S}$, one has for any $f \in L^2(\partial\Omega)$,

$$\lim_{\substack{Y \to X \\ Y \in \Gamma_\pm(X)}} \frac{\partial \mathcal{S}f}{\partial n}(Y) = \mp\frac{1}{2}f(X) + \mathcal{D}^*f(X), \quad \text{for a.e. } X \in \partial\Omega,$$

where $\mathcal{D}^*$ is the formal transpose of $\mathcal{D}$ in (11.2.2), and

$$\lim_{\substack{Y \to X \\ Y \in \Gamma_-(X)}} (\nabla_T \mathcal{S}f)(Y) = \lim_{\substack{Y \to X \\ Y \in \Gamma_+(X)}} (\nabla_T \mathcal{S}f)(Y),$$

where $\nabla_T := \nabla - n(\partial/\partial n)$ is the usual tangential gradient.

(iv) $\mathcal{D}\mathcal{S} = -\mathcal{S}\mathcal{D}^*$ on $L^2(\partial\Omega)$.

(v) If $f \in L^2(\partial\Omega)$, then $\mathcal{S}f$ satisfies the Sommerfeld radiation condition in $\mathbb{R}^3 \setminus \overline{\Omega}$.

Note that "left-handed" versions of the acoustic layer potential operators can be defined as well (i.e., formal expressions as (11.2.1) and (11.2.3), respectively, except that $f(Y)$ appears in the leftmost part of the integrand), and that these versions satisfy similar results to the one discussed above.

11.3   Inverting Quaternionic-Valued Boundary Operators

This section is devoted to proving the Theorem 11.1.1 stated in the introduction. Our main tools (cf. [Ve], [Br]) are going to be the Green formula and a certain Rellich type identity for the solutions of $(\triangle - a)u = 0$.

Assume first that u is smooth up to the boundary of Ω. Integrating by parts, we get

$$\iint_\Omega \langle \nabla u, \nabla \bar{u} \rangle + \iint_\Omega u a \bar{u} = \int_{\partial\Omega} u \frac{\partial \bar{u}}{\partial n} \, dS.$$

In particular, using Lemma 11.2.1, we obtain

$$\iint_\Omega |\nabla u|^2 + |u|^2 \leq C \left| \int_{\partial\Omega} u \frac{\partial \bar{u}}{\partial n} \, dS \right|. \tag{11.3.1}$$

A similar estimate holds in Ω_- provided u decays at infinity.

Also, if θ is a smooth, compactly supported vector field in $\mathbb{R}^3$, a direct calculation yields

$$\mathrm{div} \left\{ \frac{1}{2} \langle \nabla \bar{u}, \nabla u \rangle \theta - \langle \nabla \bar{u}, \theta \rangle \nabla u \right\}$$

$$= \frac{1}{2} \sum_{j,k} \{ (\partial_j \bar{u})(\partial_j \partial_k u)\theta_k - (\partial_j \partial_k \bar{u})(\partial_j u)\theta_k \}$$

$$+ \mathcal{O}(|\nabla u|^2) + \mathcal{O}(|u||\nabla u|).$$

Consequently,

$$\mathrm{div} \left\{ \frac{1}{2} \mathrm{Re} \, \langle \nabla \bar{u}, \nabla u \rangle \theta - \mathrm{Re} \, \langle \nabla \bar{u}, \theta \rangle \nabla u \right\} = \mathcal{O}(|\nabla u|^2) + \mathcal{O}(|u||\nabla u|). \tag{11.3.2}$$

LEMMA 11.3.1

If u is a quaternionic-valued function satisfying either $(\triangle - a)u = 0$ in Ω_+ and $(\nabla u)^* \in L^2(\partial\Omega)$, or $(\triangle - a)u = 0$ in Ω_-, $(\nabla u)^* \in L^2(\partial\Omega)$, and the

Sommerfeld radiation condition in Ω_-, then

$$\int_{\partial\Omega} \left|\frac{\partial u}{\partial n}\right|^2 + |u|^2 \, dS \approx \int_{\partial\Omega} |\nabla_T u|^2 + |u|^2 \, dS. \qquad (11.3.3)$$

A similar estimate for the solutions of $u(\triangle - a) = 0$ holds as well.

PROOF   Assuming first that u is smooth up to the boundary, Gauss' divergence theorem, (11.3.1) and (11.3.2) yield

$$\left|\frac{1}{2}\int_{\partial\Omega} \operatorname{Re}\langle\nabla\bar{u}, \nabla u\rangle\langle\theta, n\rangle \, dS - \int_{\partial\Omega} \operatorname{Re}\langle\nabla\bar{u}, \theta\rangle \ \langle\nabla u, n\rangle \, dS\right|$$

$$\leq C\left|\int_{\partial\Omega} u\frac{\partial\bar{u}}{\partial n} \, dS\right|.$$

If we now specialize θ such that $\langle\theta, n\rangle \geq C > 0$ on $\partial\Omega$ (see, e.g., [Gr]), then a simple analysis of the normal and tangential components of ∇u immediately gives (11.3.3).

Using this, an approximation procedure as in Lemma 11.2.4, and Lebesgue's dominated convergence theorem, the general case easily follows. ∎

Next we are ready to present the proof of Theorem 11.1.1. The reasoning is standard by now ([Ve], [DK]), hence we only sketch its main points.

PROOF   [**Proof of Theorem 11.1.1**] It suffices to prove the theorem only for $s = 0, 1$, as the general case then follows by interpolation. Assume first that $s = 0$. Let $f \in L^2(\partial\Omega)$ and set $u_\pm := \mathcal{S}f$ in $\Omega_\pm$. Using (11.3.1) for $u_\pm$ and the fact that $\nabla_T u_- = \nabla_T u_+$ a.e. on $\partial\Omega$ (Theorem 11.2.6), we get, on account of the jump-relations for the normal derivative of the single layer potential, that

$$\|(-\tfrac{1}{2}I + D^*)f\|_{L^2} \approx \|(\tfrac{1}{2}I + D^*)f\|_{L^2} + \|\operatorname{Comp}(f)\|_{L^2},$$

where Comp stands for a (generic) compact operator on $L^2(\partial\Omega)$. In particular, by the triangle inequality,

$$\|f\| \leq C\|(\pm\tfrac{1}{2}I + D^*)f\|_{L^2} + \|\operatorname{Comp}(f)\|_{L^2},$$

uniformly for $f \in L^2(\partial\Omega)$. With this at hand, essentially the same reasoning as in [Ve, p. 585-588] completes the proof of the invertibility of $\pm\tfrac{1}{2}I + D$ on $L^2(\partial\Omega)$.

On turning our attention to the case $s = 1$, note that, by (iv) of Theorem 11.2.6, it suffices to prove the invertibility of $\mathcal{S}$ between $L^2(\partial\Omega)$ and

$H^1(\partial\Omega)$. To this end, the key element ([DK], [Ve]) is the *a priori* estimate

$$\|f\|_{L^2} \leq C\|\mathcal{S}f\|_{H^1} + \|\operatorname{Comp}(f)\|_{L^2}$$

uniformly for $f \in L^2(\partial\Omega)$ so that $\pm\frac{1}{2} + D^*$ have closed ranges on $L^2(\partial\Omega)$. Since

$$\|\mathcal{S}f\|_{H^1} \geq \|\nabla_T(\mathcal{S}f)\|_{L^2} \approx \|(\partial\mathcal{S}f)/(\partial n)\|_{L^2},$$

the conclusion follows as before.   ∎

11.4   Finite Element Spaces

The results in this section are of an abstract nature. We get started by defining the localized spaces alluded to in the introduction in precise terms.

DEFINITION 11.4.1   *Call X a finite element space localized space at the scale $\varrho > 0$ (or simply a localized space) for the Lipschitz domain $\Omega \subset \mathbb{R}^3$ if:*

1. *X is a closed vector subspace of $L^\infty(\partial\Omega)$;*

2. *There exists a finite partition $\{E^j\}_{j\in J}$ of $\partial\Omega$ each of E^j has the diameter at most ϱ, and such that, for some $C_0 > 0$, $\varrho^2 \leq C_0 dS(E^j)$ for all $j \in J$;*

3. *$a\chi_{E^j} \in X$ for any $j \in J$ and any $a \in \mathbb{H}$, where χ_{E^j} denotes the characteristic function of E^j.*

4. *For all $f \in X$, and all $j \in J$,*

$$|f(X)| \leq \frac{C_0}{dS(E^j)} \sup \left| \int_{\partial\Omega} g_1 f \, \overline{g_2} \, dS \right|, \quad X \in E^j, \qquad (11.4.1)$$

    where the supremum is taken with respect to the set of all $g_1, g_2 \in X$ which are supported in E^j and have $\|g_1\|_{L^\infty}, \|g_2\|_{L^\infty} \leq 1$.

The simplest and perhaps the most suggestive example is to think of X as being the collection of all piece-wise constant functions with respect to some finite partition of $\partial\Omega$. Another source of examples is to consider for each set E^j of a finite partition of $\partial\Omega$ a finite dimensional subspace X_j of $L^\infty(E^j)$ that contains the quaternionic constants, and finally to set

$$X := \bigoplus_{j\in J} X_j.$$

See also [AGJL]. Next, let Π denote the projection operator of $L^2(\partial\Omega)$ onto X.

LEMMA 11.4.2
Let X be a finite element space on $\partial\Omega$ localized on the scale $\varrho > 0$, and let $f \in H^s(\partial\Omega)$ for some $0 \le s \le 1$. Then

$$\|\Pi f - f\|_{L^2} \le C\varrho^s \|f\|_{H^s}.$$

PROOF Consider first the case of a Lipschitz continuous function f on $\partial\Omega$. Let P be an arbitrary point in E^j. Using (11.4.1) we have

$$
\begin{aligned}
|\Pi f(P) - f(P)| &\le \|\Pi f - f(P)\|_{L^\infty} \\
&\le \frac{C_0}{dS(E^j)} \sup \left| \int_{E^j} (f(Q) - f(P))\bar{g}(Q)\, dS(Q) \right| \\
&\le C\varrho \mathcal{M}(\nabla f)(P),
\end{aligned}
$$

where $\mathcal{M}$ is the usual Hardy-Littlewood maximal function (see, e.g., [St]). The result for $s = 1$ is then a direct consequence of the L^2-boundedness of $\mathcal{M}$. The general case is then obtained by complex interpolation between $s = 0$ and $s = 1$. ∎

Let now $\{\Omega_h\}_h$ be a sequence of approximating domains in the sense of Lemma 11.2.4 and for each $h > 0$ let X_h be a localized space at the scale $\varrho(h) > 0$. Denote by Π_h the corresponding projection operator of $L^2(\partial\Omega_h)$ onto X_h introduced in the above lemma. A remark which will be useful later on is that the norm of the operator Π_h is uniformly bounded in h.

LEMMA 11.4.3
With the above notation, if $\varrho(h) \longrightarrow 0$ as $h \longrightarrow 0$, then

$$\|\Pi_h(f \circ F_h^{-1}) - (f \circ F_h^{-1})\|_{L^2} \longrightarrow 0, \quad as \quad h \longrightarrow 0,$$

for any $f \in L^2(\partial\Omega)$.

PROOF Let $\epsilon > 0$ be given and take a continuous function g such that $\|f - g\|_{L^2} \le \epsilon$. As $0 \le \text{diam}(E_h^j) \le \varrho(h) \longrightarrow 0$, it is possible then to find $g_h \in X_h$ with $\|g_h - (g \circ F_h^{-1})\|_{L^2} \le \epsilon$ if h is small enough. Consequently,

$$
\begin{aligned}
\|\Pi_h(f \circ F_h^{-1}) \; &- (f \circ F_h^{-1})\|_{L^2} \le \|\Pi_h(f \circ F_h^{-1} - g \circ F_h^{-1})\|_{L^2} \\
&+ \|\Pi_h(g \circ F_h^{-1} - g_h)\|_{L^2} + \|g_h - (g \circ F_h^{-1})\|_{L^2} \\
&+ \|g \circ F_h^{-1} - f \circ F_h^{-1}\|_{L^2} \le C\epsilon
\end{aligned}
$$

if h is small enough. The proof is therefore complete. ∎

Following [AGJL], we make the following definition.

DEFINITION 11.4.4   Let Ω be a Lipschitz domain and let X be a localized space on $\partial\Omega$. We say that the operator T on $L^2(\partial\Omega)$ satisfies the local approximation property with constant δ, if there are some functions $\{f_j\}_{j\in J}$ from X such that

$$\sum_{j\in J}\int_{E^j}|Tf - f_j|^2 dS \leq \delta\|f\|_{L^2}^2, \quad \text{for all } f \in L^2(\partial\Omega). \tag{11.4.2}$$

Note that any operator fulfilling (11.4.2) must be bounded on $L^p(\partial\Omega)$, and conversely, that any bounded operator on $L^2(\partial\Omega)$ satisfies (11.4.2) for some δ simply by choosing $f_j = 0$ for all j. The important facts about such operators are detailed in the next proposition.

PROPOSITION 11.4.5
Let Ω be a Lipschitz domain, X a localized space for $\partial\Omega$, and let T be a bounded linear operator on $L^2(\partial\Omega)$. Then there exists a constant $\delta_0 > 0$ such that if the operators T and T^* satisfy the local approximation property with constant $0 < \delta < \delta_0$, then for any real λ for which $\lambda - T$ is invertible on $L^2(\partial\Omega)$, the operator $\Pi(\lambda - T)$ is invertible on X.
    Moreover, the operator norm of $[\Pi(\lambda - T)]^{-1}$ is equivalent to the operator norm of $(\lambda - T)^{-1}$.

PROOF   Let $f \in X$ and set $g := \Pi(\lambda - T)f \in X$. Also, let $f_j \in X$ be the functions as in Definition 11.4.4. As $g = (\lambda f + f_j) - \Pi(Tf - f_j)$, we see that for any j and any $X \in E^j$

$$|(\lambda f + f_j)(X)| \leq \frac{C}{dS(E^j)} \sup \left|\int_{\partial\Omega}(f + f_j)\bar{w}\,dS\right|$$

$$\leq CMg(X) + \frac{C}{dS(E^j)} \sup \left|\text{Re}\int_{\partial\Omega}\Pi(Tf - f_j)\bar{w}\,dS\right|$$

$$\leq CMg(X) + \left(\frac{1}{dS(E^j)}\int_{E^j}|Tf - f_j|^2 dS\right)^{1/2}.$$

Consequently, for any j,

$$\int_{E^j}|(\lambda - T)f|^2 dS \leq \int_{E^j}|Tf - f_j|^2 dS + \int_{E^j}|\lambda f + f_j|^2 dS$$

$$\leq C\int_{E^j}(Mg)^2 dS + C\int_{E^j}|Tf - f_j|^2 dS.$$

Adding up for all $j \in J$, using the L^2–boundedness of the maximal function and the local approximation property for T, we get

$$\|f\|_{L^2}^2 \leq C\|(\lambda - T)f\|_{L^2}^2 \leq C\|g\|_{L^2}^2 + C\delta\|f\|_{L^2}^2.$$

If δ is taken sufficiently small, this implies that for some $C > 0$, $\|f\|_{L^2} \leq C\|g\|_{L^2}$, i.e., $\Pi(\lambda - T)$ is injective and has closed range. A similar reasoning for $\Pi(\lambda - T^*)$ shows that $\Pi(\lambda - T)$ has a dense range, too, and the proof is complete. ∎

11.5 Local Galerkin Estimates

We are now in a position to present the proof of the Theorem 11.1.2 stated in the introduction. This will be accomplished in a series of steps. As much as possible, we shall try to keep the notation introduced thus far. Also, we shall assume that the constant C_0 from the Definition 11.4.1 is independent of the parameter $h > 0$.

Step 1. For any fixed $f \in L^2(\partial\Omega)$,

$$(D_h(f \circ F_h^{-1})) \circ F_h \longrightarrow Df \text{ in } L^2(\partial\Omega), \text{ as } h \longrightarrow 0. \tag{11.5.1}$$

The case of D_0 has been treated in [DV]. Moreover, from Theorem 11.2.5 we know that $D_h - D_{0h}$ converges to $D - D_0$ as $h \longrightarrow 0$ in the strong operator norm, hence the conclusion follows.

Step 2. If (11.1.4) holds then, for any $\delta > 0$, D_h and D_h^* have the local approximation property with constant δ.

Here our argument goes along the lines of [AGJL]. For the brevity of notation, we shall drop the index h. The basic observation is that if we pick $P_j \in E_j$ for each j, then

$$|Df(P) - Df(P_j)| \leq C(\kappa(P_j, \gamma\varrho)\gamma\varrho + \gamma^{-1})f^*(P) \tag{11.5.2}$$

(recall that $\kappa(P, r)$ has been introduced in the introduction). In turn, by taking $f_j := Df(P_j) \in X$ and using the boundedness of the maximal operator, this is readily seen to imply the local approximation property with constant δ. Note that δ can be arbitrarily small by choosing γ appropriately large.

To see (11.5.2), we decompose f as $f = u_j + v_j$ with $u_j := f$ on $B(P_j, \gamma\varrho) \cap \partial\Omega$ and zero elsewhere on $\partial\Omega$. A well-known argument now yields

$$|Dv_j(P) - Dv_j(P_j)| \leq C\varrho \int_{\partial\Omega \setminus B(P_j, \gamma\varrho)} \frac{|f(Q)|}{|P_j - Q|^3} dS(Q) \leq \frac{C}{\gamma}Mf(P_j)$$

(see also Stein [St, p. 62-63]). Furthermore, since the domain $\Omega (= \Omega_h)$ is smooth, we have

$$|\langle X - Y, n(Y)\rangle| \leq C\kappa(Y, |X - Y|)|X - Y|^2.$$

Using this, a routine estimate shows that

$$|Du_j(P)| \leq \kappa(P_j, \gamma\varrho) \int_{B(P_j,\gamma\varrho)\cap\partial\Omega} \frac{|f(Q)|}{|P-Q|} dS(Q) \leq C\kappa(P_j, \gamma\varrho)\gamma\varrho f^*(P).$$

The arguments for D_h^* are completely analogous, and therefore omitted.

Step 3. If $f \in H^s(\partial\Omega)$, $0 \leq s \leq 1$, then

$$\|(D_h(f \circ F_h^{-1})) \circ F_h - Df\|_{L^2} \leq C h^s \|f\|_{H^s}. \qquad (11.5.3)$$

For $s = 0$ the above estimate is an immediate consequence of Theorem 11.2.5. In order to conclude by an usual interpolation argument, we only need to show that, if f is a Lipschitz continuous function on $\partial\Omega$, then (11.5.3) holds for $s = 1$.

To this effect, it will be enough to show that such an estimate is valid for D_{0h}, the singular double layer potential operator for the Laplace operator on Ω_h, and for the difference $D_h - D_{0h}$. Let $(U_j)_j$ be an atlas for $\partial\Omega$ as in Lemma 11.2.4, and let $(\theta_j)_j$ be a Lipschitz continuous partition of unity subordinated to this covering. Also, take Θ_j supported on U_j and identically 1 in an open neighborhood of the support of θ_j. Then, for $f \in H^1(\partial\Omega)$, we write

$$D_{0h}(f \circ F_h^{-1})(F_h(X)) = \sum_j D_{0h}(\theta_j f \circ F_h^{-1})(F_h(X))$$

$$= \sum_j \Theta_j(X) D_{0h}(\theta_j f \circ F_h^{-1})(F_h(X))$$

$$+ \sum_j (1 - \Theta_j)(X) D_{0h}(\theta_j f \circ F_h^{-1})(F_h(X))$$

$$=: \Sigma_1 + \Sigma_2.$$

To treat a typical term from the first sum, we compute

$$-2\pi\{\Theta_j(X)D_{0h}(\theta_j f \circ F_h^{-1})(F_h(X)) - \Theta_j(X)D_0(\theta_j f)(X)\} =$$

$$\int_{\mathbb{R}^2} \Theta_j(\varphi(x), x) \frac{\langle x - y, \nabla(\theta_j f)(y)\rangle}{|x-y|^2}$$

$$\left\{ \lambda\left(\frac{\varphi(x) - \varphi(y)}{|x-y|}\right) - \lambda\left(\frac{\varphi_h(x) - \varphi_h(y)}{|x-y|}\right) \right\} dy,$$

where λ is the odd antiderivative of the mapping $t \mapsto (1+t^2)^{-3/2}$, $t \in \mathbb{R}$.

Now, we set $\varphi_h^t := \varphi + t(\varphi_h - \varphi)$ and note that φ_h^t is a Lipschitz continuous function with Lipschitz constant $\leq CM$ (recall that M is the Lipschitz constant of the domain Ω). We get

$$\lambda\left(\frac{\varphi(x) - \varphi(y)}{|x - y|}\right) - \lambda\left(\frac{\varphi_h(x) - \varphi_h(y)}{|x - y|}\right)$$
$$= \frac{\varphi(x) - \varphi_h(x) - (\varphi(y) - \varphi_h(y))}{|x - y|} \int_0^1 \frac{|x - y|^3}{[|x - y|^2 + (\varphi_h^t(x) - \varphi_h^t(y))^2]^{3/2}} dt.$$

Summarizing, we see that the sum Σ_1 can be majorized by

$$C\|\varphi - \varphi_h\|_{L^\infty} \sum_j \left|\int_0^1 \int_{\mathbb{R}^2} \frac{\langle x - y, \nabla(\theta_j f)\rangle}{[|x - y|^2 + (\varphi_h^t(x) - \varphi_h^t(y))^2]^{3/2}} dy dt\right|,$$

and since $\|\varphi - \varphi_h\|_{L^\infty} \le C h$, using the results of Coifman, McIntosh and Meyer [CMM], we infer that this part is of the right order. Further, any term of the second sum can be treated similarly, thus we arrive at the same conclusion as before. Note that in this situation the calculations are considerably simpler for the singularities are completely avoided, and we can even obtain the point-wise estimate

$$|(1-\Theta_j)(X)D_{0h}(\theta_j f \circ F_h^{-1})(F_h(X)) - (1-\Theta_j)(X)D_0(\theta_j f)(X)| \le C h \|f\|_{H^1}.$$

This takes care of D_{0h}. On turning our attention to $D_h - D_{0h}$, it is not difficult to see that the kernel of this operator is (modulo a quaternionic multiplicative constant)

$$\langle X_h - Y_h, n_h(Y_h)\rangle \int_0^1 \frac{e^{tb|X_h - Y_h|}}{|X_h - Y_h|} dt, \quad X_h, Y_h \in \partial\Omega_h. \tag{11.5.4}$$

To estimate $\|((D_h - D_{0h})(f \circ F_h^{-1})) \circ F_h - (D - D_0)f\|_{L^2}$ we proceed as before. However, the computations are more direct since the kernel (11.5.4) together with its gradient are weakly singular.

In fact, the only difficulty occurs when we have to estimate $n_h(Y_h) - n(Y)$. This is because we cannot control this difference in the uniform norm. The idea is then to write $n_h(Y_h) - n(Y)$ locally as $\nabla\varphi(y) - \nabla\varphi_h(y)$ and then to integrate by parts in order to move the derivative away from $\varphi(y) - \varphi_h(y)$. The fact that f is regular enough, i.e., $f \in H^1(\partial\Omega)$, ensures that this integration by parts works. Moreover, the integral operator obtained in this fashion will still have a weakly singular kernel, thus, the conclusion follows.

Step 4. If $f \in H^s(\partial\Omega)$, for some $0 \le s \le 1$, then

$$\|f_h - \Pi_h(f \circ F_h^{-1})\|_{L^2} \le C(h^s + \varrho(h)^s)\|f\|_{H^s}.$$

To see this, we use the last part of Proposition 11.4.5 so that we can write

$$
\begin{aligned}
\|f_h &- \Pi_h(f \circ F_h^{-1})\|_{L^2} \\
&\leq C\|\Pi_h(\tfrac{1}{2}I + D_h)f_h - \Pi_h(\tfrac{1}{2}I + D_h)\Pi_h(f \circ F_h^{-1})\|_{L^2} \\
&= C\|\Pi_h(Df) \circ F_h^{-1} - \Pi_h D_h \Pi_h(f \circ F_h^{-1})\|_{L^2} \\
&\leq C\|(Df) \circ F_h^{-1} - D_h \Pi_h(f \circ F_h^{-1})\|_{L^2} \\
&\leq C\|(Df) \circ F_h^{-1} - D_h(f \circ F_h^{-1})\|_{L^2} \\
&\quad + \|D_h(f \circ F_h^{-1}) - D_h \Pi_h(f \circ F_h^{-1})\|_{L^2} \\
&\leq C\|(Df) \circ F_h^{-1} - D_h(f \circ F_h^{-1})\|_{L^2} \\
&\quad + \|f \circ F_h^{-1} - \Pi_h(f \circ F_h^{-1})\|_{L^2} \\
&=: I + II.
\end{aligned}
$$

Now, Step 3 shows that $I \leq Ch^s\|f\|_{H^s}$, while Lemma 11.4.2 yields $II \leq C\varrho(h)^s\|f\|_{H^s}$, hence we are done.

Step 5. Here we shall finally end the proof of Theorem 11.1.2.

Note that (1) is a consequence of Proposition 11.4.5 and of Step 2. Actually, the last part of Proposition 11.4.5 also implies (2). To see (3), we write

$$
\begin{aligned}
\|f_h \circ F_h - f\|_{L^2} &\leq C\|f_h - (f \circ F_h^{-1})\|_{L^2} \\
&\leq C\|f_h - \Pi_h(f \circ F_h^{-1})\|_{L^2} \\
&\quad + \|(f \circ F_h^{-1}) - \Pi_h(f \circ F_h^{-1})\|_{L^2} \\
&\leq C\|(Df) \circ F_h^{-1} - D_h(f \circ F_h^{-1})\|_{L^2} \\
&\quad + \|(f \circ F_h^{-1}) - \Pi_h(f \circ F_h^{-1})\|_{L^2} \\
&=: A + B.
\end{aligned}
$$

To get the last inequality, we have used the estimates derived in the previous step. At this stage, Lemma 11.4.3 gives that B tends to zero as h approaches 0. Also, Step 1 implies the same thing for A. From this, the point (3) of Theorem 11.1.2 follows.

Finally, to see (4), first note that from Step 4 we have

$$
\|f_h \circ F_h - f\|_{L^p} \leq Ch^s\|f\|_{H^s}, \tag{11.5.5}
$$

uniformly for $f \in H^s(\partial\Omega)$. Next, since Theorem 11.1.1 gives that $\pm\tfrac{1}{2}I + D$ are invertible mappings of $H^s(\partial\Omega)$ for $0 \leq s \leq 1$, we have that for $g \in H^s(\partial\Omega)$, with s as before,

$$
\|f\|_{H^s} \leq C\|g\|_{H^s},
$$

thus the Theorem 11.1.2 is completely proved. ∎

A similar argument works for D^* as well.

Bibliography

[AGJL] V. Adolfsson, M. Goldberg, B. Jawerth and H. Lennerstad, *Localized Galerkin estimates for boundary integral equations on Lipschitz domains*, SIAM J. Math. Anal. **23**, (1990), 1365–1374.

[BDS] F. Brackx, R. Delanghe and F. Sommen, *Clifford Analysis*, Pitman Advanced Publ. Program, 1982.

[Br] R. Brown, *The method of layer potentials for the heat equation in Lipschitz cylinders*, Amer. J. Math. **111**, (1989), 339–379.

[CMM] R. R. Coifman, A. McIntosh and Y. Meyer, *L'intégrale de Cauchy définie un opérateur borné sur L^2 pour les courbes lipschitziennes*, Annals of Math. **116**, (1982), 361–378.

[CK] D. Colton and R. Kress, *Integral Equation Methods in Scattering Theory*, John Wiley & Sons, New York, 1983.

[CH] R. Courant and D. Hilbert, *Methods of Mathematical Physics*, Interscience, New York, 1962.

[DK] B. E. J. Dahlberg and C. E. Kenig, *Hardy spaces and the L^p-Neumann problem for Laplace's equation in a Lipschitz domain*, Ann. of Math. **125**, (1987), 437–465.

[DV] B. E. J. Dahlberg and G. Verchota, *Galerkin methods for the boundary integral equations of elliptic equations in nonsmooth domains*, Contemp. Math. **107**, (1990), 39–60.

[Gr] P. Grisvard, *Elliptic Problems in Nonsmooth Domains*, Pitman Advanced Publishing Program, 1985.

[GS] K. Gürlebeck and W. Spröessig, *Quaternionic Analysis and Elliptic Boundary Value Problems*, Birkhauser-Verlag, Basel, 1990.

[L] Lin, N. *Galerkin methods for the boundary integral equations of the Dirichlet problem of the Laplace equation in Lipschitz domains*, preprint, 1990.

[Me] Y. Meyer, *Ondelettes et opérateurs*, Hermann, Paris, 1990.

[Mi] M. Mitrea, *Clifford Wavelets, Singular Integrals, and Hardy Spaces*, Lecture Notes in Mathematics, **1575**, Springer-Verlag, New York, 1994.

[Ne] J. Nečas, *Les méthodes directes en théorie des équations élliptique*, Academia, Prague, 1967.

[St] E. M. Stein, *Singular Integrals and Differentiability Properties of Functions*, Princeton University Press, Princeton, NJ, 1970.

[Ve]   G. Verchota, *Layer potentials and regularity for the Dirichlet problem for Laplace's equation in Lipschitz domains*, Journal of Functional Analysis **59**, (1984), 572–611.

Department of Mathematics,
University of South Carolina,
Columbia, South Carolina 92908
E-mail: bj@loki.math.scarolina.edu

Department of Mathematics,
University of Minnesota,
Minneapolis, Minnesota 55455
E-mail: mitrea@math.umn.edu

Clifford Algebras, Hardy Spaces, and Compensated Compactness

Zhijian Wu

12.1 Introduction

Perhaps the most important idea for using Clifford algebras in harmonic analysis is to extend a function of n real variables monogenically (holomorphically if $n = 1$) to a function of $n + 1$ real variables with values in a complex Clifford algebra and to use the power of Clifford analysis and the associated function theory. A great amount of work has been done by many authors along these lines. We refer the reader to [GM], [LMcS] and [Mc] and references therein. In this paper, by using Clifford algebra, we get another two factorizations for functions in the Hardy space $H^1(\mathbb{R}^n)$. We find links between our results, the main result in [CRW] and the div-curl result about the compensated compactness in [CLMS], thereby giving different proofs to those results. Applications on the regularity of certain nonlinear quantities in PDE are also indicated.

Let us outline the main motivation of this paper (the notations and terminologies will be defined later). Suppose $F = f + \sum_{j=1}^{n} \mathcal{R}_j(f)e_j \in L^p(\mathbb{R}^n)$ and $G = g + \sum_{j=1}^{n} \mathcal{R}_j(g)e_j \in L^q(\mathbb{R}^n)$ are two monogenic functions (here $\mathcal{R}_j$, $1 \leq j \leq n$ are the Riesz transforms on $\mathbb{R}^n$). Then the product FG can be written as

$$fg - \sum_{j=1}^{n} \mathcal{R}_j(f)\mathcal{R}_j(g) + \sum_{j=1}^{n} \left(f\mathcal{R}_j(g) + g\mathcal{R}_j(f)\right) e_j$$
$$+ \sum_{j<k} \left(\mathcal{R}_j(f)\mathcal{R}_k(g) - \mathcal{R}_j(g)\mathcal{R}_k(f)\right) e_{jk} .$$

Therefore, corresponding to the product of two monogenic functions, there are three natural sets of components:

$$I_0 = \{fg - \mathcal{H}(f) \cdot \mathcal{H}(g) : f \in L^p(\mathbb{R}^n), g \in L^q(\mathbb{R}^n)\} ;$$

0-8493-8481-8/96/$0.00+$.50
© 1996 by CRC Press

$$I_1 = \{f\mathcal{R}_j(g) + g\mathcal{R}_j(f) : f \in L^p(\mathbb{R}^n), g \in L^q(\mathbb{R}^n), 1 \le j \le n\};$$

$$I_2 = \{\mathcal{R}_j(f)\mathcal{R}_k(g) - \mathcal{R}_j(g)\mathcal{R}_k(f) : f \in L^p(\mathbb{R}^n), g \in L^q(\mathbb{R}^n), 1 \le j < k \le n\}.$$

If $n = 1$, the set I_2 is empty and the sets I_0 and I_1 are the same.

The main result in [CRW] says that the set $I_1(1/p + 1/q = 1)$ spans $H^1(\mathbb{R}^n)$. This motivates us to show that I_0 and I_2 also span $H^1(\mathbb{R}^n)$ (if $n > 1$, the three sets are different from one another).

The main results in this paper are as follows.

THEOREM 12.1.1
Suppose m is a positive integer, $1 < p, q < \infty$ and $\frac{1}{p} + \frac{1}{q} = 1$. Then the bilinear form

$$\left\langle fg - \sum_{j_1, j_2, \cdots, j_m = 1}^{n} \mathcal{R}_{j_1} \mathcal{R}_{j_2} \cdots \mathcal{R}_{j_m}(f) \mathcal{R}_{j_1} \mathcal{R}_{j_2} \cdots \mathcal{R}_{j_m}(g), b \right\rangle_{L^2(\mathbb{R}^n)} \quad (12.1.1)$$

is bounded on $L^p(\mathbb{R}^n) \times L^q(\mathbb{R}^n)$, if and only if b is in BMO.

THEOREM 12.1.2
Suppose b is a function on $\mathbb{R}^n$, $1 < p, q < \infty$ and $\frac{1}{p} + \frac{1}{q} = 1$. Then the bilinear forms

$$\left\langle \mathcal{R}_j(f)\mathcal{R}_k(g) - \mathcal{R}_j(g)\mathcal{R}_k(f), b \right\rangle_{L^2(\mathbb{R}^n)}, \quad 1 \le j < k \le n$$

are bounded on $L^p(\mathbb{R}^n) \times L^q(\mathbb{R}^n)$, if and only if b is in BMO.

The first consequence is immediate: together with the result of Fefferman and Stein on the duality between $H^1(\mathbb{R}^n)$ and BMO, we get two new factorizations for functions in $H^1(\mathbb{R}^n)$. More precisely, every function f in $H^1(\mathbb{R}^n)$ can be written as

$$f = \sum_{k=1}^{\infty} \left(g_k h_k - \sum_{j_1, j_2, \cdots, j_m = 1}^{n} \mathcal{R}_{j_1} \mathcal{R}_{j_2} \cdots \mathcal{R}_{j_m}(g_k) \mathcal{R}_{j_1} \mathcal{R}_{j_2} \cdots \mathcal{R}_{j_m}(h_k) \right),$$

or

$$f = \sum_{l=1}^{\infty} \sum_{1 \le j < k \le n} (\mathcal{R}_j(g_{jkl})\mathcal{R}_k(h_{jkl}) - \mathcal{R}_j(h_{jkl})\mathcal{R}_k(g_{jkl}))$$

with

$$\|f\|_{H^1} \asymp \sum_{k=1}^{\infty} \|g_k\|_{L^2(\mathbb{R}^n)} \|h_k\|_{L^2(\mathbb{R}^n)}, \text{ or}$$

$$\|f\|_{H^1} \asymp \sum_{l=1}^{\infty} \sum_{1 \le j < k \le n} \|g_{jkl}\|_{L^2(\mathbb{R}^n)} \|h_{jkl}\|_{L^2(\mathbb{R}^n)},$$

respectively. If we accept these factorizations in $H^1(\mathbb{R}^n)$, then we have new proofs for Fefferman's result.

The other consequences are that by using Theorems 12.1.1 and 12.1.2, we get new proofs of the main theorem in [CRW] and the div-curl theorem in [CLMS], and as an application, we get $H^1(\mathbb{R}^n)$ regularity for certain nonlinear quantities in PDE.

The next section contains the preliminaries. In Section 12.3, we study the components of the product of two monogenic functions and relations between these components. We complete the proofs of our main results in Section 12.4. Applications in PDE are gathered in Section 12.5.

Throughout this paper, the letter "C" denotes a positive constant which may vary at each occurrence but is independent of the essential variables. "$X \asymp Y$" means $\frac{1}{C}|X| \leq |Y| \leq C|X|$. We always assume $1 < p, q < \infty$ and $\frac{1}{p} + \frac{1}{q} = 1$. We will use $\langle \cdot, \cdot \rangle$ for the pairing in $L^2(\mathbb{R}^n)$ and $\|\cdot\|_r$ ($r \geq 1$) for the norm in $L^r(\mathbb{R}^n)$.

Acknowledgment

The author thanks Professor Alan McIntosh for his friendship and many invaluable personal communications, which made the author realize the importance of using Clifford algebra in analysis. The author also thanks Professor John Ryan and Dr. Chun Li for their help in improving this paper.

12.2 Preliminaries

Let $\mathbb{C}_{(n)}$ be a 2^n–dimensional algebra on $\mathbb{C}$ defined as follows: the standard basis of $\mathbb{C}_{(n)}$ is $\{\mathbf{e}_S\}$, where S is any subset of $\{1, 2, \cdots, n\}$ with the properties

$$\mathbf{e}_\emptyset = \mathbf{e}_0 = 1,$$

$$\mathbf{e}_j^2 = -1, \quad 1 \leq j \leq n \quad \text{and} \quad \mathbf{e}_j \mathbf{e}_k = -\mathbf{e}_k \mathbf{e}_j = \mathbf{e}_{jk}, \quad 1 \leq j < k \leq n,$$

$$\mathbf{e}_{j_1} \mathbf{e}_{j_2} \cdots \mathbf{e}_{j_s} = \mathbf{e}_S, \quad 1 \leq j_1 < j_2 < \cdots < j_s \leq n \text{ and } S = \{j_1, \cdots, j_s\},$$

$$\overline{\mathbf{e}_j} = -\mathbf{e}_j, \quad \overline{\mathbf{e}_R \mathbf{e}_S} = \overline{\mathbf{e}_S}\,\overline{\mathbf{e}_R}.$$

Suppose that $\lambda = \sum\limits_S \lambda_S \mathbf{e}_S$, $\mu = \sum\limits_S \mu_S \mathbf{e}_S \in \mathbb{C}_{(n)}$, with $\{\lambda_S\}$ and $\{\mu_S\}$ are sets of complex numbers. The product and the dot product of λ and μ are defined, respectively, by

$$\lambda\mu = \sum_{S,R} \lambda_S \mu_R \mathbf{e}_S \mathbf{e}_R \quad \text{and} \quad \lambda \cdot \mu = \sum_S \lambda_S \overline{\mu_S} \,.$$

The magnitude of λ is $|\lambda| = (\lambda \cdot \lambda)^{\frac{1}{2}}$. We observe that $|\lambda\mu| \leq C |\lambda| |\mu|$, where C is a positive constant dependent only on n.

Suppose F is a function which takes value in $\mathbb{C}_{(n)}$. We can write

$$F = \sum_S f_S \mathbf{e}_S,$$

where $\{f_S\}$, the components of F, are complex valued functions. We say a Clifford valued function F on $\mathbb{R}^n$ is in $L^r(\mathbb{R}^n)$ if its components are in $L^r(\mathbb{R}^n)$, or equivalently if $|F|$ is in $L^r(\mathbb{R}^n)$.

In this paper, we will always use the upper case letter for Clifford valued function (e.g., F, G, etc.) and lower case letter for complex valued function (e.g., f, g, etc.).

Identify a point $x = (x_1, \cdots, x_n) \in \mathbb{C}^n$ by $x = x_1 \mathbf{e}_1 + \cdots + x_n \mathbf{e}_n$ and a point $(x, y) \in \mathbb{C}^{n+1}$ by $(x, y) = y + x_1 \mathbf{e}_1 + \cdots + x_n \mathbf{e}_n$. We have $|x|^2 = x \cdot x$ and $(x, y) \cdot (x, y) = |x|^2 + |y|^2$.

Define the Dirac operator by

$$\mathcal{D} = \sum_{j=0}^{n} \frac{\partial}{\partial x_j} \mathbf{e}_j, \quad (x_0 = y).$$

Suppose $F = \sum_S f_S \mathbf{e}_S$. Then formulas for $\mathcal{D}F$ and $F\mathcal{D}$ are

$$\mathcal{D}F = \sum_{0}^{n} \sum_S \frac{\partial f_S}{\partial x_j} \mathbf{e}_j \mathbf{e}_S \quad \text{and} \quad F\mathcal{D} = \sum_{0}^{n} \sum_S \frac{\partial f_S}{\partial x_j} \mathbf{e}_S \mathbf{e}_j \,.$$

A Clifford valued function F on $\mathbb{R}^{n+1}_+$ is said to be left (or right) monogenic if $\mathcal{D}F = 0$ (or $F\mathcal{D} = 0$). We simply say F is monogenic if it is both left and right monogenic.

The Fourier transform of a function F on $\mathbb{R}^n$ is defined by

$$\hat{F}(\xi) = \int_{\mathbb{R}^n} F(x) e^{-ix \cdot \xi} dx, \qquad \xi \in \mathbb{R}^n.$$

For $1 \leq k \leq n$, the kth Riesz transform is defined by

$$\widehat{\mathcal{R}_k f}(\xi) = i \frac{\xi_k}{|\xi|} \hat{f}(\xi).$$

A standard result says (see [S]) that Riesz transforms, are bounded on $L^p(\mathbb{R}^n)$.

For $F \in L^p(\mathbb{R}^n)$, we use $F(x, y)$ to mean the harmonic extension of F onto $\mathbb{R}^{n+1}_+$. More precisely

$$F(x, y) = \mathcal{P}_y(F)(x) = p_y * F(x),$$

where $\mathcal{P}_y$ is called Poisson transform and $p_y(x) = c_n y / \left(|x|^2 + y^2 \right)^{(n+1)/2}$ is the Poisson kernel with $\hat{p}_y(\xi) = e^{-y|\xi|}$.

Suppose $F \in L^p(\mathbb{R}^n)$. It is well known that the harmonic extension of F onto $\mathbb{R}^{n+1}_+$ is left, or right monogenic if and only if

$$\xi^- \hat{F}(\xi) = 0, \quad \text{or} \quad \hat{F}(\xi) \xi^- = 0. \tag{12.2.1}$$

Here $\xi^- = \sum_{j=1}^{n} i\xi_j \mathbf{e}_j - |\xi|$. We will simply say that F ($\in L^p(\mathbb{R}^n)$) is left, or right monogenic if the first, or the second formula of (12.2.1) holds, respectively. Define $\mathcal{H} = \sum_{j=1}^{n} \mathcal{R}_j \mathbf{e}_j$ (if $n = 1$, $\mathcal{H}$ is the Hilbert transform). For any $f \in L^p(\mathbb{R}^n)$ it is easy to check, by using (12.2.1), that the function

$$F = f + \mathcal{H}(f)$$

is monogenic and is in $L^p(\mathbb{R}^n)$ with $\|F\|_p \asymp \|f\|_p$.

The Hardy space $H^r(\mathbb{R}^n)(r \geq 1)$ is defined by

$$H^r(\mathbb{R}^n) = \{ f \in L^r(\mathbb{R}^n) : \mathcal{H}(f) \in L^r(\mathbb{R}^n) \}.$$

Suppose $E = (E_1, E_2, \cdots, E_n)$ and $B = (B_1, B_2, \cdots, B_n)$ are vector fields on $\mathbb{R}^n$. We say E is divergence free, denoted by $div(E) = 0$, and B is curl free, denoted by $curl(B) = 0$, if $\sum_{j=1}^{n} \frac{\partial E_j}{\partial x_j} = 0$ and $\frac{\partial B_j}{\partial x_k} - \frac{\partial B_k}{\partial x_j} = 0$, $1 \leq j, k \leq n$, respectively, in the sense of distributions.

For easy reference, we record the following two results from [CRW] and [CLMS], respectively.

THEOREM 12.2.1 (R. Coifman, R. Rochberg and G. Weiss)
Suppose b is a function on $\mathbb{R}^n$ and $1 < p < \infty$. Then the commutators $[\mathcal{M}_b, \mathcal{R}_k], k = 1, 2, \cdots, n$ are bounded on $L^p(\mathbb{R}^n)$ if and only if b is in BMO. Moreover, every f in $H^1(\mathbb{R}^n)$ can be written as

$$f = \sum_{j=1}^{\infty} \sum_{k=1}^{n} \{ g_{jk} \mathcal{R}_k(h_{jk}) + h_{jk} \mathcal{R}_k(g_{jk}) \}$$

with

$$\|f\|_{H^1} \asymp \sum_{j=1}^{\infty} \sum_{k=1}^{n} \|g_{jk}\|_2 \|h_{jk}\|_2 .$$

THEOREM 12.2.2   (R. Coifman, P. Lions, Y. Meyer and S. Semmes)
Suppose $1 < p, q < \infty$ and $\frac{1}{p} + \frac{1}{q} = 1$. If $E \in L^p(\mathbb{R}^n)^n, B \in L^q(\mathbb{R}^n)^n$ and

$$div(E) = 0, \quad curl(B) = 0,$$

then $E \cdot B$ is in $H^1(\mathbb{R}^n)$. Moreover, every function $f \in H^1$ can be written as

$$f = \sum_{j=1}^{\infty} \lambda_j w_j \quad with \quad \sum_{j=1}^{\infty} |\lambda_j| < \infty.$$

Here $w_j \in W (\forall j \geq 1)$ and W is the subset of $H^1(\mathbb{R}^n)$ formed by the functions $w = E \cdot B$ where $E, B \in L^2(\mathbb{R}^n)^n$, $\|E\|_2, \|B\|_2 \leq 1$ and $div(E) = 0, curl(B) = 0$ in $\mathbb{R}^n$.

12.3   Components of the product of two monogenic functions

Suppose $F \in L^p(\mathbb{R}^n)$ and $G \in L^q(\mathbb{R}^n)$ are right and left monogenic functions, respectively. It is clear that the product FG is in $L^1(\mathbb{R}^n)$. The following theorem says we can expect more.

THEOREM 12.3.1
Suppose $F \in L^p(\mathbb{R}^n)$ and $G \in L^q(\mathbb{R}^n)$ are right and left monogenic functions, respectively. Then each component of FG is in $H^1(\mathbb{R}^n)$.

REMARK   By the later discussion in this section and Lemma 12.4.1 in Section 12.4, we may reduce Theorem 12.3.1 to Theorem 12.2.1. Here we will give an easy proof by using Green's theorem and the properties of monogenic functions. This method also gives a uniform approach to the "if" parts of Theorems 12.2.1, 12.1.1 and 12.1.2 and the first part of Theorem 12.2.2.

We will use the following notations frequently.

$$\nabla_x = \left(\frac{\partial}{\partial x_1}, \frac{\partial}{\partial x_2}, \cdots, \frac{\partial}{\partial x_n} \right) \quad and \quad \nabla = \left(\nabla_x, \frac{\partial}{\partial y} \right);$$

$$\triangle = \nabla \cdot \nabla = \nabla_x \cdot \nabla_x + \frac{\partial^2}{\partial y^2} = \sum_{j=1}^{n} \frac{\partial^2}{\partial x_j{}^2} + \frac{\partial^2}{\partial y^2};$$

$$\nabla\Phi \cdot \nabla\Psi = \frac{\partial\Phi}{\partial y}\frac{\partial\Psi}{\partial y} + \sum_{j=1}^{n} \frac{\partial\Phi}{\partial x_j}\frac{\partial\Psi}{\partial x_j} \quad and \quad |\nabla\Phi| = |\nabla\Phi \cdot \nabla\Phi|^{\frac{1}{2}}.$$

To prove Theorem 12.3.1, we need the following standard results.

LEMMA 12.3.2
Let $F \in C_0^\infty(\mathbb{R}^n)$. Then

$$\int_{\mathbb{R}_+^{n+1}} y \, |\nabla F(x,y)|^2 \, dxdy \leq C \, \|F\|_2^2 \, .$$

PROOF For real valued F, the lemma can be proved easily by using Green's theorem (see [S]) or the Fourier transform on the variable x. The general case therefore follows from this special case. ∎

LEMMA 12.3.3
Suppose b is a complex valued function on $\mathbb{R}^n$ and is in BMO. Then for any $G \in C_0^\infty(\mathbb{R}^n)$

$$\int_{\mathbb{R}_+^{n+1}} |G(x,y)|^2 \, |\nabla b(x,y)|^2 \, y \, dxdy \leq C \, \|b\|_{BMO}^2 \, \|G\|_2^2 \, .$$

PROOF For real valued G, the lemma is true (see for example [T]). The proof of the general case is then trivial. ∎

PROOF [**Proof of Theorem 12.3.1**] Suppose $b \in BMO$ is a complex valued function. By Fefferman's duality theorem, we only need to show that $|\langle FG, b \rangle|$ is bounded by $C \, \|b\|_{BMO} \, \|F\|_p \, \|G\|_q$ for all F, $G \in C_0^\infty(\mathbb{R}^n)$ (note that $C_0^\infty(\mathbb{R}^n)$ is dense in $L^r(\mathbb{R}^n)$). In the following, the use of Green's theorem and integration by parts, need that the associate functions in (x,y) goes to zero when $y \to \infty$ or $|x| \to \infty$. We omit the details, because it is standard (see for example [S] or [T]).

By Green's theorem, we have

$$\langle FG, b \rangle = \int_{\mathbb{R}_+^{n+1}} y \triangle \left(F(x,y)G(x,y)b(x,y) \right) \, dxdy \, .$$

Here $F(x,y)$, $G(x,y)$ and $b(x,y)$ are the harmonic extensions of F, G and b, respectively. Using the harmonicity of these functions and the product rule of derivatives, we get

$$\triangle (FGb) = \triangle (FG) \, b + 2 \nabla (FG) \cdot \nabla b = 2 \left(\nabla F \cdot \nabla G \right) b + 2 \nabla (FG) \cdot (\nabla b) \, ,$$

Therefore

$$\langle FG, b \rangle = 2 \int_{\mathbb{R}_+^{n+1}} \nabla (FG) \cdot \nabla b \, y \, dxdy + 2 \int_{\mathbb{R}_+^{n+1}} \left(\nabla F \cdot \nabla G \right) b \, y dxdy \, .$$

$$(12.3.1)$$

Since $F(x, y)$ and $G(x, y)$ are right and left monogenic on $\mathbb{R}^{n+1}_+$, we have the following identities

$$\frac{\partial F}{\partial y} = -\sum_{j=1}^{n} \frac{\partial F}{\partial x_j} \mathbf{e}_j \quad \text{and} \quad \frac{\partial G}{\partial y} = -\sum_{j=1}^{n} \mathbf{e}_j \frac{\partial G}{\partial x_j}.$$

Therefore

$$\int_{\mathbb{R}^{n+1}_+} (\nabla F \cdot \nabla G)\, b\, y dx dy = \int_{\mathbb{R}^{n+1}_+} \frac{\partial F}{\partial y} \frac{\partial G}{\partial y} b\, y dx dy$$

$$+ \int_{\mathbb{R}^{n+1}_+} (\nabla_x F \cdot \nabla_x G)\, b\, y dx dy$$

$$= \int_{\mathbb{R}^{n+1}_+} \sum_{j,k=1}^{n} \frac{\partial F}{\partial x_j} \mathbf{e}_j \mathbf{e}_k \frac{\partial G}{\partial x_k} b\, y dx dy$$

$$+ \int_{\mathbb{R}^{n+1}_+} \sum_{j=1}^{n} \frac{\partial F}{\partial x_j} \frac{\partial G}{\partial x_j} b\, y dx dy.$$

Integration by parts in x allows us to continue the above computation to

$$= -\int_{\mathbb{R}^{n+1}_+} \sum_{j,k=1}^{n} F \mathbf{e}_j \mathbf{e}_k \frac{\partial^2 G}{\partial x_k \partial x_j} b\, y dx dy + \int_{\mathbb{R}^{n+1}_+} \sum_{j=1}^{n} F \mathbf{e}_j \frac{\partial G}{\partial y} \frac{\partial b}{\partial x_j} y dx dy$$

$$- \int_{\mathbb{R}^{n+1}_+} \sum_{j=1}^{n} F \frac{\partial^2 G}{\partial x_j{}^2} b\, y dx dy - \int_{\mathbb{R}^{n+1}_+} F \sum_{j=1}^{n} \frac{\partial G}{\partial x_j} \frac{\partial b}{\partial x_j} y dx dy.$$

Note that

$$\sum_{j,k=1}^{n} \mathbf{e}_j \mathbf{e}_k \frac{\partial^2 G}{\partial x_k \partial x_j} = -\sum_{j=1}^{n} \frac{\partial^2 G}{\partial x_j{}^2} \quad \text{and} \quad \sum_{j=1}^{n} \mathbf{e}_j \frac{\partial b}{\partial x_j} = \mathcal{D}b - \frac{\partial b}{\partial y}.$$

Refining the result of the above computation, we get

$$\int_{\mathbb{R}^{n+1}_+} (\nabla F \cdot \nabla G)\, b\, y dx dy = \int_{\mathbb{R}^{n+1}_+} F(\mathcal{D}b) \frac{\partial G}{\partial y} y dx dy$$

$$- \int_{\mathbb{R}^{n+1}_+} F \nabla G \cdot \nabla b\, y dx dy.$$

Similarly we have

$$\int_{\mathbb{R}^{n+1}_+} (\nabla F \cdot \nabla G)\, b\, y dx dy = \int_{\mathbb{R}^{n+1}_+} \frac{\partial F}{\partial y} (\mathcal{D}b)\, G\, y dx dy$$

$$- \int_{\mathbb{R}^{n+1}_+} \nabla F \cdot (G \nabla b)\, y dx dy.$$

By these two formulas, together with (12.3.1), we get

$$\langle FG, b \rangle = \int_{\mathbb{R}^{n+1}_+} \left(\nabla (FG) \cdot \nabla b + F(Db)\frac{\partial G}{\partial y} + \frac{\partial F}{\partial y}(Db)G \right) y \, dx \, dy.$$

$$(12.3.2)$$

If $p = 2$, by Schwarz inequality and Lemmas 12.3.2 and 12.3.3, we get

$$|\langle FG, b \rangle| \le C \int_{\mathbb{R}^{n+1}_+} (|\nabla F| |G| + |F| |\nabla G|) |\nabla b| \, y \, dx \, dy$$

$$\le C \|b\|_{BMO} \|F\|_2 \|G\|_2.$$

For general p, we need a deep result from harmonic analysis (which involves the the Littlewood-Paley theory, see, for example, [CM] or [Chr]) which views each term on the right-hand side of (12.3.2) as a paraproduct with symbol b, and says the magnitude of these paraproducts are bounded by $C \|b\|_{BMO} \|F\|_p \|G\|_q$. The proof is complete. ∎

A direct consequence of Theorem 12.3.1 follows:

COROLLARY 12.3.4
Suppose $f \in L^p(\mathbb{R}^n)$ and $g \in L^q(\mathbb{R}^n)$ are complex valued functions. Then

(i) *$fg - \mathcal{H}(f) \cdot \mathcal{H}(g)$ is in H^1;*

(ii) *$f\mathcal{R}_j(g) + g\mathcal{R}_j(f)$, $j = 1, 2, \cdots, n$ are in H^1;*

(iii) *$\mathcal{R}_j(f)\mathcal{R}_k(g) - \mathcal{R}_k(f)\mathcal{R}_j(g)$, $j, k = 1, 2, \cdots, n$ are in H^1.*

REMARK (i) By Theorems 12.1.1, 12.1.1 and 12.1.2, we know that the three parts of Corollary 12.3.4 are equivalent.

(ii) The "if" part of Theorem 12.1.2 is just the part 3 of Corollary 12.3.4.

(iii) The "if" part of Theorem 12.1.1 is a consequence of Fefferman's duality theorem and the second part of Corollary 12.3.4, because the commutator $[\mathcal{M}_b, \mathcal{R}_k]$ can be regarded as the bilinear form

$$\langle [\mathcal{M}_b, \mathcal{R}_k](f), g \rangle = \langle f\mathcal{R}_k(g) + g\mathcal{R}_k(f), b \rangle, \quad \forall f, g \in C_0^\infty(\mathbb{R}^n).$$

$$(12.3.3)$$

PROOF [**Proof of the "if" part of Theorem 12.1.1**] Suppose b is in BMO. We want to show that the bilinear form (12.1.1) is bounded on $L^p(\mathbb{R}^n) \times L^q(\mathbb{R}^n)$. By Fefferman's duality theorem, we only need to show that for $f \in L^p(\mathbb{R}^n)$ and $g \in L^q(\mathbb{R}^n)$ the function

$$h_m = fg - \sum_{j_1, j_2, \cdots, j_m = 1}^{n} \mathcal{R}_{j_1} \mathcal{R}_{j_2} \cdots \mathcal{R}_{j_m}(f) \mathcal{R}_{j_1} \mathcal{R}_{j_2} \cdots \mathcal{R}_{j_m}(g)$$

is in $H^1(\mathbb{R}^n)$. If $m = 1$, then by the first part of Corollary 12.3.4, we have $h_1 \in H^1(\mathbb{R}^n)$. By induction, suppose h_l is in $H^1(\mathbb{R}^n)$. Then

$$
h_{l+1} = h_l + \sum_{j_1, j_2, \cdots, j_l = 1}^{n} \{ f_{j_1, j_2, \cdots, j_l} g_{j_1, j_2, \cdots, j_l}
$$

$$
- \sum_{k=1}^{n} \mathcal{R}_k \left(f_{j_1, j_2, \cdots, j_l} \right) \mathcal{R}_k \left(g_{j_1, j_2, \cdots, j_l} \right) \},
$$

where $f_{j_1, j_2, \cdots, j_l} = \mathcal{R}_{j_1} \mathcal{R}_{j_2} \cdots \mathcal{R}_{j_l}(f)$ and $g_{j_1, j_2, \cdots, j_l} = \mathcal{R}_{j_1} \mathcal{R}_{j_2} \cdots \mathcal{R}_{j_l}(g)$ are clearly in $L^p(\mathbb{R}^n)$ and $L^q(\mathbb{R}^n)$, respectively. By the first part of Corollary 12.3.4, we know each of the terms in $\{\ \}$ of the above formula is in $H^1(\mathbb{R}^n)$. Therefore h_{l+1} is in $H^1(\mathbb{R}^n)$. ∎

To see that the first part of Theorem 12.2.2 is also a consequence of Corollary 12.3.4, we study the relation between the set I_2 and the quantity $E \cdot B$ in Theorem B. We note first that $\mathcal{R}_j(f)\mathcal{R}_k(g) - \mathcal{R}_k(f)\mathcal{R}_j(g)$ can be written as $E \cdot B$ with $div(E) = 0$ and $curl(B) = 0$. In fact let $\mathcal{E}_{jk}$ be the operator defined by $\mathcal{E}_{jk}(f) = \mathcal{R}_k(f)\mathbf{e}_j - \mathcal{R}_j(f)\mathbf{e}_k$. Then

$$
\widehat{div(\mathcal{E}_{jk})}(\xi) = i\xi_j \widehat{\mathcal{R}_k(f)}(\xi) - i\xi_k \widehat{\mathcal{R}_j(f)}(\xi) = -\frac{\xi_j \xi_k}{|\xi|} \hat{f}(\xi) + \frac{\xi_k \xi_j}{|\xi|} \hat{f}(\xi) = 0.
$$

Therefore $div\left(\mathcal{E}_{jk}(f) \right) = 0$. Let $B = \mathcal{H}(g)$. Then

$$
\mathcal{R}_j(f)\mathcal{R}_k(g) - \mathcal{R}_k(f)\mathcal{R}_j(g) = \mathcal{E}_{jk}(f) \cdot B.
$$

That $curl(B) = 0$ follows by the use of the Fourier transform:

$$
\widehat{\frac{\partial B_j}{\partial x_k}}(\xi) - \widehat{\frac{\partial B_k}{\partial x_j}}(\xi) = i\xi_k \widehat{\mathcal{R}_j(g)}(\xi) - i\xi_j \widehat{\mathcal{R}_k(g)}(\xi) = 0, \quad 1 \le j, k \le n.
$$

The following theorem establishes a relation between $div(E) = 0$ and the operators $\mathcal{E}_{jk}$.

THEOREM 12.3.5
Suppose $E = \sum_{j=1}^{n} E_j \mathbf{e}_j \in L^p(\mathbb{R}^n)$ and $div(E) = 0$. Then E can be represented as

$$
E = \sum_{j<k} \mathcal{E}_{jk}(f_{jk}),
$$

where $f_{jk} = \mathcal{R}_j(E_k) - \mathcal{R}_k(E_j)$. Moreover

$$
\|E\|_p^p \asymp \sum_{j<k} \|f_{jk}\|_p^p \asymp \sum_{j<k} \|\mathcal{E}_{jk}(f_{jk})\|_p^p.
$$

And if $p = 2$, the first "$\asymp$" sign above can be replaced by the "$=$" sign.

PROOF We first note that $div(E) = 0$ implies $\sum_{j=1}^n \mathcal{R}_j(E_j) = 0$. In fact

$$\sum_{j=1}^n \widehat{\mathcal{R}_j(E_j)}(\xi) = i\sum_{j=1}^n \frac{\xi_j}{|\xi|}\hat{E}_j(\xi) = \frac{1}{|\xi|}\widehat{div(E)}(\xi)\,.$$

A straightforward computation yields

$$\sum_{j<k}\mathcal{E}_{jk}(f_{jk}) = \sum_{j<k}\{\mathcal{R}_k(f_{jk})\mathbf{e}_j + \mathcal{R}_j(f_{kj})\mathbf{e}_k\}$$

$$= \sum_{j,k=1}^n \mathcal{R}_k(f_{jk})\mathbf{e}_j$$

$$= \sum_{j,k=1}^n \left(\mathcal{R}_k\mathcal{R}_j(E_k) - \mathcal{R}_k\mathcal{R}_k(E_j)\right)\mathbf{e}_j$$

$$= \mathcal{H}\left(\sum_{k=1}^n \mathcal{R}_k(E_k)\right) - \left(\sum_{k=1}^n \mathcal{R}_k\mathcal{R}_k\right)(E)$$

$$= E\,.$$

Here we use the identity $\sum_{k=1}^n \mathcal{R}_k\mathcal{R}_k = -I$ to get the last equal sign. Therefore

$$\|E\|_p^p \le C\sum_{j<k}\|\mathcal{E}_{jk}(f_{jk})\|_p^p$$

$$\le C\sum_{j<k}\|f_{jk}\|_p^p \le C\sum_{j<k}\left(\|E_j\|_p^p + \|E_k\|_p^p\right) \le C\|E\|_p^p\,.$$

If $p = 2$, we have

$$\sum_{j<k}\|f_{jk}\|_2^2 = \frac{1}{2}\sum_{j,k=1}^n \|\mathcal{R}_j(E_k) - \mathcal{R}_k(E_j)\|_2^2$$

$$= \frac{1}{2}\sum_{j,k=1}^n \{\|\mathcal{R}_j(E_k)\|_2^2 + \|\mathcal{R}_k(E_j)\|_2^2\} - \sum_{j,k=1}^n \langle \mathcal{R}_j(E_k), \mathcal{R}_k(E_j)\rangle$$

$$= \sum_{k=1}^n \|E_k\|_2^2 - \sum_{j,k=1}^n \langle \mathcal{R}_k(E_k), \mathcal{R}_j(E_j)\rangle\,.$$

Since $div(E) = 0$ implies $\sum_{j=1}^n \mathcal{R}_j(E_j) = 0$, we get

$$\sum_{j<k}\|f_{jk}\|_2^2 = \sum_{k=1}^n \|E_k\|_2^2 = \|E\|_2^2\,.$$

The proof is complete. ∎

REMARK The geometric property of Theorem 12.3.5 is the following: $div(E) = 0$ means that $\hat{E}(\xi)$ is perpendicular to ξ. Therefore, $\hat{E}(\xi)$ can be represented as

$$\hat{E}(\xi) = \sum_{j,k=1}^{n} a_{jk}(\xi) \left(\xi_k \mathbf{e}_j - \xi_j \mathbf{e}_k \right).$$

The following corollary shows that the first part of Theorem 12.2.2 is a consequence of the third part of Corollary 12.3.4.

COROLLARY 12.3.6
Suppose $E \in L^p(\mathbb{R}^n)^n$, $B \in L^q(\mathbb{R}^n)^n$ and $div(E) = 0$ and $curl(B) = 0$. Then $E \cdot B$ can be written as a finite linear combination of functions from the set I_2.

PROOF Since $curl(B) = 0$, there is a potential function $\tilde{g}$ (in fact $\hat{\tilde{g}}(\xi) = i \sum_{j=1}^{n} \frac{\xi_j}{|\xi|^2} \hat{B}_j(\xi)$) such that $\nabla_x \tilde{g} = B$. Let $g = (-\Delta)^{\frac{1}{2}} \tilde{g}$ (i.e., $\hat{g}(\xi) = |\xi| \hat{\tilde{g}}(\xi)$), then g is in $L^q(\mathbb{R}^n)$ and $\mathcal{H}(g) = B$. Together with the result in Theorem 12.3.5, we have

$$E \cdot B = \sum_{j<k} \mathcal{E}_{jk} \cdot \mathcal{H}(g) = \sum_{j<k} \mathcal{R}_k(f_{jk}) \mathcal{R}_j(g) - \mathcal{R}_j(f_{jk}) \mathcal{R}_k(g).$$

This is the desired result. ∎

It is easy to check that functions in I_0 or I_2 can always be represented as a finite linear combinations of functions in I_1. In fact

$$\mathcal{R}_j(f)\mathcal{R}_k(g) - \mathcal{R}_k(f)\mathcal{R}_j(g)$$
$$= \{ (\mathcal{R}_j(f)) \, \mathcal{R}_k(g) + g\mathcal{R}_k (\mathcal{R}_j(f)) \}$$
$$- \{ g\mathcal{R}_j (\mathcal{R}_k(f)) + \mathcal{R}_k(f)\mathcal{R}_j(g) \};$$

and by the identity $\sum_{k=1}^{n} \mathcal{R}_k \mathcal{R}_k = -I$, it is easy to get

$$fg - \mathcal{H}(f) \cdot \mathcal{H}(g) = -\sum_{j=1}^{n} \{ f\mathcal{R}_j (\mathcal{R}_j(g)) + \mathcal{R}_j(g)\mathcal{R}_j(f) \}. \qquad (12.3.4)$$

We will see in the following that functions in I_1 can also be represented as a finite linear combinations of functions in I_0 and I_2.

THEOREM 12.3.7
Suppose $1 \le k \le n, u \in L^p(\mathbb{R}^n)$ and $v \in L^q(\mathbb{R}^n)$. Then there are functions f, g and vector fields E, B with $div(E) = 0, curl(B) = 0$ such that

$$u\mathcal{R}_k(v) + v\mathcal{R}_k(u) = fg - \mathcal{H}(f) \cdot \mathcal{H}(g) + E \cdot B.$$

Moreover

$$\|f\|_p, \|B\|_p \le C \|u\|_p \quad and \quad \|g\|_q, \|E\|_q \le C \|v\|_q.$$

PROOF Let $f = u$, $g = \mathcal{R}_k(v)$ and $B = \mathcal{H}(u)$, $E = v\mathbf{e}_k + \mathcal{H}(g)$. Using the Fourier transform, it is easy to check that $div(E) = 0$ and $curl(B) = 0$. It is also clear that

$$u\mathcal{R}_k(v) + v\mathcal{R}_k(u) = fg - \mathcal{H}(f) \cdot \mathcal{H}(g) + (v\mathbf{e}_k + \mathcal{H}(g)) \cdot \mathcal{H}(f).$$

The rest is trivial. ∎

REMARK Now we can give a new proof to Theorem 12.2.1 by using Theorems 12.1.1 and 12.2.2. In fact if $b \in BMO$, by Theorem 12.3.7, Corollary 12.3.6 and Theorems 12.1.1 and 12.1.2, we have that $f\mathcal{R}_k(g) + g\mathcal{R}_k(f)$ is in $H^1(\mathbb{R}^n)$ for any $f \in L^p(\mathbb{R}^n)$ and $g \in L^q(\mathbb{R}^n)$. Therefore, by formula (12.3.3), the commutator $[\mathcal{M}_b, \mathcal{R}_k]$ is bounded on $L^p(\mathbb{R}^n)$. On the other hand, by formulas (12.3.4) and (12.3.3), the boundedness of the commutators $[\mathcal{M}_b, \mathcal{R}_k]$, $1 \le k \le n$ on $L^p(\mathbb{R}^n)$ implies the boundedness of the bilinear form $\langle fg - \mathcal{H}(f) \cdot \mathcal{H}(g), b \rangle$ on $L^p(\mathbb{R}^n) \times L^q(\mathbb{R}^n)$. Therefore by Theorem 12.1.1 ($m = 1$), b is in BMO.

12.4 Proof of the main theorems

We prove the "only if" parts of our main theorems in this section.

Note first that the "only if" part of Theorem 12.1.2, can be reduced to the same part of Theorem 12.1.1. To see this, it is enough to check the following identity

$$fg - \sum_{j,k=1}^{n} \mathcal{R}_j\mathcal{R}_k(f)\mathcal{R}_j\mathcal{R}_k(g) = \sum_{j,k=1}^{n} \{\mathcal{R}_j\left(\mathcal{R}_j(f)\right)\mathcal{R}_k\left(\mathcal{R}_k(g)\right)$$
$$- \mathcal{R}_j\left(\mathcal{R}_k(g)\right)\mathcal{R}_k\left(\mathcal{R}_j(f)\right)\}.$$

Suppose $F \in L^p(\mathbb{R}^n)$ and $G \in L^q(\mathbb{R}^n)$. Then $FG \in L^1(\mathbb{R}^n)$. This is trivial but it is also the best that we can hope for. Theorem 12.3.1

says that by assuming in further that F and G are right and left mono-genic, respectively, the trivial $L^1(\mathbb{R}^n)$ regularity of FG can be improved by $H^1(\mathbb{R}^n)$. This compensation (made by the monogenicity of F and G) can be expressed more clearly by using Fourier transforms. This fact is the key motivation of the proofs of the main results.

For a complex valued function b on $\mathbb{R}^n$, consider the bilinear form $\langle FG , b \rangle$ on $L^p(\mathbb{R}^n) \times L^q(\mathbb{R}^n)$. By Plancherel's formula, we have formally

$$\langle FG ; b \rangle = \int_{\mathbb{R}^n} \int_{\mathbb{R}^n} \hat{F}(\eta)\hat{b}(\xi - \eta)\hat{G}(-\xi) \, d\eta d\xi .$$

Suppose further that F and G are right and left monogenic, respectively, we have, by (12.2.1), $\hat{F}(\eta)\eta^- = 0$ and $\xi^- \hat{G}(\xi) = 0$. Let $\xi^+ = -\sum_{j=1}^n i\xi_j e_j - |\xi|$. Then $\xi^+ + \xi^- = -2|\xi|$ and $(-\xi)^+ = \xi^-$. Therefore,

$$\hat{F}(\eta)\hat{G}(-\xi) = \hat{F}(\eta)\frac{\eta^+(-\xi)^+}{4|\eta||\xi|}\hat{G}(-\xi)$$

$$= \hat{F}(\eta)\frac{\eta^+(\xi)^-}{4|\eta||\xi|}\hat{G}(-\xi) .$$

So

$$\langle FG , b \rangle = \int_{\mathbb{R}^n} \int_{\mathbb{R}^n} \hat{F}(\eta)\hat{b}(\xi - \eta)\frac{\eta^+\xi^-}{4|\eta| lth\xi}\hat{G}(-\xi) \, d\eta d\xi . \qquad (12.4.1)$$

Here the function $\frac{\eta^+\xi^-}{4|\eta||\xi|}$ is called the kernel of the bilinear form (12.4.1), which contains the information of the cancellation in $\langle FG , b \rangle$. The bilinear form (12.4.1) is also a special case of the paracommutators studied in [JP] (see also [CM]). In order to have a better look at the cancellation in $\frac{\eta^+\xi^-}{4|\eta||\xi|}$ we introduce the following lemma.

LEMMA 12.4.1
Suppose $\xi, \eta \neq 0$. *Then*

$$\frac{\eta^+\xi^-}{4|\eta||\xi|} = \frac{1}{4}\left(1 - \frac{\xi \cdot \eta}{|\xi||\eta|}\right) - \frac{i}{4}\sum_{j=1}^n \left(\frac{\xi_j}{|\xi|} - \frac{\eta_j}{|\eta|}\right)e_j + \frac{1}{4}\sum_{j<k}\left(\frac{\eta_j\xi_k - \eta_k\xi_j}{|\xi||\eta|}e_{jk}\right) .$$

PROOF   This is straightforward.   ∎

We now turn to Janson and Peetre's work on paracommutators. For a complex valued function b defined on $\mathbb{R}^n$, the paracommutator with sym-bol b and kernel $A(\xi, \eta)$ is the operator defined by the bilinear form on $C_0^\infty(\mathbb{R}^n) \times C_0^\infty(\mathbb{R}^n)$

$$\langle T_b(A)f , g \rangle = \int_{\mathbb{R}^n} \int_{\mathbb{R}^n} \hat{f}(\eta)\hat{b}(\xi - \eta)A(\xi, \eta)\hat{g}(-\xi) \, d\eta d\xi . \qquad (12.4.2)$$

Janson and Peetre's theory says, in short, that the behavior of the para-commutator $T_b(A)$ (bounded on $L^p(\mathbb{R}^n)$, etc.) depends on the smoothness of the symbol b and certain regularities of the kernel $A(\xi, \eta)$. (As an application of the results in [JP], together with Lemma 12.4.1, one can get another proof for Theorem 12.3.1.) We need some notations and definitions for those regularities of the kernel $A(\xi, \eta)$.

For $U, V \subseteq \mathbb{R}^n$, the set of Schur multipliers on $U \times V$ is denoted by $M(U \times V)$. This set contains all the functions $\phi \in L^\infty(U \times V)$ that admits the representation

$$\phi(\xi, \eta) = \int_X \alpha(\xi, x)\beta(\eta, x)\, d\mu(x) \qquad (12.4.3)$$

for some σ-finite measure space (X, μ) and measurable functions α on $U \times X$ and β on $V \times X$ with

$$\int_X \|\alpha(\cdot, x)\|_{L^\infty(U)} \|\beta(\cdot, x)\|_{L^\infty(V)}\, d\mu(x) < \infty.$$

The norm of the Schur multiplier ϕ, denoted by $\|\phi\|_{M(U \times V)}$, is defined by the minimum of the left-hand side of the above inequality over all representations (12.4.3) (see [JP]).

DEFINITION 12.4.2 *The kernel $A(\xi, \eta)$ is said to have the conditions :*

A0   *Homogeneity, if $A(r\xi, r\eta) = A(\xi, \eta)$ for all $r \neq 0$ and $\xi, \eta \in \mathbb{R}^n$.*

A1   *Boundedness, if $\|A\|_{M(\mathbb{R}^n \times \mathbb{R}^n)} < \infty$.*

A3   *Zero on the diagonal, if there exist $\gamma, \delta > 0$ such that $\|A\|_{M(B \times B)} \leq C\,(r/|\xi_0|)^\gamma$, where $B = Ball(\xi_0, r) = \{\xi : |\xi - \xi_0| < r\}$ and $0 < r < \delta\,|\xi_0|$.*

A5   *Nondegeneracy, if for every $\xi_0 \neq 0$ there exist $\delta > 0$ and $\eta_0 \in \mathbb{R}^n$ such that $\|1/A\|_{M(U \times V)} \leq C$, where $U = \{\xi : |\xi/|\xi| - \xi_0/|\xi_0|| < \delta, |\xi| > |\xi_0|\}$ and $V = Ball(\eta_0, \delta\,|\xi_0|)$.*

The following theorem was proved in [JP, Theorem 10.1].

THEOREM 12.4.3   (Theorem C, Janson and Peetre)
Suppose that A satisfies $A0, A1, A3, A5$. Then the boundedness of the bi-linear form (12.4.2) on $L^p(\mathbb{R}^n) \times L^q(\mathbb{R}^n)$ implies that b is in BMO. More precisely

$$\|b\|_{BMO} \leq C\,\|T_b(A)\|_{(L^p(\mathbb{R}^n), L^p(\mathbb{R}^n))}\,.$$

REMARK   The proof of Theorem C in [JP] is for the case $p = 2$. For general p, the proof is similar (see [L]).

The proof of the "only if" parts of the main results is based on Theorem C.

Using the fact

$$\mathcal{R}_{j_1}\mathcal{R}_{j_2}\widehat{\cdots\mathcal{R}_{j_m}}(\phi)(\xi) = (i)^m \frac{\xi_{j_1}\xi_{j_2}\cdots\xi_{j_m}}{|\xi|^m}\hat{\phi}(\xi)$$

and Plancherel's formula, we can write

$$\left\langle fg - \sum_{j_1,j_2,\cdots,j_m=1}^{n} \mathcal{R}_{j_1}\mathcal{R}_{j_2}\cdots\mathcal{R}_{j_m}(f)\mathcal{R}_{j_1}\mathcal{R}_{j_2}\cdots\mathcal{R}_{j_m}(g)\,, b\right\rangle$$

$$= \int_{\mathbb{R}^n}\int_{\mathbb{R}^n} \hat{f}(\eta)A_m(\xi,\eta)\hat{b}(\xi-\eta)\hat{g}(-\xi)\,d\eta d\xi\,,$$

where the kernel $A_m(\xi,\eta)$ is

$$A_m(\xi,\eta) = 1 - \sum_{j_1,j_2,\cdots,j_m=1}^{n} \frac{\eta_{j_1}\eta_{j_2}\cdots\eta_{j_m}\xi_{j_1}\xi_{j_2}\cdots\xi_{j_m}}{|\eta|^m|\xi|^m} = 1 - \left(\frac{\xi\cdot\eta}{|\xi||\eta|}\right)^m.$$

We need to verify that this kernel satisfies the assumptions of Theorem C. It is clear that $A_m(\xi,\eta)$ is homogeneous. For independent interests, we gather the other properties of $A_m(\xi,\eta)$ in the following lemmas.

LEMMA 12.4.4   (Boundedness)
For $m = 1, 2, \cdots$, we have the estimates

$$\|A_m\|_{M(\mathbb{R}^n\times\mathbb{R}^n)} \leq 1 + n^m.$$

PROOF   In fact, let $\alpha(\xi,j) = \beta(\xi,j) = \xi_j/|\xi|$, then

$$\frac{\xi\cdot\eta}{|\xi||\eta|} = \sum_{j=1}^{n}\alpha(\xi,j)\beta(\eta,j).$$

This implies

$$\left\|\frac{\xi\cdot\eta}{|\xi||\eta|}\right\|_{M(\mathbb{R}^n\times\mathbb{R}^n)} \leq \sum_{j=1}^{n}\|\alpha(\cdot,j)\|_{L^\infty(\mathbb{R}^n)}\|\beta(\cdot,j)\|_{L^\infty(\mathbb{R}^n)} = n. \quad (12.4.4)$$

Therefore

$$\|A_m\|_{M(\mathbb{R}^n\times\mathbb{R}^n)} \leq 1 + \left\|\frac{\xi\cdot\eta}{|\xi||\eta|}\right\|_{M(\mathbb{R}^n\times\mathbb{R}^n)}^m \leq 1 + n^m.$$

Here we have used the following easy facts of Schur multipliers:

$$\|\phi + \psi\|_{M(U\times V)} \leq \|\phi\|_{M(U\times V)} + \|\psi\|_{M(U\times V)}\,,$$

$$\|\phi\psi\|_{M(U\times V)} \leq \|\phi\|_{M(U\times V)}\|\psi\|_{M(U\times V)}\,.$$

The proof is complete. ∎

REMARK The first part of the proof above can be generalized as

$$\left\|\sum_{j=1}^{m} \phi_j(\xi)\psi_j(\eta)\right\|_{M(U\times V)} \tag{12.4.5}$$

$$\leq m \left(\sup_{1\leq j\leq m} \|\phi_j\|_{L^\infty(U)}\right)\left(\sup_{1\leq j\leq m} \|\psi_j\|_{L^\infty(V)}\right).$$

LEMMA 12.4.5 (Zero on the diagonal)
If $B = Ball(\xi_0, r)$ with $0 < r < \frac{1}{2}|\xi_0|$, then

$$\|A_m\|_{M(B\times B)} \leq C\left(\frac{r}{|\xi_0|}\right)^2.$$

PROOF Note that

$$A_m(\xi,\eta) = \left(1 - \frac{\xi\cdot\eta}{|\xi||\eta|}\right)\left(1 + \frac{\xi\cdot\eta}{|\xi||\eta|} + \cdots + \left(\frac{\xi\cdot\eta}{|\xi||\eta|}\right)^{m-1}\right).$$

By formula (12.4.4) we have

$$\|A_m\|_{M(B\times B)} \leq mn^{m-1}\|A_1\|_{M(B\times B)}.$$

Therefore we only need to show that the lemma is true for $A_1(\xi,\eta)$.
For $|\xi_0| \neq 0$ and $\xi, \eta \in B$, write

$$A_1(\xi,\eta) = \frac{1}{2}\left|\frac{\xi}{|\xi|} - \frac{\eta}{|\eta|}\right|^2$$

$$= \frac{1}{2}\left|\left(\frac{\xi}{|\xi|} - \frac{\xi_0}{|\xi_0|}\right) - \left(\frac{\eta}{|\eta|} - \frac{\xi_0}{|\xi_0|}\right)\right|^2$$

$$= \frac{1}{2}\left|\frac{\xi}{|\xi|} - \frac{\xi_0}{|\xi_0|}\right|^2 + \frac{1}{2}\left|\frac{\eta}{|\eta|} - \frac{\xi_0}{|\xi_0|}\right|^2 - \left(\frac{\xi}{|\xi|} - \frac{\xi_0}{|\xi_0|}\right)\cdot\left(\frac{\eta}{|\eta|} - \frac{\xi_0}{|\xi_0|}\right).$$

Note that

$$\left|\frac{\xi_j}{|\xi|} - \frac{\xi_{0j}}{|\xi_0|}\right| \leq \left|\frac{\xi}{|\xi|} - \frac{\xi_0}{|\xi_0|}\right| = \frac{||\xi_0|(\xi-\xi_0) + (|\xi_0|-|\xi|)\xi_0|}{|\xi||\xi_0|} \leq 2\frac{|\xi-\xi_0|}{|\xi|}.$$

Together with (12.4.5), this is enough to derive the desired result. ∎

LEMMA 12.4.6 (Nondegeneracy)
Suppose $|\xi_0| = |\eta_0| \neq 0$ and $\xi_0 \cdot \eta_0 = 0$. For $\delta = \frac{1}{8n}$, let

$$U = \{\xi \in \mathbb{R}^n : |\xi/|\xi| - \xi_0/|\xi_0|| < \delta\} \quad and \quad V = Ball(\eta_0, \delta|\xi_0|).$$

Then $\frac{1}{A_m(\xi,\eta)}$ is in $M(U \times V)$. More precisely

$$\left\| \frac{1}{A_m(\xi,\eta)} \right\|_{M(U \times V)} \leq 2$$

PROOF   Suppose $\eta \in V$. Then $\frac{1}{2}|\xi_0| < |\eta| < \frac{3}{2}|\xi_0|$. Write

$$\frac{\xi \cdot \eta}{|\xi||\eta|} = \frac{\xi \cdot (\eta - \eta_0)}{|\xi||\eta|} + \left(\frac{\xi}{|\xi|} - \frac{\xi_0}{|\xi_0|} \right) \cdot \frac{\eta_0}{|\eta|}.$$

Therefore, if $\xi \in U$ and $\eta \in V$, then

$$\left| \frac{\xi \cdot \eta}{|\xi||\eta|} \right| \leq 2\delta + 2\delta = 4\delta < 1,$$

and

$$\left\| \frac{\xi \cdot \eta}{|\xi||\eta|} \right\|_{M(U \times V)} \leq 2n\delta + 2n\delta = \frac{1}{2}.$$

Place

$$\frac{1}{A_m(\xi,\eta)} = \sum_{j=0}^{\infty} \left(\frac{\xi \cdot \eta}{|\xi||\eta|} \right)^{jm}.$$

We therefore have

$$\left\| \frac{1}{A_m(\xi,\eta)} \right\|_{M(U \times V)} \leq \sum_{j=0}^{\infty} \left(\frac{1}{2} \right)^{jm}.$$

This is enough to obtain the desired result.   ∎

PROOF [**Proof of the "only if" part of Theorem 12.1.1**] Suppose the bilinear form (12.1.1) is bounded on $L^p(\mathbb{R}^n) \times L^q(\mathbb{R}^n)$. We want to show b is in BMO. By our previous discussion, we know that the kernel of the bilinear form (12.1.1) is

$$A_m(\xi,\eta) = 1 - \left(\frac{\xi \cdot \eta}{|\xi||\eta|} \right)^m.$$

The desired result is then a consequence of Theorem C and Lemmas 12.4.2, 12.4.3 and 12.4.4.   ∎

We end this section by deriving the factorization theorems for $H^1(\mathbb{R}^n)$.

THEOREM 12.4.7
Suppose f is in $H^1(\mathbb{R}^n)$. Then f can be written as

$$f = \sum_{k=1}^{\infty} \left(g_k h_k - \sum_{j_1, j_2, \cdots, j_m = 1}^{n} \mathcal{R}_{j_1} \mathcal{R}_{j_2} \cdots \mathcal{R}_{j_m}(g_k) \mathcal{R}_{j_1} \mathcal{R}_{j_2} \cdots \mathcal{R}_{j_m}(h_k) \right),$$

(12.4.6)

with

$$\|f\|_{H^1} \asymp \sum_{k=1}^{\infty} \|g_k\|_p \|h_k\|_q.$$

(12.4.7)

PROOF Suppose f has expression (12.4.6) with a finite right-hand side of (12.4.7). By Theorem 12.1.1 and Fefferman's duality theorem, we have $f \in H^1(\mathbb{R}^n)$.

To see that every $H^1(\mathbb{R}^n)$ function f can be written in the form (12.4.6), we consider the Banach space H of all functions f in $L^1(\mathbb{R}^n)$ which admits the decomposition (12.4.6) and is normed by

$$\|f\| = \inf \sum_{k=1}^{\infty} \|g_k\|_p \|h_k\|_q.$$

It is clear that H is a subspace of $H^1(\mathbb{R}^n)$ (as we just showed). By Theorem 12.1.1, we have

$$\|b\|_{BMO} \asymp \sup\{|\langle f, b \rangle| : \|f\| = 1\}.$$

This shows that $\|\cdot\|$ and the $H^1(\mathbb{R}^n)$ norm are equivalent and $H^* = BMO$. Therefore $H = H^1(\mathbb{R}^n)$. ∎

THEOREM 12.4.8
Suppose f is in $H^1(\mathbb{R}^n)$. Then f can be written as

$$f = \sum_{l=1}^{\infty} \sum_{1 \leq j < k \leq n} (\mathcal{R}_j(g_{jkl}) \mathcal{R}_k(h_{jkl}) - \mathcal{R}_j(h_{jkl}) \mathcal{R}_k(g_{jkl}))$$

with

$$\|f\|_{H^1} \asymp \sum_{l=1}^{\infty} \sum_{1 \leq j < k \leq n} \|g_{jkl}\|_p \|h_{jkl}\|_q.$$

PROOF The proof is the same as the proof of Theorem 12.4.5. ∎

12.5   Applications in compensated compactness

In the theory of compensated compactness, improving the regularity of nonlinear quantities in PDE is very important. For example (an example studied in [CLMS])

$$u \triangle_x u + |\nabla_x u|^2 \in H^1(\mathbb{R}^n)$$

if $u \in L^p(\mathbb{R}^n)$, $\triangle_x u \in L^q(\mathbb{R}^n)$ and $\frac{2n}{n-2} \le p \le \infty$.

Denote by $W^{k,r}(\mathbb{R}^n)$, $k = 1, 2$, the usual Sobolev spaces.

The following theorem is about the regularity of more symmetric elliptic equations (compared to the above example).

THEOREM 12.5.1
Suppose $u \in W^{1,p}(\mathbb{R}^n)$ and $v \in W^{1,q}(\mathbb{R}^n)$. Then

$$\nabla_x u \cdot \nabla_x v - (-\triangle_x)^{\frac{1}{2}} u (-\triangle_x)^{\frac{1}{2}} v$$

is in $H^1(\mathbb{R}^n)$.

PROOF Let $f = (-\triangle_x)^{\frac{1}{2}} u$ and $g = (-\triangle_x)^{\frac{1}{2}} v$. Then $f \in L^p(\mathbb{R}^n)$ and $g \in L^q(\mathbb{R}^n)$. And

$$\nabla_x u \cdot \nabla_x v - (-\triangle_x)^{\frac{1}{2}} u (-\triangle_x)^{\frac{1}{2}} v = -(fg - \mathcal{H}(f) \cdot \mathcal{H}(g)) .$$

The desired result is then a consequence of Theorem 12.1.1 ($m = 1$).   ∎

The following theorem is about the Von Karman equation (also called the Hessian matrix). The same result was proved in [CLMS] and [G]. We list it here because it is a direct consequence of Theorem 1.

THEOREM 12.5.2
Suppose $u \in W^{2,p}(\mathbb{R}^n)$ and $v \in W^{2,q}(\mathbb{R}^n)$. Then

$$\triangle_x u \triangle_x v - \sum_{j,k=1}^{n} \frac{\partial^2 u}{\partial x_j \partial x_k} \frac{\partial^2 v}{\partial x_j \partial x_k}$$

is in $H^1(\mathbb{R}^n)$.

PROOF Let $f = -\triangle_x u$ and $g = -\triangle_x v$. Then $f \in L^p(\mathbb{R}^n)$, $g \in L^q(\mathbb{R}^n)$ and $\mathcal{R}_j \mathcal{R}_k(f) = \frac{\partial^2 u}{\partial x_j \partial x_k}$, $\mathcal{R}_j \mathcal{R}_k(g) = \frac{\partial^2 v}{\partial x_j \partial x_k}$. Therefore

$$\triangle_x u \triangle_x v - \sum_{j,k=1}^{n} \frac{\partial^2 u}{\partial x_j \partial x_k} \frac{\partial^2 v}{\partial x_j \partial x_k} = fg - \sum_{j,k=1}^{n} \mathcal{R}_j \mathcal{R}_k(f) \mathcal{R}_j \mathcal{R}_k(g) .$$

Applying Theorem 12.1.1 with $m = 2$, we get the desired result.   ∎

Like Theorems 12.5.1 and 12.5.2, there are higher-order PDE quantities corresponding to Theorem 12.1.1. We do not intend to write them down.

THEOREM 12.5.3 (Final Conjecture)
Suppose $f \in H^1(\mathbb{R}^n)$. Then there are functions $g, h \in L^2(\mathbb{R}^n)$ so that

$$f = gh - \sum_{j=1}^{n} \mathcal{R}_j(g)\mathcal{R}_j(h).$$

For $n = 1$ this conjecture is clearly true.

Bibliography

[Ch]   S. Chanillo, *Sobolev inequalities involving divergence free maps*, Comm. P. D. E. **16** (1991), 1969-1994.

[Chr]   M. Christ, *Lectures on Singular Integral Operators*, AMS, Providence, RI, 1990.

[CG]   R. Coifman and L. Grafakos, *Hardy Spaces Estimates for Multilinear Operators, I* Revista Mat. Iberoamericana **8** (1992), 45–67.

[CLMS] R. Coifman, P.L. Lions, Y. Meyer and S. Semmes, *Compensated compactness and Hardy spaces*, preprint, 1991.

[CM]   R. Coifman and Y. Meyer, *Au delá des operatéurs pseudo-differentiels*, Astérique, Société Mathématique de France, 1978.

[CRW]   R. Coifman, R. Rochberg and G. Weiss, *Factorization theorems for Hardy spaces in several variables*, Ann. Math. **103**, (1976), 611-635.

[G]   L. Grafakos, *Hardy Spaces Estimates for Multilinear Operators, II*, Revista Mat. Iberoamericana **8**, (1992), 69–92.

[GM]   J. Gilbert & M. Murray, *Clifford Algebras and Dirac operators in Harmonic Analysis*, C.U.P., Cambridge, 1991.

[JP]   S. Janson & J. Peetre, *Paracommutators–Boundedness and Schatten-Von Neumann Properties*, Trans. Amer. Math. Soc. **305**, (1988), 467–504.

[L]   C. Li, *Boundedness of paracommutators on L^p spaces*, Acta Math. Sinica, New Series **6**, (1990), 131–147.

[LMcS] C. Li, A. McIntosh, & S. Semmes, *Convolution singular integrals on Lipschitz surfaces*, J. of Amer. Math. Soc. **5**, (1992), 455–481.

[LMcWZ] C.Li, A. McIntosh, Z. Wu and K. Zhang, *Compensated compactness, paracommutators and Hardy spaces*, J. Funct. Anal. (to appear).

[Mc]   A. McIntosh, *Clifford algebras and the higher dimensional Cauchy integral*, Approximation Theory and Function Spaces, Banach Center Publications, Warsaw, Poland **22**, (1989), 253–267.

[S]   E. M. Stein, *Singular Integrals and Differentiability Properties of Functions*, Princeton University Press, Princeton, NJ, 1970.

[T]   A. Torchinsky, *Real–Variable Methods in Harmonic Analysis*, Academic Press, New York, 1986.

Department of Mathematics,
University of Alabama,
Tuscaloosa, Alabama 35487
E-mail: zwu@mathdept.as.ua.edu

13

Frame Decompositions of Form-Valued Hardy Spaces

J.E. Gilbert, J.A. Hogan, and J.D. Lakey

13.1 Introduction

Wavelets have appeared in the theory of Hardy H^p-spaces for a long time, though the precise connection has been recognized only in the past few years. For example, when Hardy spaces consist of functions analytic in the upper half-plane, Meyer [M1, p. 22] reinterprets Lusin's representation

$$F(z) = \int\int_{\mathbb{R}^2_+} (z - \overline{\zeta})^{-2}\alpha(\zeta)dudv, \quad (\zeta = u + iv, v > 0) \qquad (13.1.1)$$

of such functions as a wavelet-type synthesis where $\alpha(\zeta)$ plays the role of the coefficient function and $(z - \overline{\zeta})^{-2}$ that of analyzing wavelets; there is no unique choice of α, but there is a natural choice $\alpha(\zeta) = v\partial F/\partial v$ linking coefficient norms with the Lusin area function. By taking boundary values in (13.1.1), so that Hardy spaces are thought of as subspaces of $L^p(\mathbb{R}^n)$, we arrive at the prototypical wavelet representation of analytic signals first studied by Grossmann and Morlet [GrM]. On the other hand, the idea of an orthonormal wavelet basis having desirable function-theoretic properties can be traced to Strömberg's discovery [Str] of explicit unconditional bases for Hardy spaces, where the number of vanishing moments of the analyzing wavelet is dictated by p whenever $p \leq 1$. This last case deals with *real* Hardy theory, *i.e.*, with real-valued functions which can be interpreted as the *real part* of the boundary values of functions analytic in the upper half-plane. The prototypical Calderón reproducing formula for real-valued functions can also be obtained from (13.1.1) by taking real parts and the natural choice of coefficient function α. More generally, Meyer [M2] (cf., [AEPT]) has shown that the *atomic* decompositions obtained earlier for these real Hardy spaces can be obtained from expansions in terms of compactly supported wavelets. Such a decomposition is possible even when

0-8493-8481-8/96/$0.00+$.50
© 1996 by CRC Press

orthonormal wavelet expansions are replaced by decompositions in terms of compactly supported wavelet frames. Actually, the molecular version of these decompositions dates to work by Coifman and Rochberg, cf. also, [CGT]. In short, wavelets provide an efficient set of building blocks for Hardy spaces whether these spaces are comprised of analytic functions or of the real parts of their boundary values.

Hardy spaces may be studied in a variety of settings: analytic functions may be replaced by solutions of generalized Cauchy–Riemann systems, Euclidean space may be replaced by more general domains [CKS] or Riemannian manifolds, while atomic decompositions on $\mathbb{R}$ can be replaced by corresponding decompositions on much more general homogeneous spaces [CW]. In particular, one may consider H^p theory associated with first-order differential operators of the *Dirac type* [GM2]. For such operators Cauchy integral theorems can be established, and when boundary regularity is available, as it usually is, it is possible to establish a reproducing formula for boundary values leading to a real H^p theory.

In this paper we shall develop a theory of wavelet type frames for boundary values of solutions of one particular first-order system, the Hodge-deRham (d, d^*) system (cf. [SW]) on $\mathbb{R}^{n+1}_+$. The aim will be to set the stage for a theory applicable to the whole family of first order systems associated with the group of rigid motions of Euclidean space ([DGK]); for simplicity, we shall ignore the subtleties introduced when values of $p \leq 1$. Most significantly perhaps, we shall ignore virtually all connections with group representation theory despite the fact that the deepest understanding does not emerge until connections with representations of the transformation group of Euclidean space generated by translations, dilations and rotations are explored. This will be done in a later paper.

Complex-valued functions are replaced by r-forms on open subsets of $\mathbb{R}^{n+1}$, *i.e.*, functions taking values in the *exterior* product $\Lambda_r(\mathbb{R}^{n+1})$, while their real-valued boundary values will now be (r-1)-forms and so have values in the *exterior* product $\Lambda_{r-1}(\mathbb{R}^n)$. For $1 < p < \infty$, the boundary values are divergence free, or *massless*. We show that such r-forms may decomposed, in a sense, into superpositions of translates and dilates of a fixed massless vector field. Orthonormal bases and Riesz bases of divergence free vector fields for the *massless Sobolev space* have already been studied by Battle and Federbush [BF], and by Lemarié [Le]. While our work in a sense parallels theirs, in the same way that the theory of wavelet frames parallels that of wavelet orthonormal bases, our goal is to develop wavelet type tools that will be available for Hardy space theory in a variety of settings — in particular ones where orthonormal wavelet bases are not available.

In two dimensions, divergence free wavelets are easy to construct because if ϕ is any reasonable real valued function on $\mathbb{R}^2$, the function

$$\vec{\phi} = (-R_2\phi, R_1\phi),$$

obtained from the respective Riesz transforms $R_1\phi, R_2\phi$ of ϕ will have divergence zero. Since Riesz transforms commute with translations and dilations they map wavelets to wavelets. Consequently, if ψ_λ is an orthonormal wavelet basis for $L^2(\mathbb{R}^2)$, the corresponding $\vec{\psi}_\lambda$ will form an orthonormal basis for the *massless Sobolev space* of divergence free vector fields with scalar product $(-\Delta\vec{f}, \vec{g})$.

In higher dimensions divergence free wavelets no longer arise automatically, and a differentiation/cancellation condition must be built in. The Battle-Federbush construction involves solving a minimization problem in gauge field theory. Lemarié's construction on the other hand produces *biorthogonal* bases of divergence free wavelets and is based on the fact that certain biorthogonal wavelets come in *dual pairs* that are related to one another by differentiation. Vector wavelets satisfying differential constraints are then constructed by taking tensor products in an appropriate manner. All such divergence free wavelets can then be used to study solutions of the Navier-Stokes equation, for instance.

Our goal here is to produce the corresponding *wavelet frames* for massless r-forms that are most suitably viewed in terms of boundary values of Clifford analytic functions. Section 13.2 contains some background about affine frames and properties obtained in [GHL] for scalar-valued frames. In Section 13.3 we review basic facts about the deRham complex and discuss Hardy spaces of Clifford analytic functions as well as associated boundary-value theory, while in Section 13.4 algebraic properties of solutions of the Hodge-deRham (d, d^*) system are discussed. This leads into a study of the Cauchy-Szegö reproducing kernel for the Hardy spaces contained in Section 13.5. In Section 13.6 we present a theory of discrete frames for real Hardy spaces of forms, and, finally, in Section 13.7 we list some directions for further investigation.

13.2 Dilatation Frames

Our approach deals simultaneously with discrete and *continuous* frames, cf., [T1, AAG]. Let $\mathcal{B}$ be a Banach space (or linear metrizable space) with dual $\mathcal{B}'$.

DEFINITION 13.2.1 A collection of pairs $\{e_\alpha, f_\alpha\}_{\alpha \in A}$ indexed by a measurable space $\{A, \mu\}$ is said to be a bi–frame for $\mathcal{B}$ if there exist positive constants A and B so that the inequalities

$$A\|h\|_\mathcal{B} \leq \|\int_A (h, f_\alpha)e_\alpha d\mu(\alpha)\|_\mathcal{B} \leq B\|h\|_\mathcal{B}$$

hold for all $h \in \mathcal{B}$. In other words, the frame operator

$$S : h \longrightarrow \int_A (h, f_\alpha) e_\alpha d\mu(\alpha)$$

is bounded and invertible on $\mathcal{B}$.

One important way of generating frames occurs when $A = G$ is a locally compact group, μ is a natural measure on A, say Haar measure, and the bi-frame elements arise from the action of a representation $\pi : G \longrightarrow \mathcal{B}$ of the group on some function space $\mathcal{B}$. In particular, if the operator

$$h \longrightarrow \int_G (h, \pi^*(g)\phi)\pi^*(g)\psi d\mu(g) \qquad (13.2.1)$$

is bounded and invertible on $\mathcal{B}$ then the collection $\{\pi^*(g)\phi, \pi^*(g)\psi\}_{g \in G}$ is said to be a *frame of coherent states* for $\mathcal{B}$ with respect to $\{\pi, G\}$. More general frames of coherent states may be obtained when the group G is suitably replaced by some homogeneous space, cf., [AAG, HL].

Our terminology differs in two important ways from some commonly accepted usage, cf., [HW, D1, D2, DGM, FG]. First of all, frames studied by Daubechies in [D1, D2] are *discrete frames*, that is, the set A in Definition (13.2.1) is countable and $d\mu(\alpha)$ is a counting measure. Secondly, Definition (13.2.1) requires only the boundedness and invertibility of the frame operator S: no mention is made at the outset of an auxiliary coefficient space, though such spaces often appear later as a consequence of the theory. The Banach frames studied by Feichtinger and Gröchenig and others postulate the existence of such coefficient spaces. Our motive for this approach is to have a unified setting for studying frames of coherent states associated to the dilatation group of $\mathbb{R}^n$ for vector-valued Hardy spaces, as well as associated discrete *approximating* frames.

Several transformation groups are possible candidates for the group in (13.2.2). One of the most geometrically significant groups that can be considered in this context is the group of conformal transformations of $\mathbb{R}^n$, but for the sake of comparison to so-called affine frame theory we shall restrict attention to the subgroup H of *dilatations* generated by translations and dilations. We can identify H with the semidirect product

$$H = \mathbb{R}^n \, \circledS \, \mathbb{R}_+^* = \left\{ (x, s) : x \in \mathbb{R}^n \,, s \in \mathbb{R}_+^* \right\} \qquad (13.2.2)$$

of n-dimensional Euclidean space and the multiplicative group $\mathbb{R}_+^*$ of positive real numbers (when $n = 1$ H is the group of sense-preserving affine transformations). Group multiplication in H is then given by

$$w \cdot z = (y, t) \cdot (x, s) = (sy + x, st) \,, \qquad (x, y \in \mathbb{R}^n, s, t \in \mathbb{R}_+^*). \qquad (13.2.3)$$

the group identity is $(0,1)$, and $(x,s)^{-1} = (-x/s, 1/s)$ is the inverse of (x,s). In turn, this means that H can be identified with the upper half-space $\Omega = \mathbb{R}_+^{n+1}$ in $\mathbb{R}^{n+1}$. Furthermore, the action of H on $\mathbb{R}^n$ extends to a natural action on Ω which reduces to group multiplication when H is identified with Ω. All of these identifications will be used repeatedly and without comment in the future.

We then consider the irreducible unitary representation

$$\pi : H \to \mathcal{L}(L^2(\mathbb{R}^n)); \quad \pi(g)\psi(v) = t^{\frac{n}{2}}\psi(v.z) \quad (z = (x,t)).$$

Right Haar measure on H is $d\mu(z) = t^{-(n+1)}dxdt$, so that the operator in (13.2.2) takes the specific form

$$f \mapsto \int_H (f, \phi_{(z)})\psi_{(z)}d\mu(z) \tag{13.2.4}$$

$$= \int_0^\infty \frac{1}{t^n} \int_{\mathbb{R}^n} \psi(\frac{x-y}{t}) \int_{\mathbb{R}^n} f(v)\overline{\phi(\frac{y-v}{t})}dvdy\frac{dt}{t^{n+1}}$$

$$= \int_0^\infty \psi_t * \phi_t^* * f\frac{dt}{t}$$

Here $\psi_{(z)}(v) = \pi^*(z)\psi(v)$ with $z \in H$, whereas $\psi_t(v) = t^{-n}\psi(v/t)$ and $\phi^*(v) = \overline{\phi(-v)}$.

The condition that the operator in (13.2.5) is bounded and invertible on the Hilbert space $L^2(\mathbb{R}^n)$ is then the Fourier multiplier condition

$$0 < A_1 \le \left| \int_0^\infty \hat{\phi}(t\xi)\overline{\hat{\psi}(-t\xi)}\frac{dt}{t} \right| \le A_2 < \infty \tag{13.2.5}$$

In fact, this condition ensures that the operator remains bounded and invertible on all spaces $L^p(\mathbb{R}^n)$ when $1 < p < \infty$ (see Stein, [St, p. 50]) and convergence holds in the sense of distributions when suitably interpreted. In particular, it follows that the functions $\{\phi_{(z)}, \psi_{(z)}\}_{z \in H}$ form frames of *dilatation coherent states* for the Lebesgue spaces. When both ϕ, ψ are chosen in such a way that the multiplier in formula (13.2.6) equals one almost everywhere, (13.2.5) is the *Calderón reproducing formula* (see [FJW] for a discussion of the Calderón reproducing formula and its variants).

The frames of dilatation coherent states provide effective tools, especially for the study of Calderón-Zygmund operators. However, there are applications, for example to signal processing, where one requires discrete versions of affine frames, e.g., [PW]. One reason is that such frames are useful in numerical analysis. But we can also think of discrete frames as Riemann approximations to the Calderón reproducing formula. This provides a new point of view on approximations to the identity and this is the way we shall think about such frames here. In the remainder of the section we summarize some results of [GHL] that illustrate our point of view, and that carry over to the form-valued case discussed in the subsequent sections. We start with a notion of Riemann sum over a (uniform) partition of the dilatation group, as well as appropriate notions of approximation.

For fixed $r > 1$ and $s > 0$ we begin with a *fundamental tile*

$$\Delta = \{(y,t) \in H : 0 \le y_j < s, 1 \le t < r, (1 \le j \le n)\}. \qquad (13.2.6)$$

When

$$\lambda_{km} = \left(\frac{sk}{r^m}, \frac{1}{r^m}\right) \quad (k \in \mathbb{Z}^n, m \in \mathbb{Z}) \qquad (13.2.7)$$

it follows that

$$\Omega = \bigcup_{k,m} \Delta \cdot \lambda_{km}$$

defines a tiling of Ω into tiles having equal right Haar measure $|\Delta|$.

The Discrete Approximation operator $\mathcal{D}$ below can then be regarded as a type of Riemann sum approximation to the corresponding operator $\mathcal{E}$ associated with this tiling.

DEFINITION 13.2.2   *For each pair of functions $\phi = \phi(x)$ and $\psi = \psi(x)$ in $L^1 \cap L^2(\mathbb{R}^n)$ the operator*

$$\mathcal{D} : f \longrightarrow \frac{1}{|\Delta|} \sum_{k,m} r^{mn} \psi(r^m x - sk) \left(\int_{\mathbb{R}^n} \overline{\phi(r^m y - sk)} f(y) dy\right) \quad (r > 1, s > 0)$$

will be called the Discrete Approximation to

$$\mathcal{E} : f \longrightarrow \lim_{\epsilon \to 0} \int_\epsilon^{1/\epsilon} \frac{1}{t^{2n}} \left(\int_{\mathbb{R}^n} \left(\int_{\mathbb{R}^n} \psi(\frac{x-v}{t}) \overline{\phi(\frac{y-v}{t})} dv\right) f(y) dy\right) \frac{dt}{t}$$

with mesh (r,s).

Justification of the term *approximation* requires a way of saying that the operator $\mathcal{D} - \mathcal{E}$ is small. The two necessary ingredients are a notion of well-behavedness for the functions ϕ, ψ, and a notion of small mesh size. In turn, we are able to apply the Calderón-Zygmund theory to show that the discrete approximations are really frame operators.

Well-behavedness of ϕ, ψ is measured in a function space norm where smoothness and decay are prescribed in the following way.

DEFINITION 13.2.3   *For fixed $\delta > 0$ denote by $\mathcal{S}_\delta(\mathbb{R}^n)$ the functions f on $\mathbb{R}^n$ having derivatives up to order $[\delta] + 1$ such that the Banach norm*

$$\|f\|_\delta = \sup_{|\alpha| \le [\delta]+1} \left\{ \sup_x (1 + |x|)^{n+\delta+|\alpha|} |\partial^\alpha f(x)| \right\}$$

is finite. $\mathcal{S}_\delta^{(0)}(\mathbb{R}^n)$ *denotes the closed subspace of functions for which*

$$\int_{\mathbb{R}^n} v^\alpha \psi(v) dv = 0$$

for any multi-index α *such that* $|\alpha| \le [\delta]$.

Example 1
When

$$P_t(x) = c_n \frac{t}{(|x|^2 + t^2)^{\frac{n+1}{2}}}; \quad c_n = \frac{\Gamma(\frac{n+1}{2})}{\pi^{\frac{n+1}{2}}}$$

denotes the usual Poisson kernel, the function

$$\phi(x) = \frac{\partial P_t}{\partial t}(x)\Big|_{t=1}$$

belongs to $\mathcal{S}_\delta^{(0)}(\mathbb{R}^n)$ provided $0 < \delta < 1$.

On the other hand, if R_j denotes the j-th Riesz transform operator of convolution with the kernel $c_n x_j / |x|^{n+1}$, then for the same ϕ

$$R_j \phi(x) = \frac{\partial P}{\partial x_j}(x) = -(n+1)c_n \frac{x_j}{(1+|x|^2)^{\frac{n+3}{2}}}$$

belongs to $\mathcal{S}_\delta \cap \mathcal{S}_\gamma^{(0)}(\mathbb{R}^n)$ for any $0 < \delta < 2$, and $0 < \gamma < 1$, cf., Theorem (13.6.2) below. ⬚

Application of the Cotlar-Stein Lemma and basic convolution estimates for $\mathcal{S}_\delta^{(0)}$, together with Calderón-Zygmund theory, provides the following result.

THEOREM 13.2.4
(i) *Let* $\phi, \psi \in \mathcal{S}_\delta^{(0)}(\mathbb{R}^n), \delta > 0$. *Then the operators* $\mathcal{D}$ *and* $\mathcal{E}$ *defined above are bounded on* $L^2(\mathbb{R}^n)$ *with operator norms a multiple of* $\|\phi\|_\delta$ *and* $\|\psi\|_\delta$ *respectively.*
(ii) *Let* $\phi, \psi \in \mathcal{S}_\delta^{(0)} \cap \mathcal{S}_\gamma(\mathbb{R}^n)$ *for some* $\delta > 0$ *and* $\gamma > 1$. *Then* $\mathcal{D}$ *and* $\mathcal{E} - \mathcal{D}$ *are standard Calderón-Zygmund operators with CZ-norms satisfying*

$$\|\mathcal{D}\|_{CZ} \le \text{const.} \|\phi\|_\delta \|\psi\|_\delta; \quad \|\mathcal{E} - \mathcal{D}\|_{CZ} \le C_\Delta \|\phi\|_\delta \|\psi\|_\delta.$$

The constant C_Δ *can be made arbitrarily small by taking the mesh size* $\max\{\log r, s\}$ *to be sufficiently small.*
(iii) *If* (13.2.6) *holds, then there is a sufficiently small mesh size depending on* $p, 1 < p < \infty$, *such that* $\mathcal{D}$ *is bounded and invertible on* $L^p(\mathbb{R}^n)$.

Furthermore, the functions

$$\phi_{km}(x) = \pi^*(\lambda_{km})\phi(x), \quad \psi_{km}(x) = \pi^*(\lambda_{km})\psi(x)$$

then form a dilatation bi-frame for $L^p(\mathbb{R}^n)$.

13.3   The deRham complex and Hardy spaces of forms

Each of the several ways of thinking of the classical Hardy spaces has a natural counterpart in a wide variety of settings. In this section the spaces are comprised of solutions of the Hodge-deRham (d, d^*)-system of operators on forms defined on the upper half-space Ω. It will be helpful to recall some of the well-known, though not necessarily familiar, ideas associated with these operators (see also [Fo; GM1 pp. 49–65]).

The *exterior algebra*

$$\Lambda_*(\mathbb{R}^n) = \oplus \sum_{r=0}^{n} \Lambda_r(\mathbb{R}^n), \quad \Lambda_0 = \mathbb{R}, \quad \Lambda_1 = \mathbb{R}^n, \ldots \qquad (13.3.1)$$

associated with $\mathbb{R}^n$ is the algebra of linear combinations of wedge products $v_1 \wedge \ldots \wedge v_r$ of vectors in $\mathbb{R}^n$. This *exterior product* is r-linear and alternating, i.e., $v \wedge w = -w \wedge v$. The algebra $\Lambda_*(\mathbb{R}^n)$ also becomes a Hilbert space by linear extension of the inner product on $\mathbb{R}^n$ to forms of fixed degree via

$$(v_1 \wedge \ldots \wedge v_r, w_1 \wedge \ldots \wedge w_r) = \det[v_k \cdot w_l]$$

and we shall use the usual $|\cdot|$ to denote the corresponding norm. Now fix a basis $\{e_1, \ldots, e_n\}$ for $\mathbb{R}^n$. The *exterior* and *interior multiplication operators* on $\Lambda_*(\mathbb{R}^n)$ are the linear operators defined by

$$\mu_j : \Lambda_r \longrightarrow \Lambda_{r+1}; \quad \mu_j(1) = e_j, \mu_j(e_k) = e_j \wedge e_k, \ldots \qquad (13.3.2)$$

and

$$\mu_j^* : \Lambda_r \longrightarrow \Lambda_{r-1}; \quad \mu_j^*(e_k) = \delta_{jk}, \mu_j^*(e_k \wedge e_l) = \delta_{jk}e_l - \delta_{jl}e_k, \ldots \quad (13.3.3)$$

respectively. Direct computation shows that:

$$\mu_j\mu_k + \mu_k\mu_j = 0; \quad \mu_j^*\mu_k^* + \mu_k^*\mu_j^* = 0; \quad \mu_j^*\mu_k + \mu_k\mu_j^* = \delta_{jk}I. \qquad (13.3.4)$$

An *r-form* is defined to be a vector–valued function

$$\sum_{i_1,\ldots,i_r} f_{i_1,\ldots,i_r} e_{i_1} \wedge \ldots \wedge e_{i_r}, \qquad (13.3.5)$$

where $f_{i_1,\ldots,i_r}$ is a (real-valued) function on $\mathbb{R}^n$. We shall denote by $L^p(\mathbb{R}^n, \Lambda_*(\mathbb{R}^n))$ the usual Lebesgue space of all p-integrable forms on $\mathbb{R}^n$.

On such forms the *exterior* and *interior Riesz transform* operators

$$\mathcal{R} = \sum_{k=1}^{n} \mu_k R_k; \quad \mathcal{R}^* = -\sum_{k=1}^{n} \mu_k^* R_k, \tag{13.3.6}$$

where R_k denote the usual Riesz transforms, define bounded operators such that

$$\int_{\mathbb{R}^n} (\mathcal{R}f(x), g(x))dx = \int_{\mathbb{R}^n} (f(x), \mathcal{R}^* g(x))dx \tag{13.3.7}$$

with respect to the inner product on Λ_*; furthermore,

$$\mathcal{R}\mathcal{R} = 0, \quad \mathcal{R}^*\mathcal{R}^* = 0; \quad \mathcal{R}\mathcal{R}^* + \mathcal{R}^*\mathcal{R} = I \tag{13.3.8}$$

because of (13.3.3) and the fact that $\sum_k R_k^2 = -I$.

THEOREM 13.3.1
The operators $\mathcal{R}\mathcal{R}^*$ and $\mathcal{R}^*\mathcal{R}$ are projections on $L^p(\mathbb{R}^n, \Lambda_r)$ for each $r = 0, 1, \ldots, n$ such that

(i)$\mathcal{R}\mathcal{R}^*$ *projects* $L^p(\mathbb{R}^n, \Lambda_r)$ *onto* $\{f \in L^p(\mathbb{R}^n, \Lambda_r) : \mathcal{R}f = 0\}$,
(ii)$\mathcal{R}^*\mathcal{R}$ *projects* $L^p(\mathbb{R}^n, \Lambda_r)$ *onto* $\{f \in L^p(\mathbb{R}^n, \Lambda_r) : \mathcal{R}^*f = 0\}$.

PROOF In view of (13.3.7),

$$(\mathcal{R}^*\mathcal{R})^2 = (\mathcal{R}^*\mathcal{R}); \quad (\mathcal{R}\mathcal{R}^*)^2 = (\mathcal{R}\mathcal{R}^*);$$
$$(\mathcal{R}^*\mathcal{R})(\mathcal{R}\mathcal{R}^*) \quad = \quad 0 = (\mathcal{R}\mathcal{R}^*)(\mathcal{R}^*\mathcal{R}).$$

The result now follows immediately from the boundedness of $\mathcal{R}$ and $\mathcal{R}^*$.
∎

Corresponding spaces of forms can be defined on the upper half-space $\Omega = \mathbb{R}_+^{n+1}$. Let e_0 be the *inward* unit normal to this half-space, so that $\{e_0, \ldots, e_n\}$ is now a basis for the ambient space $\mathbb{R}^{n+1}$, and let $\mathcal{I} = \mu_0$ and $\mathcal{I}^* = \mu_0^*$ be the exterior and interior multiplication operators associated with this new element. Clearly,

$$\mathcal{I}\mathcal{I}^* + \mathcal{I}^*\mathcal{I} = I; \quad (\mathcal{I}\mathcal{I}^*)^2 = \mathcal{I}\mathcal{I}^*; \quad (\mathcal{I}^*\mathcal{I})^2 = \mathcal{I}^*\mathcal{I}. \tag{13.3.9}$$

The Hodge-deRham (d, d^*)-operators on smooth forms on any open set in Ω are defined by

$$d = \sum_{k=1}^{n} \mu_k \frac{\partial}{\partial x_k} + \mu_0 \frac{\partial}{\partial t}; \quad d^* = \left[\sum_{k=1}^{n} \mu_k^* \frac{\partial}{\partial x_k} + \mu_0^* \frac{\partial}{\partial t} \right]. \tag{13.3.10}$$

In view of (13.3.3) therefore

$$dd^* + d^*d = \Delta, \tag{13.3.11}$$

so any solution of the first-order system $dF = 0, d^*F = 0$ will have harmonic components. In other words (d, d^*) is a Generalized Cauchy-Riemann system in the sense of Stein-Weiss (see [SW]). In fact, the following local characterization of solutions is a simple consequence of the Poincaré lemma.

THEOREM 13.3.2   Local Characterization
The solutions of the Hodge-deRham system

$$dF = 0; \quad d^*F = 0$$

on r-forms are characterized as being locally of the form $F = d\Phi$ where Φ is an $(r-1)$-form such that (i) Φ has harmonic components, and (ii) $d^*\Phi = 0$.

The relation between differentiation and Riesz transforms will play a fundamental role below, e.g., (13.5.6).

DEFINITION 13.3.3   *A form $g : \mathbb{R}^n \longrightarrow \Lambda_*(\mathbb{R}^n)$ is said to be massless if $\mathcal{R}^*g = 0$.*

For the case of one-forms, massless is the same as divergence free.
   Hardy spaces of solutions of the (d, d^*)-system on Ω are defined in the traditional way.

DEFINITION 13.3.4   *For each $p > 0$, the Hardy space $H^p(\Omega, \Lambda_r(\mathbb{R}^{n+1}))$ consists of all r-form solutions, $1 \leq r \leq n$, of $dF = 0$, $d^*F = 0$ for which*

$$\|F\|_{H^p} = \sup_{s>0}\left(\int_{\mathbb{R}^n} |F(x,s)|^p dx\right)^{\frac{1}{p}}$$

is finite.

For convenience the notation $H^p_{\rho_r}(\Omega)$ will be used instead of $H^p(\Omega, \Lambda_r (\mathbb{R}^{n+1}))$, emphasizing the fact that such a Hardy space is the one canonically associated with the irreducible representation of SO_{n+1} on $\Lambda_r(\mathbb{R}^{n+1})$, that is, the one having signature $\rho_r = (1, 1, \ldots, 1, 0 \ldots, 0)$ with 1 in the first r entries. When $p \geq 1$ these Hardy spaces are Banach spaces. As Stein-Weiss showed, each F in $H^p_{\rho_r}(\Omega)$ with $p > n/(n+1)$ has boundary values $F_+ = \lim_{t\rightarrow 0+} F(x,t)$ almost everywhere on $\mathbb{R}^n$; furthermore, F_+ belongs to $L^p(\mathbb{R}^n, \Lambda_r(\mathbb{R}^{n+1}))$ and F can be recovered from these boundary values by Poisson extension

$$F(x,s) = \int_{\mathbb{R}^n} P_s(x - v)F_+(v)dv. \tag{13.3.12}$$

We shall therefore denote by

$$H^p_{\rho_r}(\mathbb{R}^n) = \left\{ F_+ : F \in H^p_{\rho_r}(\Omega) \right\} \tag{13.3.13}$$

the subspace of $L^p(\mathbb{R}^n, \Lambda_r(\mathbb{R}^{n+1}))$ consisting of all such boundary functions.

THEOREM 13.3.5
The functions in $H^p_{\rho_r}(\mathbb{R}^n)$, $1 \leq r \leq n$, $1 < p < \infty$ are precisely those of the form $F_+ = \mathcal{R}f + \mathcal{I}f$ with f a 'massless' $(r-1)$-form.

PROOF First suppose that $F_+ = \mathcal{R}f + \mathcal{I}f$ for some massless $(r-1)$-form f, and let

$$F(x,s) = \int_{\mathbb{R}^n} P_s(x - v) F_+(v) dv.$$

The condition that the L^p-norms of $F(\cdot, s)$ are uniformly bounded is just Young's inequality. On the other hand, the condition that $F(x, s)$ satisfies $dF = d^* F = 0$ is verified easily by checking the corresponding conditions on the n-dimensional Fourier transform of $F(x, s) = P_s * (\mathcal{R} + \mathcal{I})(f)(x)$. The proof of the converse, that is, that every $F_+ = \mathcal{R}f + \mathcal{I}f$ for a massless f in L^p is contained in Section 13.5 below, in particular, see (13.5.6)(iii). ∎

It is instructive to recognize the relationship between Theorem (13.3.18) and the local characterization of Theorem (13.3.13). A simple computation shows that for any form $g : \mathbb{R}^n \to \Lambda_*(\mathbb{R}^{n+1})$ one has

$$P_t * (\mathcal{R} + \mathcal{I})g = d \circ S(g)$$

where S is the *single layer potential operator* of convolution with the kernel

$$-\frac{1}{(n-1)\omega_{n+1}} \frac{1}{(|x - y|^2 + t^2)^{\frac{n-1}{2}}}.$$

Therefore $F(x,t) = P_t * (\mathcal{R} + \mathcal{I})f = d \circ S(f) = d\Phi$ where f is as in Theorem (13.3.18) and Φ is as in Theorem (13.3.13). Of course, d is not one-to-one, so one cannot conclude that $Sf = \Phi$. The point is that f provides the link between the theory of forms that are analytic upstairs and thus satisfy an overdetermined system of equations, and the boundary theory where a preferred component generates solutions.

In view of Theorem (13.3.18) we define the *real H^p-space*

$$\text{Re } H^p_{\rho_r}(\mathbb{R}^n) = \left\{ f : \mathcal{R}^* f = 0; \quad F_+ = \mathcal{R}f + \mathcal{I}f; \quad F \in H^p_{\rho_r}(\Omega) \right\}. \tag{13.3.14}$$

13.4　Projections on $H^p_{\rho_r}$

The purpose of this section is to describe algebraic conditions related to overdeterminedness of solutions of the (d, d^*)- system, rephrasing Theorem (13.3.18) in terms of the range of projections associated with the operators $\mathcal{I}, \mathcal{I}^*$ and the Riesz transform operators $\mathcal{R}, \mathcal{R}^*$.

LEMMA 13.4.1
The operators $(\mathcal{I} \pm \mathcal{R})$ and $(\mathcal{I} \pm \mathcal{R})^*$ provide decompositions of the identity in the sense that

$$\frac{1}{2}\left\{ (\mathcal{I} + \mathcal{R})(\mathcal{I} + \mathcal{R})^* + (\mathcal{I} + \mathcal{R})^*(\mathcal{I} + \mathcal{R}) \right\} = I$$

and

$$\frac{1}{2}\left\{ (\mathcal{I} - \mathcal{R})(\mathcal{I} - \mathcal{R})^* + (\mathcal{I} - \mathcal{R})^*(\mathcal{I} - \mathcal{R}) \right\} = I$$

on $L^p(\mathbb{R}^n, \Lambda_*(\mathbb{R}^{n+1}))$.

PROOF　The identities are immediate consequences of (13.3.3) and the anticommutation properties

$$\mathcal{I}\mathcal{R} + \mathcal{R}\mathcal{I} = \mathcal{I}\mathcal{R}^* + \mathcal{R}^*\mathcal{I} = \mathcal{I}^*\mathcal{R} + \mathcal{R}\mathcal{I}^* = \mathcal{I}^*\mathcal{R}^* + \mathcal{R}^*\mathcal{I}^* = 0. \qquad (13.4.1)$$

of $\mathcal{I}, \mathcal{I}^*, \mathcal{R}$, and $\mathcal{R}^*$.　∎

To relate these operators to overdeterminedness of the (d, d^*)-system, let

$$F = (\mathcal{I}^*\mathcal{I})F + (\mathcal{I}\mathcal{I}^*)F \equiv F_{\text{Tan}} + \mathcal{I}F_{\text{Nor}} \qquad (13.4.2)$$

be the orthogonal decomposition of an r-form into *tangential* and *normal* components, cf., (13.3.10). Notice that F_{Tan} is an r-form containing no e_0-terms, while F_{Nor} is an $(r-1)$-form also containing no e_0 terms.

LEMMA 13.4.2
The operators

$$\mathcal{P} = \frac{1}{2}(\mathcal{I} + \mathcal{R})(\mathcal{I} + \mathcal{R})^*, \quad \mathcal{Q} = \frac{1}{2}(\mathcal{I} - \mathcal{R})^*(\mathcal{I} - \mathcal{R})$$

are commuting projections on $L^p(\mathbb{R}^n, \Lambda_*(\mathbb{R}^{n+1}))$ such that

$$\mathcal{P}^2 = \mathcal{P}; \quad \mathcal{Q}^2 = \mathcal{Q}; \quad \mathcal{P}\mathcal{Q} = \mathcal{Q}\mathcal{P}.$$

The range of $\mathcal{P}$ is the subspace of forms satisfying $F_{\text{Tan}} = \mathcal{R}F_{\text{Nor}}$; whereas the range of $\mathcal{Q}$ is the subspace of forms satisfying $F_{\text{Nor}} = \mathcal{R}^*F_{\text{Tan}}$.

PROOF The boundedness of $\mathcal{P}$ and $\mathcal{Q}$ are clear, and the identities for $\mathcal{P}$ and $\mathcal{Q}$ are simple algebraic manipulations using (13.3.7) and (13.3.10). To identify the range of $\mathcal{P}$ and $\mathcal{Q}$, equate tangential and normal components of F with $\mathcal{P}F$ and F with $\mathcal{Q}F$ respectively. We leave the details to the reader. ∎

THEOREM 13.4.3
A form $F \in L^p(\mathbb{R}^n, \Lambda_r(\mathbb{R}^{n+1}))$, $1 < p < \infty$, *satisfies* $\mathcal{Q}\mathcal{P}F = F$ *if and only if* $F = \mathcal{R}f + \mathcal{I}f$ *for some* f *in* $\mathrm{Re}\, H^p_{\rho r}(\mathbb{R}^n)$; *in particular,* $\mathcal{Q}\mathcal{P}$ *is an orthogonal projection from* $L^p(\mathbb{R}^n, \Lambda_r(\mathbb{R}^{n+1}))$ *onto* $H^p_{\rho r}(\mathbb{R}^n)$ *for each* $1 < p < \infty$ *and* $1 \le r \le n$.

PROOF Suppose first that $F = \mathcal{R}f + \mathcal{I}f$ where $\mathcal{R}^*f = 0$. Since f has no e_0-component,

$$\mathcal{P}F = \frac{1}{2}(\mathcal{I}+\mathcal{R})(\mathcal{I}^*\mathcal{I}-\mathcal{I}\mathcal{R}^* - \mathcal{R}\mathcal{I}^* + \mathcal{R}^*\mathcal{R})f = \frac{1}{2}(\mathcal{I}+\mathcal{R})(\mathcal{I}^*\mathcal{I}+\mathcal{R}^*\mathcal{R})f = F.$$

Similarly,

$$\begin{aligned} \mathcal{Q}F = \mathcal{Q}(\mathcal{R}+\mathcal{I})f &= \tfrac{1}{2}(\mathcal{I}^* - \mathcal{R}^*)(\mathcal{I} - \mathcal{R})(\mathcal{R}+\mathcal{I})f \\ &= (\mathcal{I}^* - \mathcal{R}^*)\mathcal{I}\mathcal{R}f = (\mathcal{R}+\mathcal{I})f = F \end{aligned}$$

again because of the hypotheses on f. Thus $\mathcal{P}\mathcal{Q}F = F$, showing that $H^p_{\rho r}(\mathbb{R}^n)$ lies in the range of $\mathcal{P}\mathcal{Q}$.

Conversely, suppose $\mathcal{P}\mathcal{Q}F = F$. By the commutativity of $\mathcal{P}$ and $\mathcal{Q}$, the conditions $\mathcal{P}F = F$ and $\mathcal{Q}F = F$ then hold simultaneously. Consequently, by (13.4.4) $F_{\mathrm{Tan}} = \mathcal{R}F_{\mathrm{Nor}}$ and $\mathcal{R}^*F_{\mathrm{Nor}} = 0$. Thus $F = \mathcal{R}f + \mathcal{I}f$ with $f = F_{\mathrm{Nor}}$. The remainder of the theorem now follows from (13.3.18). ∎

13.5 Lusin's representation for $H^p_{\rho r}$

Solutions of the (d, d^*)-system have many of the properties of analytic functions of one complex variable — as they should since a complex-valued function $F = U + iV$ satisfies $\bar{\partial}F = 0$ if and only if the 1-*form* $F = Ve_0 + Ue_1$ satisfies $dF = 0$, $d^*F = 0$. This analogy carries over to Lusin's representation of analytic functions. In higher dimensions the Dirac operator

$$D = \sum_{j=1}^{n}(\mu_j - \mu_j^*)\frac{\partial}{\partial x_j} + (\mu_0 - \mu_0^*)\frac{\partial}{\partial t} = d - d^*. \tag{13.5.1}$$

on $\Lambda_r(\mathbb{R}^{n+1})$-valued functions plays the role of the Cauchy-Riemann $\bar{\partial}$-operator. In particular, any r-form solution of $DF = (d - d^*)F = 0$ must also be a solution of

$$dF = 0, \quad d^*F = 0$$

separately, since dF is an $(r+1)$-form and d^*F is an $(r-1)$-form whenever F is an r-form. Thus on $\Lambda_r(\mathbb{R}^{n+1})$-valued functions the Dirac operator reduces to the (d, d^*)-system, a fact first observed by Marcel Riesz. This is the simplest way in which Hodge-deRham theory can be subsumed within the theory of Dirac operators. It is well known that there is a Cauchy integral theory associated with (13.5.1) (cf. for instance, [GM1]). Let

$$\Gamma(x,t) = \frac{1}{(|x|^2 + t^2)^{\frac{1}{2}(n-1)}} = \frac{1}{|z|^{n-1}} \quad (z = (x,t))$$

be the fundamental solution of the Laplacian on $\mathbb{R}^{n+1}$ and set

$$C(z) = D\Gamma = (1-n)\left\{ \sum_{j=1}^{n}(\mu_j - \mu_j^*)\frac{x_j}{|z|^{n+1}} + (\mu_0 - \mu_0^*)\frac{t}{|z|^{n+1}} \right\}. \quad (13.5.2)$$

A simple application of Stokes' theorem shows that in any reasonable domain M in Ω a Cauchy integral formula

$$F(z) = \frac{1}{\omega_{n+1}} \quad \int_{\partial M} \qquad\qquad\qquad\qquad (13.5.3)$$

$$\left(\sum_{j=0}^{n}(\mu_j - \mu_j^*)\frac{z_j - w_j}{|z - w|^{n+1}} \right)\eta(w)F(w)\,dS(w) \quad (z \in M).$$

holds for any solution of $DF = 0$; here

$$\eta(w) = \sum_{j=0}^{n}(\mu_j - \mu_j^*)\eta_j(w)$$

is the unit outer normal to ∂M at w, $dS(w)$ is the scalar element of surface area on ∂M, and ω_n is the surface area of the unit sphere in $\mathbb{R}^n$. But, if F is an r-form, we can omit from the integrand any term which is not Λ_r-valued, and so produce a Cauchy integral formula *specific to r-forms*:

$$F(z) = \qquad\qquad\qquad\qquad\qquad\qquad\qquad\qquad (13.5.4)$$

$$-\int_{\partial M}\sum_{j,k}\frac{z_j - w_j}{|z - w|^{n+1}}\eta_k(w)\big\{\delta_{jk} - (\mu_k\mu_j^* - \mu_j\mu_k^*)\big\}F(w)dS(w),$$

a formula first given for 1-forms by Paul Weiss in 1946 (cf., [GM1, p. 244]). But when ∂M is a hyperplane $\mathbb{R}^n + te_0$ of height t above $\mathbb{R}^n$, then e_0 is its unit inward normal, and only $\eta_0(w) \neq 0$. Hence

$$
\begin{aligned}
F(v, s+t) &= \int_{\mathbb{R}^n} \frac{s}{(|x-v|^2+s^2)^{\frac{1}{2}(n+1)}} F(v, t)\, dt \\
&\quad - e_0 \wedge \sum_{j=1}^n \left\{ \int_{\mathbb{R}^n} \frac{x_j - v_j}{(|x-v|^2+s^2)^{\frac{1}{2}(n+1)}} \left(\mu_j^* F(v, t) \right) dv \right\} \\
&\quad + \sum_{j=1}^n e_j \wedge \left\{ \int_{\mathbb{R}^n} \frac{x_j - v_j}{(|x-v|^2+s^2)^{\frac{1}{2}(n+1)}} \mu_0^* F(v, t)\, dv \right\}.
\end{aligned}
$$

This further simplifies when F is split into its tangential and normal components, for then

$$
e_0 \wedge (\mu_j^* F) = e_0 \wedge (\mu_j^* F_{\mathrm{Tan}}) \ , \quad e_j \wedge (\mu_0^* F) = e_j \wedge F_{\mathrm{Nor}} \quad (1 \le j \le n).
$$

Consequently, the Cauchy integral over the hyperplane $\mathbb{R}^n + t e_0$ is given by

$$
F(x, s+t) = \tfrac{1}{2} P_s * \left\{ F(., t) + \mathcal{R} F_{\mathrm{Nor}}(., t) + \mathcal{I} \mathcal{R}^* F_{\mathrm{Tan}}(., t) \right\}, \tag{13.5.5}
$$

expressing $2F$ as the harmonic extension of F and the Riesz transform operators $\mathcal{R}, \mathcal{R}^*$ applied, respectively, to the normal and tangential components of F.

At this point the proof of Theorem (13.3.18) can be completed:

$$
\begin{aligned}
F(x, s) &= \lim_{t \to 0} F(x, s+t) \tag{13.5.6} \\
&= \tfrac{1}{2} P_s * \left\{ F_+(.) + \mathcal{R} F_{\mathrm{Nor},+}(.) + \mathcal{I} \mathcal{R}^* F_{\mathrm{Tan},+}(.) \right\},
\end{aligned}
$$

and so

$$
F_+(x) = \lim_{s \to 0} F(x, s) = \tfrac{1}{2} \left\{ F_+(x) + \mathcal{R} F_{\mathrm{Nor},+}(x) + \mathcal{I} \mathcal{R}^* F_{\mathrm{Tan},+}(x) \right\} \tag{13.5.7}
$$

(both limits certainly exist when $1 < p < \infty$). Equating tangential and normal components, we deduce that $F_{\mathrm{Tan}} = \mathcal{R} F_{\mathrm{Nor}}$, while $F_{\mathrm{Nor}} = \mathcal{R}^* F_{\mathrm{Tan}}$; hence automatically, $\mathcal{R}^* F_{\mathrm{Nor}} = 0$ and $\mathcal{R} F_{\mathrm{Tan}} = 0$ by Theorem (13.3.9).

The Lusin representation for r-form solutions of the (d, d^*)-system and the associated wavelet decompositions now follow easily. For any F having harmonic components,

$$
F(x, s+t) = P_s * F(., t) = P_{s+t} * F_+ \ ; \tag{13.5.8}
$$

consequently, each $F \in H^p_{\rho r}(\Omega)$ can be written as

$$
F(x, s+t) = P_{s+t} * (\mathcal{P} \mathcal{Q} F_+) \tag{13.5.9}
$$

After integration by parts twice we thus obtain the representation

$$
F(x, s) = 4 \int_0^\infty t \frac{\partial P_t}{\partial t} * \frac{\partial P_{s+t}}{\partial t} * (\mathcal{P} \mathcal{Q} F_+)(x)\, dt \tag{13.5.10}
$$

$$= 4 \int_0^\infty t \frac{\partial P_{s+t}}{\partial t} * \left(\mathcal{P} \mathcal{Q} \frac{\partial F}{\partial t} \right) (x, t) \, dt$$

$$= 4 \int_0^\infty \int_{\mathbb{R}^n} t \left(\mathcal{P} \mathcal{Q} \frac{\partial P_{s+t}}{\partial t} \right) (x - y) \frac{\partial F}{\partial t} (y, t) \, dy dt.$$

Hence for $F \in H_{p_r}^p(\Omega)$ the analogue of Lusin's representation becomes

$$F(x, s) = \int_\Omega K(z, \zeta) \alpha(\zeta) \, d\zeta \tag{13.5.11}$$

where

$$K(z, \zeta) = \left(\mathcal{P} \mathcal{Q} \frac{\partial P_{s+t}}{\partial t} \right) (x - y), \quad \begin{cases} z &= (x, s), \\ \zeta &= (y, t), \end{cases}$$

and $d\zeta$ is Lebesgue measure on Ω, the natural choice of α being $\alpha(\zeta) = t \partial F / \partial t$. It is important to note that if $\mathcal{P}\alpha = \alpha$ or $\mathcal{Q}\alpha = \alpha$, as is the case for the natural choice of α, then $\mathcal{Q}$ or $\mathcal{P}$, respectively, can be omitted from the kernel $K = K(z, \zeta)$ in Lusin's representation. For instance,

$$F(x, s) = 4 \int_0^\infty \mathcal{Q} \left(t \frac{\partial P_t}{\partial t} * t \frac{\partial P_t}{\partial t} * P_s \right) * F_+(x) \frac{dt}{t}$$

from which the prototypical Calderòn reproducing formula

$$F_+ = 2 \int_0^\infty t [(\mathcal{I} - \mathcal{R}) \frac{\partial P_t}{\partial t}]^* * [(\mathcal{I} - \mathcal{R}) \frac{\partial P_t}{\partial t}] * F_+ \, dt \tag{13.5.12}$$

follows, letting $s \to 0+$. This last representation can clearly be extended to scalar-valued functions other than $t \partial P_t / \partial t$; indeed, the representation

$$F_+ = \tfrac{1}{2} \int_0^\infty [(\mathcal{I} - \mathcal{R}) \psi]_t^* * [(\mathcal{I} - \mathcal{R}) \phi]_t * F_+ \frac{dt}{t} \tag{13.5.13}$$

holds for any pair of functions ϕ, ψ in $L^2(\mathbb{R}^n)$ such that

$$\int_0^\infty \hat{\psi}(t\xi) \overline{\hat{\phi}(t\xi)} \frac{dt}{t} = 1 \quad (|\xi| = 1). \tag{13.5.14}$$

Such a representation can be interpreted as a Continuous Wavelet transform for F_+ in the Hardy space $H_{p_r}^p(\mathbb{R}^n)$. There is an entirely analogous representation for such F_+ when $\mathcal{Q}$ is replaced by $\mathcal{P}$. Taking normal components in (13.5.11), we finally obtain the "real" Calderón reproducing formula

$$f = \int_0^\infty [\mathcal{R}\psi]_t^* * [\mathcal{R}\phi]_t * f \frac{dt}{t} \tag{13.5.15}$$

for 'massless' f in $L^p(\mathbb{R}^n)$. Of course, if one is interested solely in 'massless' forms this formula is easily derived without any reference to analytic forms.

13.6 Discrete frames: the case of one-forms

In this section we show how one can obtain discrete frames after replacing the reproducing formula (13.5.12) by a discrete approximation. We shall only consider the one-form case to make the presentation simpler, even though the same methods apply for the case of r-forms. If f is a 'massless' one-form, that is, a divergence-free vector field in $L^p(\mathbb{R}^n, \mathbb{R}^n)$, then

$$
\begin{aligned}
\sum_{j=1}^n \mu_j^* \int_0^\infty \; & [R_j\phi]_t^* * [\mathcal{R}\phi]_t * f \frac{dt}{t} \\
&= \sum_{j=1}^n \mu_j^* \sum_{k=1}^n \mu_k \int_0^\infty [R_j\phi]_t^* * [R_k\phi]_t * f \frac{dt}{t} \\
&= \sum_{j=1}^n \mu_j^* \sum_{k=1}^n \mu_k \sum_{l=1}^n \left\{ \int_0^\infty [R_j\phi]_t^* * [R_k\phi]_t * f_l \frac{dt}{t} \right\} e_l.
\end{aligned}
$$

We may approximate the last integral, replacing it by a Riemann sum over a sufficiently dense dilatation lattice Λ just as in Corollary (13.2.19):

$$
\mathcal{D}f = \sum_{j=1}^n \mu_j^* \sum_{k=1}^n \mu_k \sum_{l=1}^n \left\{ \sum_{\lambda \in \Lambda} (f_l, \pi^*(\lambda) R_k \phi) \pi^*(\lambda) R_j \phi \right\} e_l. \tag{13.6.1}
$$

This expression approximates f, a massless vector field, by a sum of massless vector "wavelets". That is, for each λ, the λ-th term in the sum, call it h_λ, satisfies $\mathcal{R}^* h_\lambda = 0$:

$$
\begin{aligned}
h_\lambda &= \sum_{j=1}^n \mu_j^* \sum_{k=1}^n \mu_k \sum_{l=1}^n \left\{ (f_l, \pi^*(\lambda) R_k \phi) \pi^*(\lambda) R_j \phi \right\} e_k \wedge e_l \\
&= \sum_{k,l=1}^n \{(f_l, \pi^*(\lambda) R_k \phi)\} \pi^*(\lambda)(R_k \phi e_l - R_l \phi e_k).
\end{aligned}
$$

Applying $\mathcal{R}^*$ to this last expression yields

$$
\sum_{k,l=1}^n \{(f_l, \pi^*(\lambda) R_k \phi)\}[R_l \pi^*(\lambda) R_k \phi - R_k \pi^*(\lambda) R_l \phi] = 0
$$

since $\pi(\lambda)$ commutes with the action of the Riesz transforms. This shows that, at least in a formal sense, we may approximate massless vector fields by linear superpositions of massless vector wavelets arising naturally from the action of a dilatation lattice on $\mathbb{R}^n$. We may *expand* f in terms of a Neumann expansion once we show that the corresponding discrete frame operators are continuous and invertible on $H_{\rho_1}^p(\mathbb{R}^n)$. For this purpose we can apply the approximation machinery outlined in Section 13.2 componentwise.

THEOREM 13.6.1
Given $1 < p < \infty$. Let Λ be the dilatation lattice of points $\lambda = \lambda_{km} = (sk/r^m, 1/r^m)$, where $m \in \mathbb{Z}$, $k \in \mathbb{Z}^n$, and $r > 1$ and $s > 0$ are fixed.

Define the corresponding lattice frame operator

$$\mathcal{T}_\Lambda f(x) = \sum_{j=1}^n \mu_j^* \sum_{k=1}^n \mu_k \sum_{l=1}^n \left\{ \sum_{\lambda \in \Lambda} (f_l, \pi^*(\lambda) R_k \phi) \pi^*(\lambda) R_j \phi \right\} e_l$$

where $\phi = \partial/\partial t P_t|_{t=1}$. *If* Λ *has sufficiently small mesh size, then* $\mathcal{T}_\Lambda$ *is bounded and invertible on* Re $H_{\rho_1}^p(\mathbb{R}^n)$. *Furthermore,* $\mathcal{T}_\Lambda(f)$ *converges to* f *in* Re $H_{\rho_1}^p(\mathbb{R}^n)$ *as the mesh size of* Λ *tends to zero.*

PROOF In view of the remarks above, it suffices that the (j, k, l)-component

$$\int_0^\infty [R_j \phi]_t^* [R_k \phi]_t * f_l(\cdot) \frac{dt}{t}$$

can be approximated in $L^p(\mathbb{R}^n)$ by the corresponding lattice sum

$$\sum_{\lambda \in \Lambda} (f_l, \pi^*(\lambda) R_k \phi) \pi^*(\lambda) R_j \phi,$$

because one may then apply Minkowski's theorem for the finite sum over the (j, k, l) components to obtain the L^p-approximation. But this approximation is an immediate consequence of Theorem (13.2.18) since the functions $R_j \phi$ belong to $\mathcal{S}_\delta^{(0)} \cap \mathcal{S}_\gamma(\mathbb{R}^n)$ whenever $0 < \delta < 1$ and $1 < \gamma < 2$. ∎

13.7 Concluding remarks and problems

We have studied massless r–form wavelet frames in terms of the real H^p-theory associated with the Hodge-deRham (d, d^*)-system and, in particular, obtained a reproducing formula for Re $H^p(\mathbb{R}^n)$ when $1 < p < \infty$.

(i) A more subtle problem is that of obtaining an explicit boundary value theory for solutions $F : \Omega \longrightarrow \Lambda_*(\mathbb{R}^{n+1})$ of $DF = 0$ where $D = d - d^*$ is the Dirac operator. The difficulty stems from the fact that the projection F_r of a solution F of $DF = 0$ onto forms of degree r is not, in general, itself a solution of $DF_r = 0$, so that one no longer has simple descriptions of the associated space Re $H^p(\mathbb{R}^n, \Lambda_*(\mathbb{R}^{n+1}))$ as in Theorem (13.3.17) and Lemma (13.4.4). For example, the Cauchy integral formula (13.5.6) generalizes to

$$F(x, s+t) = \tfrac{1}{2} P_s * \left\{ F(\cdot, t) + \mathcal{R} F_{\mathrm{Nor}}(\cdot, t) \right.$$
$$\left. + \mathcal{I} \mathcal{R}^* F_{\mathrm{Tan}}(\cdot, t) + \mathcal{R} \mathcal{I} F_{\mathrm{Tan}}(\cdot, t) + \mathcal{R}^* F_{\mathrm{Nor}}(\cdot, t) \right\}.$$

But if $F_{\mathrm{Tan}}^{(r)}$ denotes the degree r part of F_{Tan} (similarly for $F_{\mathrm{Nor}}^{(r)}$), the general conditions relating tangential and normal components are

$$F_{\mathrm{Tan}}^{(r)} = \mathcal{R}(F_{\mathrm{Nor}}^{(r-1)}) + \mathcal{R}^*(F_{\mathrm{Nor}}^{(r+1)}); \quad F_{\mathrm{Nor}}^{(r)} = \mathcal{R}^*(F_{\mathrm{Tan}}^{(r+1)}) - \mathcal{R}(F_{\mathrm{Tan}}^{(r-1)}).$$

Replacing the upper half space Ω by a general Lipschitz domain in $\mathbb{R}^{n+1}$ is clearly of interest also.

(ii) Even when we restrict to forms of fixed degree, there are important ways of generalizing (13.5.11). For example, one can look for general conditions on forms $\Phi = \sum_{j=1}^{n} \phi_j e_j$ and $\Psi = \sum_{k=1}^{n} \psi_j e_j$ so that

$$f(\cdot) = \int_0^\infty \Psi_t^* * \Phi_t * f \frac{dt}{t}$$

holds. Here (13.5.11) is the special case when $\Phi = \Psi = \mathcal{R}\phi$. It is at this point that the representation theoretic ideas will play an interesting role. It would be useful, for example, to have Φ and/or Ψ having compact support (in the sense that each component does). Such compactly supported forms could be used to fashion atomic decompositions for Re $H_{\rho_r}^p(\mathbb{R}^n)$.

(iii) One would also like to develop a theory of form-valued H^p–spaces where $p < 1$ and, in fact, p is smaller than the critical index of harmonicity. In this case the boundary-value forms would need to satisfy an additional cancellation condition in addition to the massless condition; for example, each component must satisfy a certain vanishing moment condition. The first step in developing such a theory would be to determine exactly what this cancellation condition should be.

Bibliography

[AEPT] J. Aguirre, M. Escobedo, J. C. Peral, and Ph. Tchamitchian, *Basis of wavelets and atomic decompositions of $H^1(\mathbb{R}^n)$ and $H^1(\mathbb{R}^n \times \mathbb{R}^n)$*, Proc. Amer. Math. Soc. **111** (1991), 683–693.

[AAG] S. Ali, J.-P. Antoine, and J.-P. Gazeau, *Square integrability of group representations on homogeneous spaces. I. Reproducing triples and frames; II. Coherent and quasi-coherent states. The case of the Poincaré group*, Ann. Inst. H. Poincaré Phys. Théor. **55** (1991), 829–855; 856–890.

[BF] G. Battle and P. Federbush, *Divergence-free vector wavelets*, Michigan Math. J. **40**, (1993), 181–195.

[CZ] A.P. Calderón and A. Zygmund, *Algebras of certain singular integral operators*, Amer. J. Math. **78**, (1956), 310–320.

[CKS]   D.C. Chang, S. Krantz and E. Stein, H^p-theory on a smooth do-
        main in $\mathbb{R}^N$ and elliptic boundary value problems, J. Funct. Anal.
        114, (1993), 286–347.

[CGT]   J.-A. Chao, J.E. Gilbert and P.A. Tomas, Molecular decomposi-
        tions in H^p-theory, Suppl. Rend. Circ. Mat. Palermo **1**, (1981),
        115–119.

[Ch]   F. M. Christ, Lectures on Singular Integral Operators, CBMS Re-
        gional Conf. Ser. in Math., No. 77, AMS, Providence, RI, 1990.

[CW]   R. Coifman and G. Weiss, Extensions of Hardy spaces and their
        use in analysis, Bull. Amer. Math. Soc. **83**, (1977) 569–645.

[D1]   I. Daubechies, The wavelet transform, time-frequency localization
        and signal analysis, IEEE Trans. Inform. Theory **36**, (1990), 961–
        1005.

[D2]   I. Daubechies, Ten Lectures on Wavelets, CBMS-NSF Regional
        Conf. Ser. in Appl. Math., SIAM, Philadelphia, 1992.

[DGM]   I. Daubechies, A. Grossmann and Y. Meyer, Painless nonorthog-
        onal expansions, J. Math. Phys. **27**, (1986), 1271–1283.

[DGK]   K.M. Davis, J.E. Gilbert and R.A. Kunze, Elliptic differential
        operators in harmonic analysis, I. Generalized Cauchy-Riemann
        systems, Amer. J. Math. **113**, (1991), 75–116.

[FG]   H. Feichtinger and K.H. Gröchenig, Non-orthogonal wavelet and
        Gabor expansions, and group representations, in "Wavelets and
        Their Applications," (M.B. Ruskai, et al., eds.), Jones and
        Bartlett, Boston, 1992, 353–375.

[Fo]   G.B. Folland, Harmonic analysis of the deRham complex on the
        sphere, J. Reine Angew. Math. **398**, (1989), 130–143.

[FJW]   M. Frazier, B. Jawerth and G. Weiss, Littlewood–Paley Theory
        and the Study of Function Spaces, CBMS Regional Conf. Ser. in
        Math., No. 79, AMS, Providence, RI, 1991.

[GHL]   J. Gilbert, J. Hogan, and J. Lakey, Maximal theorems for operators
        of Cotlar type and frames, preprint.

[GM1]   J. E. Gilbert and M.A.M. Murray, Clifford algebras and Dirac op-
        erators in harmonic analysis, Cambridge University Press, Cam-
        bridge, 1991.

[GM2]   J. E. Gilbert and M.A.M. Murray, H^p-theory on Euclidean space
        and the Dirac operator, Rev. Mat. Iberoam. **4**, (1988), 253–289.

[GMP]   A. Grossmann, J. Morlet, and T. Paul, Transforms associated to
        square integrable group representations. II. Examples, Ann. Inst.
        H. Poincaré Phys. Théor. **45**, (1986) 293–309.

[GrM] A. Grossmann and J. Morlet, *Decompostions of Hardy functions into square integrable wavelets of constant shape*, SIAM J. Math Anal. **15** (1984), 723-736.

[HW] C. Heil and D. Walnut, *Continuous and discrete wavelet transforms*, SIAM Review **31**, (1989), 628–666.

[Le] P. G. Lemarié-Rieusset, *Ondelettes vecteurs á divergence nulle*, C. R. Acad. Sci. Paris **313**, (1991), 213–216.

[M1] Y. Meyer, *Wavelets: Algorithms and Applications*, SIAM, Philadelphia, 1993.

[M2] Y. Meyer, *Ondelettes et Opérateurs*, Hermann, Paris, 1990.

[PW] R. Priebe and G. Wilson, *Application of "matched" wavelets to identification of metallic transients*, in Time-Frequency and Time-Scale Analysis, Proc. IEEE-SP Int. Symp., Victoria, Oct. 1992, IEEE, Piscataway, NJ, 349–252.

[St] E. Stein, *Singular Integrals and Differentiability Properties of Functions*, Princeton University Press, Princeton, NJ, 1970.

[SW] E.M. Stein and G. Weiss, *Generalization of the Cauchy-Riemann equations and representations of the rotation group*, Amer. J. Math. **90**, (1968), 163–196.

[Str] J.-O. Strömberg, *A modified Franklin system and higher-order spline systems on $\mathbb{R}^n$ as unconditional bases for Hardy spaces*, in Conference on Harmonic Analysis in Honor of Antoni Zygmund, pp.475–494, Wadsworth, Belmont, 1983.

[T1] B. Torresani, *Time-frequency representations: wavelet packets and optimal decompositions*, Ann. Inst. H. Poincaré Phys. Théor. **56**, (1992), 215–234.

[T2] B. Torresani, *Wavelets and representations of the affine Weyl–Heisenberg group*, J. Math. Phys. **32**, (1991), 1273–1279.

Department of Mathematics,
University of Texas,
Austin, Texas 78712
E-mail: gilbert@math.utexas.edu
E-mail: lakey@math.utexas.edu

School of Mathematics, Physics, Computing and Electronics,
Macquarie University,
North Ryde, New South Wales 2109,
AUSTRALIA
E-mail: jeffh@macadam.mpce.mq.edu.au

14

Applications of Clifford Analysis to Inverse Scattering for the Linear Hierarchy in Several Space Dimensions

Daniel B. Dix

ABSTRACT The Inverse Scattering Transform (IST) is a nonlinear analog of the Fourier transform which has been used to solve certain nonlinear evolution equations in the same way that the Fourier transform can be used to solve linear partial differential evolution equations. The current understanding of the IST is based on the complex inhomogeneous Cauchy-Riemann equation. This works fine in one space dimension, and has been extended (with some modifications) to two space dimensions, but there are fundamental difficulties which have yet to be overcome in order to have substantial higher-dimensional applications of the method. This situation is to be contrasted with that for the Fourier transform, which works equally well in all space dimensions. Leading researchers in [1] have called the problem of finding a satisfactory extension of IST to several space dimensions the most important open problem in this field. Here it will be shown how prototype forward and inverse scattering transforms (which reduce to the forward and inverse Fourier transforms) for the linear hierarchy in an arbitrary number of space dimensions can be constructed based on the inhomogeneous Dirac equation and Clifford analysis.

14.1 Introduction

In this work we explore the realm of Clifford analysis as a possible means of generalizing the inverse scattering method to an arbitrary number of space dimensions. The current approaches to the inverse scattering method (see Beals and Coifman [2], Ablowitz and Clarkson [1]) in more than two space dimensions do not possess as strong a resemblance to the Fourier method of

---

1991 *Mathematics subject Classification.* 35E15, 35Q58, 15A66.

solving the initial-value problem for linear evolutionary partial differential equations as one would like. So we present here a version of the inverse scattering method which works for the hierarchy of such linear equations in arbitrarily many space dimensions using the inhomogeneous Dirac equation and Clifford analysis. These results do not solve any new nonlinear equations, but they do indicate that the connection between Clifford analysis and inverse scattering theory is worthy of further investigation. In addition to the self-explanatory section titles, we remark that in section 14.3 we show how to embed complex-valued function theory into Clifford-valued function theory in the one-dimensional case. In section 14.4 we include a succinct but general description of inverse scattering theory (more properly, the *d-bar* method).

14.2   Review of Some Clifford Analysis

Good general references are Brackx, Delanghe and Sommen [4], Gilbert and Murray [9], and Li, McIntosh and Qian [12].

Let $\mathbb{C}_{(n)}$ denote the 2^n-dimensional $\mathbb{C}$-vector space (equipped with a multiplicative structure to be described below) spanned by $\{e_S\}_{S \subset \{1,\dots,n\}}$. Let e_0 denote $e_\emptyset$ and let e_j denote $e_{\{j\}}$, $j = 1, \dots, n$. Define an associative multiplication of $\pm$ basis elements by the rules: $e_{i_1} e_{i_2} \dots e_{i_k} = e_{\{i_1, i_2, \dots, i_k\}}$ whenever $1 \le i_1 < i_2 < \dots < i_k \le n$, e_0 is the multiplicative unit, $e_j^2 = -e_0$, $j = 1, \dots, n$, and $e_j e_k = -e_k e_j$ for $1 \le j \ne k \le n$. By requiring that multiplication in $\mathbb{C}_{(n)}$ agree with the multiplication just defined on $\pm$ basis elements, distribute over addition and commute with scalar multiplication, we see that $\mathbb{C}_{(n)}$ becomes an associative $\mathbb{C}$-algebra with unit. It is called the complex Clifford algebra of index n.

Suppose $\Omega \subset \mathbb{R}^{n+1}$ is an open subset and $f : \Omega \to \mathbb{C}_{(n)}$ is a differentiable function. So $f = \sum_{S \subset \{1,\dots,n\}} f_S e_S$ where $f_S : \Omega \to \mathbb{C}$ is a differentiable function. Define the Dirac operator by the rule $\mathcal{D} = \sum_{j=0}^n \partial_{k_j} e_j$ where $(k_0, k_1, \dots, k_n)$ are the coordinates in $\mathbb{R}^{n+1}$. The action of the Dirac operator $\mathcal{D}$ on the left and right of a function f is defined by

$$\mathcal{D}f = \sum_S \sum_{j=0}^n \frac{\partial f_S}{\partial k_j} e_j e_S, \qquad f\mathcal{D} = \sum_S \sum_{j=0}^n \frac{\partial f_S}{\partial k_j} e_S e_j.$$

We say f is *left* (resp. *right*) *monogenic* in Ω if $\mathcal{D}f = 0$ (resp. $f\mathcal{D} = 0$) in Ω. When $n > 1$ the pointwise product of monogenic functions is *not* necessarily monogenic.

Define the Cauchy kernel by

$$C(k_0, k_1, \ldots, k_n) = \frac{1}{\sigma_n} \frac{k_0 e_0 - \sum_{j=1}^{n} k_j e_j}{(\sum_{j=0}^{n} k_j^2)^{(n+1)/2}}, \quad \text{where} \quad \sigma_n = \frac{2\pi^{(n+1)/2}}{\Gamma((n+1)/2)}.$$

It is true that $\mathcal{D}C = C\mathcal{D} = \delta$ as ordinary vector-valued distributions on $\mathbb{R}^{n+1}$. If we allow our test functions to be Clifford algebra-valued, then we can define left (resp. right) distributions to be continuous $\mathbb{C}_{(n)}$-linear functions from $C_c^\infty(\Omega, \mathbb{C}_{(n)}) \to \mathbb{C}_{(n)}$, where these linear spaces are considered as right (resp. left) $\mathbb{C}_{(n)}$-modules (with the usual topologies). We denote the action of a left (resp. right) distribution F on the test function ϕ by $\langle F, \phi \rangle$ (resp. $\langle \phi, F \rangle$). The Dirac operator acts on a left (resp. right) distribution F by the rule $\langle F\mathcal{D}, \phi \rangle = -\langle F, \mathcal{D}\phi \rangle$ (resp. $\langle \phi, \mathcal{D}F \rangle = -\langle \phi\mathcal{D}, F \rangle$). The equation $C\mathcal{D} = \delta$ (resp. $\mathcal{D}C = \delta$) holds in the sense of left (resp. right) distributions.

Suppose that Σ is a smooth n-dimensional oriented manifold (smoothly) embedded in $\mathbb{R}^{n+1}$. A typical example is where $\Sigma = \partial\Omega$, where Ω is an open subset of $\mathbb{R}^{n+1}$ whose boundary $\partial\Omega$ is smooth. The orientation of Ω is derived from that of $\mathbb{R}^{n+1}$, and $\partial\Omega$ is endowed with the positive orientation relative to Ω. In the general situation Σ is also equipped with the Riemannian metric inherited from $\mathbb{R}^{n+1}$. Thus Σ, an oriented Riemannian manifold, possesses a canonical volume form, dV. Furthermore, since Σ has codimension one in $\mathbb{R}^{n+1}$ and is oriented, only one of the two possible unit normal vectors N at a point in Σ has the property that if $(y_1, \ldots, y_n)$ is a positively oriented frame of vectors tangent to Σ at that same point then $(N, y_1, \ldots, y_n)$ is a positively oriented frame in $\mathbb{R}^{n+1}$. Thus there is a canonical unit normal vector field $N : \Sigma \to \mathbb{R}^{n+1}$ on Σ. Let $N_j : \Sigma \to \mathbb{R}$ denote the components of the vector N, $j = 0, 1, \ldots, n$; locally they can be written in terms of the component functions of an orientation-preserving coordinate patch. As is well known (Edwards [6], page 363), the volume form dV can be expressed explicitly in terms of the components N_j: $dV = \sum_{j=0}^{n}(-1)^j N_j\, dk_0 \wedge \ldots \wedge \widehat{dk_j} \wedge \ldots \wedge dk_n$. Define the $\mathbb{C}_{(n)}$-valued n-form $\omega = \sum_{j=0}^{n}(-1)^j e_j dk_0 \wedge \ldots \wedge \widehat{dk_j} \wedge \ldots \wedge dk_n$. It is also well known (Edwards [6], page 381) that this n-form ω agrees in value with the $\mathbb{C}_{(n)}$-valued n-form $\sum_{j=0}^{n} N_j e_j\, dV$ when evaluated at any n-tuple of vectors tangent to Σ at a point.

Suppose $f : \Sigma \to \mathbb{C}_{(n)}$ is a Hölder continuous (order α, $0 < \alpha < 1$) function. Assume for simplicity that f has compact support. Then we define the left and right Cauchy integrals of f by the formulae

$$F^L(k) = \int_{k' \in \Sigma} C(k - k') \left(\sum_{j=0}^{n} N_j(k') e_j \right) f(k')\, dV(k')$$

$$= \int_{\Sigma} C(k - k')\omega(k')f(k')$$

$$F^R(k) = \int_{k' \in \Sigma} f(k') \left(\sum_{j=0}^{n} N_j(k') e_j \right) C(k - k') \, dV(k')$$

$$= \int_{\Sigma} f(k') \omega(k') C(k - k'),$$

where $k \notin \text{supp}(f)$. F^L (resp. F^R) is left (resp. right) monogenic in $\mathbb{R}^{n+1} \setminus \text{supp}(f)$. Under the conditions we have stated, the boundary values

$$F_{\pm}(k) = \lim_{\epsilon \to 0^+} F(k \pm \epsilon N(k))$$

exist at each $k \in \Sigma$, and define Hölder continuous functions on Σ. These boundary values satisfy the Plemelj relations (Iftimie [11], Mitrea [13])

$$F_+^L(k) - F_-^L(k) = f(k) \qquad F_+^L(k) + F_-^L(k) =$$

$$-2 \lim_{\epsilon \to 0^+} \int_{\Sigma} C_\epsilon(k - k') \omega(k') f(k'),$$

$$F_+^R(k) - F_-^R(k) = f(k) \qquad F_+^R(k) + F_-^R(k) =$$

$$-2 \lim_{\epsilon \to 0^+} \int_{\Sigma} f(k') \omega(k') C_\epsilon(k - k'),$$

for all $k \in \Sigma$, where (in the following expression $\|k\|$ denotes the Euclidean norm of $k \in \mathbb{R}^{n+1}$)

$$C_\epsilon(k) = \begin{cases} C(k) & \|k\| > \epsilon \\ 0 & \|k\| \le \epsilon \end{cases}.$$

We define the right (resp. left) distribution $\omega|_\Sigma f$ (resp. $f\omega|_\Sigma$) in the obvious way

$$\langle \phi, \omega|_\Sigma f \rangle = \int_{\Sigma} \phi \, \omega f \qquad \langle f\omega|_\Sigma, \phi \rangle = \int_{\Sigma} f \omega \phi.$$

Then as right and left distributions we have the formulae

$$\mathcal{D} F^L = \omega|_\Sigma (F_+^L - F_-^L) \qquad F^R \mathcal{D} = (F_+^R - F_-^R) \omega|_\Sigma.$$

Because of Painlevé's theorem (Theorem 10.6, page 64, Brackx, Delanghe and Sommen [4]), these formulae hold under less restrictive conditions on F^L, F^R, namely whenever F^L (resp. F^R) is left (resp. right) monogenic on $\Omega \setminus \Sigma$, where Ω is some open subset of $\mathbb{R}^{n+1}$ intersecting Σ, such that the boundary values $F_{\pm}^L$ (resp. $F_{\pm}^R$) exist and are Hölder continuous on $\Sigma \cap \Omega$.

Let S_h denote the group of all bijections of the set $\mathbf{h} = \{1, \ldots, h\}$ where $h \ge 1$ is an integer. Let $\mathbf{n}^{\mathbf{h}}$ denote the set of all mappings $l : \{1, \ldots, h\} \to \{1, \ldots, n\}$. S_h acts on the set $\mathbf{n}^{\mathbf{h}}$ on the right by the rule $(l, \nu) \mapsto l \circ \nu$. An orbit of this action is called a *combination* of h elements from the set $\mathbf{n} = \{1, \ldots, n\}$. If $l \in \mathbf{n}^{\mathbf{h}}$ then let $(l) = lS_h$ denote the orbit, or combination, associated to l. The elements of this orbit will be called the *distinguishable permutations* of the combination (l). The set of all combinations will be denoted by $\mathbf{n}^{\mathbf{h}}/S_h$.

For $j = 1, \ldots, n$ define the monogenic functions $Z_j : \mathbb{R}^{n+1} \to \mathbb{C}_{(n)} : k \mapsto k_j e_0 - k_0 e_j$. When $h = 0$ define $V_0(k) = e_0$ for all $k \in \mathbb{R}^{n+1}$. When $h \geq 1$ is an integer, $(l) \in \mathbf{n}^{\mathbf{h}}/S_h$, and $k \in \mathbb{R}^{n+1}$, define

$$V_{(l)}(k) = \tfrac{1}{h!} \sum_{l' \in (l)} Z_{l'(1)}(k) Z_{l'(2)}(k) \cdots Z_{l'(h)}(k).$$

The functions $V_{(l)}$, where $(l) \in \mathbf{n}^{\mathbf{h}}/S_h$, form a basis of the $\mathbb{C}_{(n)}$-module of all inner spherical monogenics of order h (Brackx, Delanghe and Sommen, page 70, [4]). They are homogeneous functions of order h in the k variable. Dually, we can define for $h = 0$, $W_0(k) = C(k)$, $k \in \mathbb{R}^{n+1}$, and for all $h \geq 1$, $(l) \in \mathbf{n}^{\mathbf{h}}/S_h$

$$W_{(l)}(k) = (-1)^h \partial_{k_{l(1)}} \cdots \partial_{k_{l(h)}} C(k).$$

$W_{(l)}$ is clearly well-defined. These functions form a basis of the $\mathbb{C}_{(n)}$-module of all outer spherical monogenics of order h. They are all homogeneous of order $-(n + h)$ in k.

Using these families of functions we can give expansions of the Cauchy kernel

$$C(k - k') = \sum_{h=0}^{\infty} \left[\sum_{(l) \in \mathbf{n}^{\mathbf{h}}/S_h} W_{(l)}(k) V_{(l)}(k') \right]$$

$$= \sum_{h=0}^{\infty} \left[\sum_{(l) \in \mathbf{n}^{\mathbf{h}}/S_h} V_{(l)}(k') W_{(l)}(k) \right],$$

which converge if $\|k\| > \|k'\|$ (Brackx, Delanghe and Sommen, page 78, [4]).

14.3 Relation between Complex and Clifford Analysis in Dimension One

In a desire to give a functional calculus for singular convolution operators acting on functions defined on n dimensional Lipschitz graphs in $\mathbb{R}^{n+1}$, Li, McIntosh and Qian [12] define a Fourier transform which has a closer relationship to the ordinary n dimensional Fourier transform than another generalization proposed by Brackx, Delanghe and Sommen, [4].

The transform of Li, McIntosh and Qian sets up an invertible correspondence between a class of holomorphic $\mathbb{C}_{(n)}$-valued (Fourier) symbols (functions of $x \in \mathbb{C}^n$) and a class of monogenic $\mathbb{C}_{(n)}$-valued kernels (functions of $k \in \mathbb{R}^{n+1}$), which generalized a well-known correspondence in dimension one. This connection between holomorphic functions of n complex variables and monogenic functions of $n + 1$ real variables is evidence of the fundamental importance of their transform. Their Fourier transform of a smooth temperate test function $f : \mathbb{R}^n \to \mathbb{C}_{(n)}$ (which might possibly possess a holomorphic extension beyond $\mathbb{R}^n$ to some domain in $\mathbb{C}^n$) was defined as $\hat{f}(k) = \int_{\mathbb{R}^n} E(x, -k) f(x)\, dx$, where $k = (k_0, k_1, \ldots, k_n) \in \mathbb{R}^{n+1}$ and

$$E(x, k) = \exp\left(i \sum_{j=1}^n x_j k_j \right) \left[e_0 \cosh(k_0 \alpha) - \frac{\sinh(k_0 \alpha)}{\alpha} i \sum_{j=1}^n x_j e_j \right],$$

where

$$\alpha^2 = \sum_{j=1}^n x_j^2$$

for all $x = (x_1, \ldots, x_n) \in \mathbb{C}^n$. When $k_0 = 0$, this exponential function simplifies to the usual complex-valued exponential function times e_0. Thus we have used the usual notation $\hat{f}$ for their transform. The results of Li, McIntosh and Qian depend essentially on the fact that $E(x, k)$ is left (also right) monogenic in k for each fixed x and holomorphic in x for each fixed k. In fact, as follows from an analog of Painlevé's theorem (Theorem 10.6, page 64, Brackx, Delanghe and Sommen [4]), the function $E(x, k)$ is the unique function satisfying the following properties:

1. $E : \mathbb{C}^n \times \mathbb{R}^{n+1} \to \mathbb{C}_{(n)}$;
2. $E((x_1, \ldots, x_n), (0, k_1, \ldots, k_n)) = \exp(i \sum_{j=1}^n x_j k_j) e_0$ for all

$$(x_1, \ldots, x_n), (k_1, \ldots, k_n) \in \mathbb{R}^n;$$

3. $E(x, k)$ is a holomorphic function of $x \in \mathbb{C}^n$ (i.e., each component is a holomorphic complex-valued function) for every fixed $k \in \mathbb{R}^{n+1}$;

4. $E(x, k)$ is a left monogenic function of $k \in \mathbb{R}^{n+1}$ for every fixed $x \in \mathbb{C}^n$.

The exponential function $E(x, k)$, together with an exposition of some of its basic properties, first appeared in Sommen [16].

However, in order that their theory truly generalize results which hold when $n = 1$, there needs to be a close relationship between the $\mathbb{C}_{(1)}$-valued function $E(x, k)$ (defined for $(x, k) \in \mathbb{C} \times \mathbb{R}^2$) and the complex-valued function $e^{ix(k_1 + ik_0)}$ (defined for $(x, k_1 + ik_0) \in \mathbb{C} \times \mathbb{C}$). It is true that if in $E(x, k)$ we replace e_0, e_1 by $1, -i$, respectively, then it reduces to $e^{ix(k_1 + ik_0)}$.

Of course, for any function $g(x,k)$, $E(x,k) + g(x,k)(e_0 - ie_1)/2$ will also reduce to $e^{ix(k_1+ik_0)}$ under this substitution, so the substitution alone is not enough to explain the close relationship between the theory of $\mathbb{C}_{(1)}$-valued monogenic functions on $\mathbb{R}^2$ and the theory of $\mathbb{C}$-valued holomorphic functions on $\mathbb{C}$.

Li, McIntosh and Qian [12] have given a way of associating to every complex-valued holomorphic function $f(x)$, $x \in \mathbb{C}$, a $\mathbb{C}_{(1)}$-valued holomorphic function. Basically, if $f(x) = \sum_{j=0}^{\infty} a_j x^j$ is a convergent power series representation of f then one associates to f the $\mathbb{C}_{(1)}$-valued function

$$(\theta f)(x) = \sum_{j=0}^{\infty} a_j (xie_1)^j = \frac{f(x)+f(-x)}{2} e_0 + \frac{f(x)-f(-x)}{2} ie_1.$$

This last expression enables us to define θf for arbitrary functions f. For example, if one fixes $k_0 \in \mathbb{R}$ and considers the complex-valued holomorphic function $f(x) = e^{-xk_0}$, then $(\theta f)(x) = E(x,(k_0,0))$. The map θ defined by Li, McIntosh and Qian does not completely solve our problem however, because if $k = (k_0,k_1)$ is fixed $(k_1 \neq 0)$ and $f(x) = e^{ix(k_1+ik_0)}$ it is not true that $(\theta f)(x) = E(x,k)$. Even if this were true it would not be clear *a priori* that θf would be monogenic in the parameter k.

We would like to extend the mapping θ to include in its domain functions of both x and k in such a way that holomorphic functions of k get mapped to $\mathbb{C}_{(1)}$-valued monogenic functions of k. Furthermore, we would like our extended mapping θ to transform complex-valued function theory into $\mathbb{C}_{(1)}$-valued function theory, i.e., we want θ to be an algebra homomorphism and to have good properties relative to differentiation.

In this section we will study an extension of θ and record here for later use (see the next section) a number of formulae. In particular, the properties of θ will enable us to transform the scattering problem $m_x(x,z) - izm(x,z) = q(x)$ for the complex-valued Jost function $m(x,z)$ (which is holomorphic as a function of $z = k_1 + ik_0$) into a scattering problem for a $\mathbb{C}_{(1)}$-valued Jost function, which will be monogenic as a function of $k = (k_0,k_1) \in \mathbb{R}^2$. This will give us a formulation of inverse scattering for the linear hierarchy in one space dimension which will generalize directly to the n dimensional case.

Let $C(X,Y)$ denote the set of continuous functions from X to Y, where X and Y are topological spaces.

THEOREM 14.3.1
Suppose $D \subset \mathbb{C} \times \mathbb{C}$ is a set satisfying the condition

$$(x, k_1 + ik_0) \in D \quad implies \quad (-x, -k_1 + ik_0) \in D.$$

Define $D_1 = \{(x, (k_0, k_1)) \in \mathbb{C} \times \mathbb{R}^2 \mid (x, k_1 + ik_0) \in D\}$ and $\theta : C(D, \mathbb{C}) \to C(D_1, \mathbb{C}_{(1)})$ by the rule

$$(\theta f)(x, k) = f(x, k_1 + ik_0)\frac{e_0 + ie_1}{2} + f(-x, -k_1 + ik_0)\frac{e_0 - ie_1}{2}, \quad k = (k_0, k_1).$$

Then the mapping θ has the following properties.

1. *θ is an injective $\mathbb{C}$-algebra homomorphism (preserves multiplicative units). On the image of θ, the inverse of θ can be computed by replacing e_0, e_1 by $1, -i$, respectively.*

2. *θ relates the Cauchy-Riemann operators and the Dirac operators in the following way*

$$\theta\left(\frac{\partial f}{\partial k_1} + i\frac{\partial f}{\partial k_0}\right) = i\left(\partial_{k_0} e_0 + \partial_{k_1} e_1\right)(\theta f)$$

$$\theta\left(\frac{\partial f}{\partial k_1} - i\frac{\partial f}{\partial k_0}\right) = -i\left(\partial_{k_0} e_0 - \partial_{k_1} e_1\right)(\theta f)$$

Therefore, when D is open, θ maps functions holomorphic in z for each fixed x into functions monogenic in k for each fixed x.

3. *$\theta(\partial_x f) = ie_1\partial_x(\theta f)$, where either f is a differentiable function of the real variable x, or f is a holomorphic function of the complex variable x.*

4. *If f and g are functions of the real variable x and $(f * g)(x) = \int_{-\infty}^{\infty} f(x - y)g(y)\, dy$, then $\theta(f * g) = (\theta f) * (\theta g)$.*

5. *If f and g are functions of the complex variable $z = k_1 + ik_0$ and*

$$(f * g)(z) = \int_{-\infty}^{\infty} \int_{-\infty}^{\infty} f(z - \xi_1 - i\xi_0)g(\xi_1 + i\xi_0)\, d\xi_1 d\xi_0,$$

then $\theta(f * g) = (\theta f) * (\theta g)$, where the last $*$ denotes ordinary convolution of functions defined on $\mathbb{R}^2$.

PROOF   Some simple calculations.   ∎

Examples:

1. If $f(x, z) = e^{ixz}$ for $(x, z) \in \mathbb{C} \times \mathbb{C}$ then $(\theta f)(x, k) = E(x, k)$, where $z = k_1 + ik_0$ and $k = (k_0, k_1)$.

2. Suppose $f_\pm(x) = [1 \pm \text{sgn}(\text{Re } x)]/2 = \chi_{[0,\infty)}(\pm \text{Re } x)$. Then $(\theta f_\pm)(x) = [e_0 \pm \text{sgn}(\text{Re } x)ie_1]/2 = \chi_\pm(x)$, where $\chi_\pm$ are functions discussed by Li, McIntosh and Qian [12].

3. If $f(z) = 1/(\pi z)$, then $(\theta f)(x, k) = -2iC(k) = -i(k_0 e_0 - k_1 e_1)/[\pi(k_0^2 + k_1^2)]$. So θ maps the fundamental solution of the operator $\partial_{\bar{z}}$ into $-2i$ times the fundamental solution of the Dirac operator.

4. Note that when $n = 1$ we have that $\mathbf{n^h} = \{1\}$ for every $h \geq 1$, so $V_{(1)}(k') = (k'_1 e_0 - k'_0 e_1)^h/h!$ and $W_{(1)}(k) = (-1)^h(\partial^h_{k_1} C)(k)$. If $\xi = k'_1 + ik'_0$ then

$$\theta[(i\xi)^h] = \{\theta[i\xi]\}^h = \{-e_1(k'_1 e_0 - k'_0 e_1)\}^h = (-e_1)^h h! V_{(1)}(k').$$

Notice that if $f(z)$ is holomorphic, then θf is monogenic, and thus $\partial_{k_0}(e_0 f) = -\partial_{k_1}(e_1 f)$. Therefore $\theta[\partial^h_z f] = (ie_1)^h \partial^h_{k_1} \theta f$ for all $h \geq 1$. Thus

$$\begin{aligned}
\theta[(iz)^{-h-1}] &= i^{-h-1}(-1)^h(h!)^{-1}\theta[\partial^h_z(1/z)] \\
&= i^{-h-1}(-1)^h(h!)^{-1}(ie_1)^h \partial^h_{k_1} \theta[1/z] \\
&= -i(-e_1)^h(h!)^{-1}(-2\pi i)(\partial^h_{k_1} C)(k) \\
&= -2\pi e^h_1(h!)^{-1}W_{(1)}(k).
\end{aligned}$$

5. If $f : \mathbb{C} \to \mathbb{C}$ is a locally integrable function and $\phi : \mathbb{R}^2 \to \mathbb{C}$ is a smooth function with compact support, then

$$\begin{aligned}
\int_{-\infty}^{\infty} \int_{-\infty}^{\infty} (\theta f)(k)\phi(k) \, dk_0 dk_1 &= \int_{-\infty}^{\infty} \int_{-\infty}^{\infty} [f(k_1 + ik_0)e_+ \\
&\quad + f(-k_1 + ik_0)e_-]\phi(k_0, k_1) \, dk_0 dk_1 \\
&= \int_{-\infty}^{\infty} \int_{-\infty}^{\infty} f(k_1 + ik_0)[\phi(k_0, k_1)e_+ \\
&\quad + \phi(k_0, -k_1)e_-] \, dk_0 dk_1,
\end{aligned}$$

where $e_\pm = (e_0 \pm ie_1)/2$. Thus, when acting on functions of z alone, θ can be extended to $\mathbb{C}$-valued distributions by the rule $\langle \theta f, \phi \rangle = \langle f, \phi \rangle e_+ + \langle f, \phi^\checkmark \rangle e_-$, where $\phi^\checkmark(k_0, k_1) = \phi(k_0, -k_1)$. Notice that in that situation θf is a $\mathbb{C}_{(1)}$-valued distribution.

6. Suppose $k_1(t) + ik_0(t)$ is a parameterized curve γ in $\mathbb{C}$, where $t \in [a, b]$. Let $dz = dk_1 + idk_0$ be the differential form. We will use the same notation for the corresponding curves and differential forms in $\mathbb{R}^2$, where as usual we associate $z = k_1 + ik_0 \in \mathbb{C}$ with $(k_0, k_1) \in \mathbb{R}^2$. Suppose f is a complex-valued continuous function defined on γ. We denote by $f \, dz|_\gamma$ the distribution defined for each test function $\phi : \mathbb{C} \to \mathbb{C}$ by the rule

$$\begin{aligned}
\langle f \, dz|_\gamma, \phi \rangle &= \int_\gamma \phi(z)f(z) \, dz \\
&= \int_a^b \phi(k_1(t) + ik_0(t))f(k_1(t) + ik_0(t))[k'_1(t) + ik'_0(t)] \, dt.
\end{aligned}$$

Let $\gamma^\vee$ denote the curve parameterized by $-k_1(t) + ik_0(t)$, $f^\vee(k_1 + ik_0) = f(-k_1 + ik_0)$ for all $z \in \gamma^\vee$, and let $d\bar{z} = dk_1 - idk_0$ be the differential form. Then

$$\langle \theta(f\,dz|_\gamma), \phi \rangle = \langle f\,dz|_\gamma, \phi \rangle e_+ + \langle f\,dz|_\gamma, \phi^\vee \rangle e_-$$
$$= \langle f\,dz|_\gamma, \phi \rangle e_+ - \langle f^\vee\,d\bar{z}|_{\gamma^\vee}, \phi \rangle e_-$$
$$= \langle f\,dz|_\gamma e_+ - f^\vee\,d\bar{z}|_{\gamma^\vee} e_-, \phi \rangle.$$

So $\theta(f\,dz|_\gamma) = f\,dz|_\gamma e_+ - f^\vee\,d\bar{z}|_{\gamma^\vee} e_-$. In the special situation where $\gamma^\vee$ traces out the same curve as γ, but in the opposite direction, i.e., $\gamma^\vee = -\gamma$, we have $-d\bar{z}|_{\gamma^\vee} = d\bar{z}|_\gamma$, and thus $\theta(f\,dz|_\gamma) = (\theta f)\omega|_\gamma = \omega|_\gamma(\theta f)$, where $\theta f = fe_+ + f^\vee e_-$, and $\omega = dk_1 e_0 - dk_0 e_1$. It should be noted that if γ is the positively oriented boundary of a region in $\mathbb{C}$, then because of our conventional identification of $\mathbb{C}$ with $\mathbb{R}^2$, γ will be the *negatively* oriented boundary of the corresponding region in $\mathbb{R}^2$.

The map θ enables us to infer the properties of Cauchy integrals of certain $\mathbb{C}_{(1)}$-valued functions on certain curves in $\mathbb{R}^2$ from their known properties in the $\mathbb{C}$-valued case. As we have seen in the previous section these properties continue to hold for more general $\mathbb{C}_{(1)}$-valued functions on more general curves. Suppose γ is a curve in $\mathbb{C}$ as in example (6) above. Assume further that $|k_1'(t) + ik_0'(t)| \neq 0$ for all $t \in [a, b]$ and that γ does not intersect itself unless it forms a simple closed curve. In the latter case we also assume $k_1'(a) + ik_0'(a) = k_1'(b) + ik_0'(b)$. Suppose a $\mathbb{C}$-valued Hölder continuous (of order α, $0 < \alpha < 1$) function f is given on the curve γ. Consider the Cauchy integral

$$F(z) = \frac{1}{2\pi i} \int_\gamma \frac{f(\zeta)}{\zeta - z}\,d\zeta = \frac{1}{2\pi i} \int_a^b \frac{f(k_1(t) + ik_0(t))}{k_1(t) + ik_0(t) - z}[k_1'(t) + ik_0'(t)]\,dt,$$

for $z \in \mathbb{C} \setminus \gamma$. As is well known (Muskhelishvili [14]) F is holomorphic on $\mathbb{C} \setminus \gamma$ and has well-defined limiting values $F_\pm(\zeta_0)$ as $z \to \zeta_0$ from the left $+$ or the right $-$ side of γ at each non-endpoint ζ_0 of γ. The left or right side of γ is defined relative to the positive orientation of γ. The formulae of Plemelj state that

$$F_+(\zeta_0) - F_-(\zeta_0) = f(\zeta_0)$$
$$F_+(\zeta_0) + F_-(\zeta_0) = \frac{1}{\pi i} \, \mathrm{pv} \int_\gamma \frac{f(\zeta)}{\zeta - \zeta_0}\,d\zeta$$

at all such non-endpoints ζ_0.

Since $1/(\pi z)$ is the fundamental solution of the inhomogeneous $\partial_{\bar{z}}$ equation, i.e., $\partial_{\bar{z}}(\pi z)^{-1} = \delta$, where δ is the Dirac delta function located at $z = 0$, we can reformulate the above properties of $F(z)$ in terms of distributions on $\mathbb{C}$. Notice that the distribution F is just the convolution of $(\pi z)^{-1}$ with the distribution $(i/2)f(z)\,dz|_\gamma$. So $\partial_{\bar z}F = (i/2)f(z)\,dz|_\gamma = (i/2)(F_+ - F_-)\,dz|_\gamma$. So the distributional $\partial_{\bar z}$ derivative of F can be expressed as a distribution obtained by multiplying the jump $F_+ - F_-$ on γ by the distribution $(i/2)\,dz|_\gamma$ of the type discussed in example (5) above.

Now suppose that $\gamma^\checkmark = -\gamma$. Now we apply the mapping θ to obtain the equation $\theta F = (-2iC) * [(i/2)\omega|_\gamma \theta f] = C * (\omega|_\gamma \theta f)$. $\theta f = \theta(F_+) - \theta(F_-) = (\theta F)_- - (\theta F)_+$ provided $(\theta F)_\pm$ are defined relative to $\gamma^\checkmark$, considered now as a contour in $\mathbb{R}^2$, in the following new manner. $(\theta F)_+$ (resp. $(\theta F)_-$) is the limit of θF from the right (resp. left) of $\gamma^\checkmark$ with respect to the positive orientation of $\gamma^\checkmark$. In the case where γ is the positively oriented boundary of a bounded region in $\mathbb{C}$, $\gamma^\checkmark$ is the positively oriented boundary of the corresponding region in $\mathbb{R}^2$, and $(\theta F)_\pm(k) = \lim_{\epsilon \to 0^+}(\theta F)(k \pm \epsilon N(k))$, where $N(k)$ is the unit outward normal vector to the curve $\gamma^\checkmark$ at the point $k \in \gamma^\checkmark$. It is the agreement with this dimensionally invariant definition that motivates our change of convention. So

$$\theta F = C * \{\omega|_{\gamma^\checkmark}[(\theta F)_+ - (\theta F)_-]\}.$$

Thus θ maps the complex-valued Cauchy integral of f into the (left or right) Clifford-valued Cauchy integral of θf defined in the previous section. So $\mathcal{D}(\theta F) = \omega|_{\gamma^\checkmark}[(\theta F)_+ - (\theta F)_-]$. The same result could have been obtained by applying θ directly to the equation $\partial_{\bar z}F = (i/2)f(z)\,dz|_\gamma = (i/2)(F_+ - F_-)\,dz|_\gamma$.

A similar argument leads to a reformulation of the second Plemelj formula. The image under θ of $F_+(\zeta_0) + F_-(\zeta_0)$ is $(\theta F)_-(k) + (\theta F)_+(k)$. The right-hand side is the limit as $\epsilon \to 0^+$ of the convolution of $\chi_{|z|>\epsilon}(z)/(\pi z)$ with $if\,dz|_\gamma$ evaluated at $\zeta_0 \in \gamma$. Applying θ to the convolution, evaluating on $\gamma^\checkmark$, and taking the limit as $\epsilon \to 0^+$ we get

$$(\theta F)_-(k) + (\theta F)_+(k) = -2\lim_{\epsilon \to 0^+}\int_{\gamma^\checkmark}C_\epsilon(k - k')\omega(k')(\theta f)(k'),$$

which coincides with the statement of the second Plemelj relation given in the previous section.

So complex function theory on certain symmetric domains is included as a subset, via the map θ, of Clifford-valued function theory. Various important properties of complex Cauchy integrals have their mirror image in the Clifford case. Furthermore, the Clifford versions hold under more general circumstances, i.e., outside the image of their complex counterparts.

14.4   Review of the Linear Hierarchy in One Space Dimension

A good general survey of the theory of the forward and inverse scattering transforms as a tool to solve hierarchies of nonlinear evolutionary partial differential equations is provided by Beals and Coifman [2]. We will give an extremely brief sketch of this theory in order to provide context for our results; the interested reader is referred to Beals and Coifman for the details. There are certain nonlinear evolutionary partial differential equations for which the initial-value problem

$$\begin{aligned} Q_t &= F(D, Q) \\ Q(0) &= Q_0, \end{aligned}$$

where D denotes differentiation in the spatial variable(s) x, can be solved by the following scheme:

$$Q(0) \xrightarrow{\ M=I+G(z)[QM]\ } M(0) \xrightarrow{\ \partial_t M - T(x,V)M=0\ } V(0)$$

$$\text{nonlinear PDE} \Big\downarrow \qquad\qquad\qquad\qquad\qquad\qquad \Big\downarrow \text{linear ODE}$$

$$Q(t) \xleftarrow{\ P(D,z)M - QM=0\ } M(t) \xleftarrow{\ M=I+\mathcal{C}T(x,V)M\ } V(t)$$

We have suppressed all variables except those which appear as parameters. Q is called the potential, and is an $N \times N$ off-diagonal matrix-valued function of $x \in \mathbb{R}^n$. M is called the Jost function, and is a non-singular $N \times N$ matrix-valued function (or distribution) of both x and the complex spectral variable $z \in H \subset \mathbb{C}^n$. The set H be be discussed more fully at the beginning of section 14.5. V is called the scattering data (see below for a more detailed description). I denotes the $N \times N$ identity matrix. The upper level in the above diagram describes the forward scattering transform; the lower, the inverse scattering transform. Each transform is a succession of two invertible steps. The equation $P(D, z)M - QM = 0$ is called the scattering equation. $P(D, z)$, where $D = (-i\partial_{x_1}, \ldots, -i\partial_{x_n})$, is a linear (pseudo-)differential operator acting in the x variable and containing z as a parameter. Note that if M is known, then Q can be recovered by the formula $Q = [P(D, z)M]M^{-1}$. The equation $\partial_{\bar{z}}M - T(x, V)M = 0$ is called the *d-bar* equation. The linear operator $T(x, V)$ contains x and V as parameters, and depends explicitly on z (although this dependence has been suppressed in our notation). $T(x, V)$ maps the distribution M into a distribution with support contained in H which describes the departure from holomorphy of M. V denotes the collection of free parameters occurring in the operator $T(x, V)$ which depend on the potential Q. $T(x, V)$ has the crucial property that if M is known, then the equation $\partial_{\bar{z}}M - T(x, V)M = 0$ can be solved uniquely for V.

We are following the traditional terminology of this field, which was derived from the classical scattering theory for the Schrödinger operator $-\partial_x^2 + q(x)$, considered as an unbounded operator in $L^2(\mathbb{R})$. It would take us too far afield to describe here exactly how the quantities we will discuss are related to the principal players in scattering theory: eigenfunction expansions, Møller wave operators, scattering matrices, discrete and continuous spectra, etc. It will be sufficient to mention that in the classical case the Jost function M is simply expressed in terms of the generalized eigenfunctions of the Schrödinger operator with spectral variable z. For an ample discussion of these relationships the reader should consult Dodd, Eilbeck, Gibbon, Morris [5], Beals, Deift, Tomei [3], Ablowitz, Clarkson [1], Reed, Simon [15], Hörmander [10]. The *d-bar* formulation used by Beals Coifman, which we are describing, has several advantages over the classical scattering viewpoint, namely its elegant symmetry and simplicity, its natural way of defining and unifying the continuous and discrete components of the scattering data, and the fact that it becomes almost obvious that the Jost function and hence the potential can be uniquely recovered from the scattering data.

We return now to our description of the *d-bar* method, starting, for the sake of ease of motivation, with the inverse transform. Given V, the first stage of the inverse transform is to find a particular solution M of the equation $\partial_{\bar{z}}M = T(x, V)M$, namely the one having the property that $M(z) \to I$ as $z \to \infty$. This is the same as finding the solution of the integral equation $M = I + \mathcal{C}T(x, V)M$. The linear operator $\mathcal{C}$ denotes the Cauchy integral, i.e., convolution with $1/(\pi z)$ (when $n = 1$), the fundamental solution of the inhomogeneous $\partial_{\bar{z}}$ operator. The second stage of the inverse transform is to find Q from M. This can be done either directly using the scattering equation (as mentioned above) or, provided V decays sufficiently rapidly as $z \to \infty$, by relating Q to the large z asymptotic behavior of M. The first stage of the forward transform is if given Q, find a particular solution M of the scattering equation, namely the one which was used in the second stage of the inverse transform to find the given Q. This involves solving the integral equation $M = I + G(z)[QM]$, where the linear operator $G(z)$ is an appropriately chosen fundamental solution operator for the scattering equation, i.e., $\mu = G(z)f$ satisfies $P(D, z)\mu = f$ and $\mu(z) \to 0$ as $z \to \infty$. The second stage of the forward transform is to find V from M. This can either be done directly or by relating V to the large x asymptotics of M, which is possible when Q possesses sufficient decay as $x \to \infty$.

The vertical arrow on the left denotes the process of evolving the potential from its initial value $Q(0) = Q_0$ according to the nonlinear evolution equation to obtain its value at a subsequent time t. If the operators $G(z)$ and $T(x, \cdot)$ are properly related to each other and to the equation

$Q_t = F(D, Q)$, then the corresponding evolution of the scattering data $V(0) \mapsto V(t)$ is (somehow miraculously) that of a trivially solvable linear ordinary differential equation in t, where z is involved only as a parameter (there is no coupling between different values of z).

If we leave out the intermediate step of finding the Jost function M, then the scheme we have described is very much like the use of the forward and inverse Fourier transform to solve the initial-value problem for linear evolution equations. Even from the viewpoint of classical scattering theory, the relationship with the Fourier transform is very close since Fourier analysis decomposes a function as a linear combination of e^{izx}, which are the eigenfunctions of the differential operator $D = -i\partial_x$. In fact, the method of the scattering transforms has always been considered as a generalization of the method of Fourier transforms, because although the scattering transforms are nonlinear, their Fréchet derivatives at zero are usually closely related (if not equal) to (forward or inverse) Fourier transforms. However, the relationship is much deeper than this. The above general scheme (which is identical to that proposed by Beals and Coifman) is a direct generalization of the Fourier method in the sense that the Fourier method itself can be derived as a special case of the above scheme. We will now demonstrate this in the case $n = 1$. Define

$$Q = \begin{pmatrix} 0 & q \\ 0 & 0 \end{pmatrix}, \quad M = \begin{pmatrix} 1 & m \\ 0 & 1 \end{pmatrix}, \quad V = \begin{pmatrix} 1 & v \\ 0 & 1 \end{pmatrix}.$$

q is a function of $x \in \mathbb{R}$, m a function of x and $z \in \mathbb{C} \setminus \mathbb{R}$, v a function of $k_1 \in \mathbb{R} \subset \mathbb{C}$. Let $M_{\pm}(x, k_1) = \lim_{\epsilon \to 0^+} M(x, k_1 \pm i\epsilon)$. It is convenient to start with the inverse problem, so assume v is a temperate test function. Define $I = \mathrm{diag}(1, 1)$, $J = \mathrm{diag}(i, 0)$. Consistent with the Riemann-Hilbert problem $M_+(x, k_1) = M_-(x, k_1)e^{xk_1 J}V(k_1)e^{-xk_1 J}$ and the formula $\partial_{\bar{z}}M = (i/2)(M_+ - M_-)dk_1|_{\mathbb{R}}$ we define:

$$
\begin{aligned}
T(x, V)M &= M_-(x, k_1)e^{xk_1 J}[V(k_1) - I]e^{-xk_1 J}(i/2)dk_1|_{\mathbb{R}} \\
&= \begin{pmatrix} 1 & m_-(x, k_1) \\ 0 & 1 \end{pmatrix} \begin{pmatrix} 0 & e^{ixk_1}v(k_1) \\ 0 & 0 \end{pmatrix} (i/2)dk_1|_{\mathbb{R}} \\
&= \begin{pmatrix} 0 & e^{ixk_1}v(k_1) \\ 0 & 0 \end{pmatrix} (i/2)dk_1|_{\mathbb{R}}.
\end{aligned}
$$

So the equation $M = I + CT(x, V)M$, which is usually an integral equation for M, simplifies in this case to an explicit formula ($\mathrm{Im}(z) \neq 0$)

$$m(x, z) = \frac{1}{2\pi i} \int_{-\infty}^{\infty} \frac{e^{ix\xi}v(\xi)}{\xi - z} \, d\xi.$$

Define the scattering operator

$$P(\partial_x, z)M = M_x - zJM + MzJ$$

$$= \begin{pmatrix} 0 & m_x \\ 0 & 0 \end{pmatrix} - \begin{pmatrix} iz & izm \\ 0 & 0 \end{pmatrix} + \begin{pmatrix} iz & 0 \\ 0 & 0 \end{pmatrix}$$

$$= \begin{pmatrix} 0 & m_x - izm \\ 0 & 0 \end{pmatrix}.$$

Solving the equation $P(\partial_x, z)M = QM$ for Q we get $Q = [P(\partial_x, z)M]M^{-1}$. Because of the special form of M, this reduces to

$$q(x) = m_x - izm = \frac{1}{2\pi i} \int_{-\infty}^{\infty} i\xi \frac{e^{ix\xi}v(\xi)}{\xi - z} \, d\xi - iz \frac{1}{2\pi i} \int_{-\infty}^{\infty} \frac{e^{ix\xi}v(\xi)}{\xi - z} \, d\xi$$

$$= \frac{1}{2\pi i} \int_{-\infty}^{\infty} (i\xi - iz) \frac{e^{ix\xi}v(\xi)}{\xi - z} \, d\xi$$

$$= \frac{1}{2\pi} \int_{-\infty}^{\infty} e^{ix\xi}v(\xi) \, d\xi.$$

So with the choices of the operators $T(x, V)$ and $P(\partial_x, z)$ we have made, the inverse scattering transform $V \mapsto Q$ becomes essentially the inverse Fourier transform.

Notice that since $\xi v(\xi)$ is integrable, $m_x(x, z) \to 0$ as $z \to \infty$, and hence

$$q(x) = \lim_{z \to \infty} -izm(x, z), \quad |\operatorname{Im} z| \geq \epsilon > 0.$$

In fact, if v has compact support, then by employing the series expansion $(i\xi - iz)^{-1} = -\sum_{j=0}^{\infty}(i\xi)^j (iz)^{-j-1}$, where $|z| > |\xi|$, in the above integral representation of m we obtain the convergent series

$$m(x, z) = -\sum_{j=0}^{\infty} \frac{(\partial_x^j q)(x)}{(iz)^{j+1}}, \quad |z| > \sup\{|\xi| : \xi \in \operatorname{supp} v\}.$$

We will pretend we do not know how to invert the inverse Fourier transform, and proceed to specify the forward scattering transform using the above scheme. We need to define the operator $G(z)$ so that for every Schwarz tempered test function V the solution M' of the integral equation $M' = I + G(z)[QM']$, where $Q = [P(\partial_x, z)M]M^{-1}$ and $M = I + CT(x, V)M$, satisfies $M' = M$. Now define $g(x, z)$ to be the following fundamental solution of the equation $g_x - izg = \delta$:

$$g(x, z) = e^{ixz} \tfrac{1}{2}[\operatorname{sgn}(\operatorname{Im} z) + \operatorname{sgn}(x)] = \begin{cases} e^{ixz}\chi_{[0,\infty)}(x) & \operatorname{Im} z > 0, \\ -e^{ixz}\chi_{(-\infty,0]}(x) & \operatorname{Im} z < 0, \end{cases}$$

where $\operatorname{Im} z \neq 0$. We define

$$G(z) \begin{pmatrix} f_{11} & f_{12} \\ f_{21} & f_{22} \end{pmatrix} = \\ \begin{pmatrix} \int_{-\infty}^{x} f_{11}(y) \, dy & \int_{-\infty}^{\infty} g(x - y, z) f_{12}(y) \, dy \\ \int_{-\infty}^{\infty} g(x - y, -z) f_{21}(y) \, dy & \int_{-\infty}^{x} f_{22}(y) \, dy \end{pmatrix}.$$

If $\operatorname{Im} z \neq 0$ then the equation $M = I + G(z)[QM]$ has a unique solution which must be unit upper triangular, since in that case $QM = Q$, and the integral equation reduces to the integral representation

$$m(x, z) = \int_{-\infty}^{\infty} g(x - y, z) q(y) \, dy.$$

Repeated integration by parts in this integral representation of m gives rise to the same series for m that we saw was convergent for $|z| > \sup\{|\xi| : \xi \in \operatorname{supp} v\}$ when v has compact support. Thus our choice g of a fundamental solution was correct, since the above integral representation gives the analytic continuation of $m(x, z)$ to the domain $\operatorname{Im} z \neq 0$ of that convergent expansion when v has compact support.

Now we will continue and compute V. Note that from the way we defined $T(x, V)$ we can solve the equation $\partial_{\bar z} M = T(x, V) M$ for V to obtain:

$$V(k_1) = e^{-xk_1 J} M_-(x, k_1)^{-1} M_+(x, k_1) e^{xk_1 J}$$
$$= \begin{pmatrix} 1 & e^{-ixk_1}[m_+(x, k_1) - m_-(x, k_1)] \\ 0 & 1 \end{pmatrix}.$$

Therefore we have

$$v(k_1) = e^{-ixk_1}[m_+(x, k_1) - m_-(x, k_1)]$$
$$= \lim_{\epsilon \to 0^+} \int_{-\infty}^{\infty} e^{-i(x-y)k_1}[g(x - y, k_1 + i\epsilon)$$
$$- g(x - y, k_1 - i\epsilon)]e^{-iyk_1} q(y) \, dy$$
$$= \lim_{\epsilon \to 0^+} \int_{-\infty}^{\infty} e^{-\epsilon|x-y|} e^{-iyk_1} q(y) \, dy$$
$$= \int_{-\infty}^{\infty} e^{-iyk_1} q(y) \, dy.$$

Thus the forward scattering transform $Q \mapsto V$ is essentially the forward Fourier transform.

Notice that since $q(x)$ is integrable we have that $e^{-ixk_1} m_\pm(x, k_1) \to 0$ as $x \to \mp\infty$ and $e^{-ixk_1} m_\pm(x, k_1) \to \pm v(k_1)$ as $x \to \pm\infty$. Hence we also have the asymptotic formula

$$v(k_1) = \pm \lim_{|x| \to \infty} [e^{-ixk_1} m_\pm(x, k_1) + e^{ixk_1} m_\pm(-x, k_1)].$$

In fact, if $\pm \operatorname{Im} z \geq 0$ and there exists an $M > 0$ such that $q(x) = 0$ for all $\pm x > M$, then for all $\pm x > M$ we have the equality $v(z) = \pm e^{-ixz} m(x, z)$, since in that case the function $v(k_1)$ is the boundary value of a function $v(z)$ holomorphic in the domain $\pm \operatorname{Im} z > 0$.

The perspective that the *inhomogeneous* scalar equations $m_x - izm = q$ and $m_+(x, k_1) - m_-(x, k_1) = e^{ixk_1} v(k_1)$ lead to the forward and inverse

Fourier transforms as scattering transforms is due to Fokas [7], and Fokas and Gel'fand [8]. We have presented the theory in terms of the *homogeneous* forward and inverse scattering equations $M_x - zJM + MzJ = QM$ and $\partial_{\bar{z}} M = T(x, V)M$ in order to show how the inhomogeneous scattering equations fit in with the usual (homogeneous) formulation of the theory.

The Linear Hierarchy is the collection of all partial differential equations of the evolutionary type whose initial-value problems are solved by the above transforms. This includes all linear equations with constant coefficients of the type $q_t = p(\partial_x)q$, where $p \in \mathbb{C}[x]$ is a polynomial. This evolution of the potential is equivalent to the relation $v(k_1, t) = e^{p(ik_1)t}v(k_1, 0)$ on the scattering data side.

With an eye to a generalization of the forward and inverse scattering transform for the linear hierarchy to the context of several space dimensions (presented in the next section) we will now reformulate the one-dimensional theory. We will use the integral representations

$$m(x, z) = \int_{-\infty}^{\infty} g(x - y, z)q(y)\,dy, \quad m(x, z) = \frac{1}{2\pi i} \int_{-\infty}^{\infty} \frac{e^{ixk_1}v(k_1)}{k_1 - z}\,dk_1,$$

and their associated inhomogeneous differential equations

$$m_x(x, z) - izm(x, z) = q(x), \quad (\partial_{\bar{z}}m)(x, k_1 + ik_0) = e^{ixk_1}v(k_1)(i/2)\,dk_1|_{\mathbb{R}},$$

as the basis of our reformulation. First we will apply the mapping θ defined in the previous section to the fundamental solution $g(x, k_1 + ik_0) = e^{ixk_1 - |xk_0|}\frac{1}{2}[\text{sgn}(k_0) + \text{sgn}(x)]$ to obtain

$$(\theta g)(x, k) = e^{ixk_1 - |xk_0|}\frac{1}{2}[\text{sgn}(k_0)e_0 + \text{sgn}(x)ie_1]$$

$$= \begin{cases} E(x, k)\chi_+(x) & k_0 > 0, \\ -E(x, k)\chi_-(x) & k_0 < 0, \end{cases}$$

where $k = (k_0, k_1)$, $E(x, k)$ is the exponential function discussed previously, and $\chi_\pm(x) = [e_0 \pm \text{sgn}(x)ie_1]/2$. Applying the results of the previous section to the above integral representations we get

$$(\theta m)(x, k) = \int_{-\infty}^{\infty} (\theta g)(x - y, k)(\theta q)(y)\,dy,$$

$$(\theta m)(x, k) = \int_{-\infty}^{\infty} C(k - k_1')e^{ixk_1'}(\theta v)(k_1')\,dk_1'.$$

These provide particular solutions to the following inhomogeneous differential equations

$$ie_1\partial_x(\theta m)(x, k) + (k_0 e_0 + k_1 e_1)\ (\theta m)(x, k) = (\theta q)(x),$$

$$\mathcal{D}(\theta m)(x, k) = e^{ixk_1}(\theta v)(k_1)\,dk_1|_{\mathbb{R}}.$$

One can also use θ to transform the series expansion for m (see example (4) from section 14.3) and the asymptotic formulae for v and q into the Clifford context. All these formulae will be generalized to n dimensions in the next section. As suggested by these formulae, it will turn out (as a consequence of the results in the next section) that these equations define forward and inverse Fourier transforms for general $\mathbb{C}_{(1)}$-valued potentials q and scattering data v, not just for those of the form θq and θv for $\mathbb{C}$-valued q and v. So if one speaks of a "scalar-valued" potential in the one-dimensional context one usually means a $\mathbb{C}_{(1)}$-valued potential of the form $q(x)e_0$, where q is complex-valued, rather than one of the form θq. Of course these two notions coincide when $q(x) = q(-x)$.

14.5 The Linear Hierarchy in Several Space Dimensions

The scheme described in the previous section based on the $\partial_{\bar{z}}$ equation works well for a variety of nonlinear hierarchies in one space dimension, and for the Kadomtsev-Petviashvili, Davey-Stewartson, and Ishimori hierarchies in two space dimensions. In the one-dimensional case there is a single complex spectral parameter z, and $H = \mathbb{C}$. The scattering data are supported on a rather sparse subset of H. In n space variables, $n \geq 2$, the n complex spectral parameters are constrained to lie on an $n-1$ dimensional variety H in $\mathbb{C}^n$ (in the scheme for scalar problems described by Beals and Coifman [2]), and because of the overdetermined nature of the inhomogeneous $\partial_{\bar{z}}$ equation (for $n \geq 3$) the types of evolutions of the scattering data which preserve the solvability conditions are highly restricted. When $n = 2$, Fokas and Gel'fand [8] have given a "scattering theory" for the linear hierarchy based, again, on an inhomogeneous $\partial_{\bar{z}}$ equation, where $H = \mathbb{C}$. But for $n \geq 3$ there is to the author's knowledge no way of understanding the Fourier method as a special instance of (not merely formally analogous to) the $\partial_{\bar{z}}$ method. This is not surprising in light of the fact that t-evolutions of the Fourier transforms of solutions of linear equations with constant coefficients are not constrained except in regard to forward well-posedness of the initial-value problem. Besides, the "dimensional asymmetry" of the inhomogeneous $\partial_{\bar{z}}$ equation creates doubt about whether it is the correct approach to a multidimensional theory. This author accepts as axiomatic that any reasonable generalization of scattering transforms from one to several space dimensions should have the property that the Fourier transform and the linear hierarchy should be included as a special case. We are not able at the present time to give such a general theory, but we will examine here a particular way in which the theory of the linear hierarchy can be obtained from something analogous to a forward and inverse scattering theory. In so doing we will draw attention to a possible setting within which such a general theory could conceivably be constructed.

The "scattering theory" we propose for the linear hierarchy is most easily motivated starting with the inverse problem. We suppose for simplicity that the scattering data $v : \mathbb{R}^n \to \mathbb{C}_{(n)}$ is a smooth function with compact support. We define the Jost function $m(x, k)$ for $k_0 \neq 0$ by the integral representation

$$m(x, k) = \int_{\mathbb{R}^n} C(k - k') w(k') \exp\left(i \sum_{j=1}^{n} x_j k_j' \right) v(k'),$$

where we denote by $\mathbb{R}^n$ the hyperplane in $\mathbb{R}^{n+1}$ where $k_0' = 0$. This is a particular solution of the inhomogeneous differential equation

$$\mathcal{D}m = w|_{\mathbb{R}^n} \exp\left(i \sum_{j=1}^{n} x_j k_j' \right) v(k').$$

Let $D = (D_1, \ldots, D_n) = (-i\partial_{x_1}, \ldots, -i\partial_{x_n})$. Define the Fourier multiplier operator (with a Clifford algebra-valued symbol)

$$P(D, k) = \frac{1}{(2\pi)^n} C(k - D)^{-1}$$

$$= \frac{\sigma_n}{(2\pi)^n} \left| k_0^2 + \sum_{j=1}^{n} (k_j - D_j)^2 \right|^{\frac{n-1}{2}} \left[i \sum_{j=1}^{n} \partial_{x_j} e_j + \sum_{j=0}^{n} k_j e_j \right].$$

This is a differential operator when n is odd, but when n is even, the nonlocal part must be defined using the Fourier integral theorem, i.e.,

$$\left[\left| k_0^2 + \sum_{j=1}^{n} (k_j - D_j)^2 \right|^{\frac{n-1}{2}} f \right](x)$$

$$= \frac{1}{(2\pi)^n} \int_{\mathbb{R}^n} \exp\left(i \sum_{j=1}^{n} x_j k_j' \right) \left| k_0^2 + \sum_{j=1}^{n} (k_j - k_j')^2 \right|^{\frac{n-1}{2}} \hat{f}(k') \, dk'.$$

If we apply this operator to the Jost function we obtain

$$P(D, k)m = \frac{1}{(2\pi)^n} C(k - D)^{-1} \int_{\mathbb{R}^n} C(k - k') w(k') \exp\left(i \sum_{j=1}^{n} x_j k_j' \right) v(k')$$

$$= \frac{1}{(2\pi)^n} \int_{\mathbb{R}^n} C(k - k')^{-1} C(k - k') w(k') \exp\left(i \sum_{j=1}^{n} x_j k_j' \right) v(k')$$

$$= \frac{1}{(2\pi)^n} \int_{\mathbb{R}^n} \exp\left(i \sum_{j=1}^{n} x_j k_j' \right) v(k') \, dk_1' \wedge \ldots \wedge dk_n'.$$

We call this function (which is independent of k) the potential $q(x)$. The mapping $v \mapsto q$ is clearly the inverse Fourier transform.

The reader might feel that when n is even our method is somewhat circular, since in order to recover q from m we apply an operator which is only defined using the very transforms our method is supposed to be defining. Fortunately, as in the one-dimensional case, there is an alternate method of obtaining q from m which does not require us to be able to apply the operator $P(D, k)$. We now proceed to describe this alternate method. Using the expansion of the Cauchy kernel discussed in section 1 we get a series expansion for $m(x, k)$

$$
m(x, k) = \sum_{h=0}^{\infty} \left\{ \sum_{(l) \in \mathbf{n}^h / S_h} W_{(l)}(k) \int_{\mathbb{R}^n} V_{(l)}(k') \exp\left(i \sum_{j=1}^{n} x_j k_j' \right) v(k') dk' \right\}
$$

$$
= (2\pi)^n \sum_{h=0}^{\infty} \left\{ \sum_{(l) \in \mathbf{n}^h / S_h} W_{(l)}(k)[V_{(l)}(D)q](x) \right\}
$$

which converges when $\|k\| > \sup\{\|k'\| : k' \in \operatorname{supp} v\}$. Even if $\operatorname{supp} v$ is no longer compact, but v remains rapidly decaying, the series continues to be asymptotically valid as $\|k\| \to \infty$. Thus we obtain the alternate expression for the potential

$$
q(x) = \lim_{\|k\| \to \infty} \frac{1}{(2\pi)^n} C(k)^{-1} m(x, k).
$$

Nonlocal scattering operators are certainly not unprecedented in the known realm of nonlinear hierarchies, so the role that the (sometimes) nonlocal operator $P(D, k)$ might play in any (future) theory of nonlinear multi-dimensional hierarchies is not *a priori* a restricted one.

The forward scattering problem commences by seeking an integral representation of the Jost function in terms of the potential, expressing the "monogenic continuation" of the above series expansion. This involves first of all the determination of the fundamental solution g of the equation $P(D, k)g = \delta$:

$$
g(x, k) = \int_{\mathbb{R}^n} C(k - k') \exp\left(i \sum_{j=1}^{n} x_j k_j' \right) dk_1' \ldots dk_n'
$$

$$
= \exp\left(i \sum_{j=1}^{n} x_j k_j \right) \cdot
$$

$$
\cdot \int_{\mathbb{R}^n} C(k_0, k_1'', \ldots, k_n'') \exp\left(-i \sum_{j=1}^{n} x_j k_j'' \right) dk_1'' \ldots dk_n''
$$

$$= \exp\left(i \sum_{j=1}^{n} x_j k_j - |k_0| \cdot \|x\| \right) \tfrac{1}{2} \left[\operatorname{sgn}(k_0) e_0 + \|x\|^{-1} i \sum_{j=1}^{n} x_j e_j \right]$$

$$= \begin{cases} E(x,k)\chi_+(x) & k_0 > 0, \\ -E(x,k)\chi_-(x) & k_0 < 0, \end{cases}$$

where $E(x,k)$ is the exponential function defined in section 14.2, and

$$\chi_\pm(x) = \tfrac{1}{2}\left[e_0 \pm \|x\|^{-1} i \sum_{j=1}^{n} x_j e_j \right].$$

In terms of the potential, the Jost function is defined by the usual expression

$$m(x,k) = \int_{\mathbb{R}^n} g(x-y,k) q(y)\, dy.$$

m is clearly left monogenic for $k_0 \neq 0$, so $\mathcal{D}m = \omega|_{\mathbb{R}^n}(m_+ - m_-)$. Thus we define the scattering data to be

$$v(k') = \exp\left(-i \sum_{j=1}^{n} x_j k_j' \right) [m_+(x,k') - m_-(x,k')]$$

$$= \int_{\mathbb{R}^n} \exp\left(-i \sum_{j=1}^{n} (x_j - y_j) k_j' \right) [g_+(x-y,k') - g_-(x-y,k')] \cdot$$

$$\cdot \exp\left(-i \sum_{j=1}^{n} y_j k_j' \right) q(y)\, dy$$

$$= \lim_{\epsilon \to 0+} \int_{\mathbb{R}^n} e^{-\epsilon\|x-y\|} \exp\left(-i \sum_{j=1}^{n} y_j k_j' \right) q(y)\, dy$$

$$= \int_{\mathbb{R}^n} \exp\left(-i \sum_{j=1}^{n} y_j k_j' \right) q(y)\, dy.$$

So the forward scattering transform $q \mapsto v$ is the Fourier transform.

Suppose $x = ax'$ where $a > 0$ and $x' \in \mathbb{R}^n$ satisfies $\|x'\| = 1$. Since q is integrable we obtain for $k_0 = 0$ the asymptotic result

$$\lim_{a \to \infty} \exp\left(-ia \sum_{j=1}^{n} x_j' k_j \right) m_\pm(ax', k)$$

$$= \int_{\mathbb{R}^n} \exp\left(-i \sum_{j=1}^{n} y_j k_j \right) \cdot$$

$$\cdot \tfrac{1}{2} \left[\pm e_0 + \|ax' - y\|^{-1} i \sum_{j=1}^{n} (ax_j' - y_j) e_j \right] q(y)\, dy$$

$$= \pm \chi_\pm(x')v(k).$$

In fact, as can be easily seen, v can be recovered from the asymptotic behavior of m:

$$v(k_1,\ldots,k_n) = \pm \lim_{\|x\|\to\infty} \cdot$$

$$\cdot \left\{ \exp\left(-i\sum_{j=1}^n x_j k_j\right) m_\pm(x,k) + \exp\left(i\sum_{j=1}^n x_j k_j\right) m_\pm(-x,k)\right\}.$$

Notice that since $\chi_\pm^2 = \chi_\pm$, $\chi_+ + \chi_- = e_0$, $\chi_+\chi_- = \chi_-\chi_+ = 0$, and $\chi_\pm(-x) = \chi_\mp(x)$, we have that if q is "one-sided" in the sense that $\chi_\pm q = 0$ then $m(0,k) = 0$ for $\pm k_0 < 0$. Hence we have the equality

$$v(k') = \pm m_\pm(0,k'),$$

i.e., $v(k')$ is the boundary value of a left monogenic function defined for $\pm k_0 > 0$. The results of Li, McIntosh and Qian [12] showed that if in addition to being "one-sided", $q(x)$ is the restriction to $\mathbb{R}^n$ of a bounded holomorphic function defined on certain "wedge-shaped" domains in $\mathbb{C}^n$, then its Fourier transform v has a monogenic extension to a "sector" in $\mathbb{R}^{n+1}$ properly containing the half-space $\pm k_0 > 0$.

Suppose $p(D) = \sum_{|\alpha|\leq N} D^\alpha A_\alpha$ is a polynomial in $D = (D_1,\ldots,D_n)$ with coefficients $A_\alpha \in \mathbb{C}_{(n)}$, where α denotes a multi-index. Let $qp(D)$ denote the action of the operator $p(D)$ on q, i.e., $qp(D) = \sum_{|\alpha|\leq N} D^\alpha q A_\alpha$. Then the linear evolution equation for the potential $q_t = qp(D)$ is clearly equivalent to the scattering data evolution $v(k',t) = v(k',0)e^{tp(k')}$. The exponential $e^{tp(k')}$ is in general $\mathbb{C}_{(n)}$-valued, and can be computed explicitly in principle, since $\mathbb{C}_{(n)}$ can be thought of as an algebra of matrices (Gilbert and Murray [9]).

The noncommutative multiplicative structure of $\mathbb{C}_{(n)}$ gives rise to various difficulties when one tries to incorporate the above scheme for the linear hierarchy, based on *inhomogeneous* equations for m, into a more general scheme for solving various nonlinear hierarchies, based on *homogeneous* equations for M, in a manner analogous to what we have presented in section 14.4. It remains to be seen whether or not these difficulties can be resolved or bypassed. Given the large degree of structural similarity between the n-dimensional and the one-dimensional cases, and the fact that our scheme works for the linear hierarchy in the same way for all space dimensions, this author still feels hopeful about the prospects.

Acknowledgment

The author would like to thank M. Mitrea and J. Ryan for interesting discussions and references on Clifford analysis. The author would like to thank L. Y. Sung for making him aware of the preprint [8].

Bibliography

[1] M. J. Ablowitz and P. A. Clarkson, *Solitons, Nonlinear Evolution Equations and Inverse Scattering*, London Math. Soc. Lect. Note Series, no. 149, Cambridge University Press, Cambridge, 1991.

[2] R. Beals and R. R. Coifman, *Linear Spectral Problems, Non-linear Equations and the $\bar{\partial}$-Method*, Inverse Problems **5**, (1989), 87–130.

[3] R. Beals, P. Deift and C. Tomei, *Direct and Inverse Scattering on the Line*, American Mathematical Society, Providence, RI, 1988.

[4] F. Brackx, R. Delanghe and F. Sommen, *Clifford Analysis*, Pitman Publishing, Marshfield, MA, 1982.

[5] R. K. Dodd, J. C. Eilbeck, J. D. Gibbon and H. C. Morris, *Solitons and Nonlinear Wave Equations*, Academic Press, London, 1982.

[6] C. H. Edwards, *Advanced Calculus of Several Variables*, Academic Press, New York, 1973.

[7] A. S. Fokas, NSF-CBMS Regional Research Conference on Nonlinear Dispersive Wave Systems, Lecture at University of Central Florida, March 1991.

[8] A. S. Fokas and I. M. Gel'fand *Integrability of Linear and Nonlinear Evolution Equations*, Institute for Nonlinear Studies, (preprint #217), August 1993.

[9] J. E. Gilbert and A. M. Murray, *Clifford Algebras and Dirac Operators in Harmonic Analysis*, Cambridge University Press, New York, 1991.

[10] L. Hörmander, *The Analysis of Linear Partial Differential Operators II*, Springer-Verlag, New York, 1983.

[11] V. Iftimie, *Fonctions hypercomplexes*, Bull. Math. Ry. Soc. Roumanie **9**, (1965), 279–332.

[12] C. Li, A. McIntosh and T. Qian, *Clifford Algebras, Fourier Transforms, and Singular Convolution Operators on Lipschitz Surfaces*, Macquarie Mathematics Reports no. 91-087, Macquarie University, New South Wales, Australia, 1991.

[13] M. Mitrea, *Hardy Spaces and Clifford Algebras*, preprint.

[14] N. I. Muskhelishvili, *Singular Integral Equations*, Noordhoff, Gronin-
gen, 1953.

[15] M. Reed and B. Simon, *Methods of Modern Mathematical Physics III:
Scattering Theory*, Academic Press, New York, 1979.

[16] F. Sommen, *Microfunctions with Values in a Clifford Algebra, II*. Scientific Papers of the College of Arts and Sciences, University of Tokyo
36, (1986), 15–37.

Department of Mathematics,
University of South Carolina,
Columbia, South Carolina 29208
E-mail: dix@bigcheese.math.scarolina.edu

15

On Riemann-Hilbert Problems for Nonhomogeneous Dirac Equations in a Half Space of $\mathbf{R}^m (m \geq 2)$

Zhenyuan Xu and Chiping Zhou

ABSTRACT In this paper, Riemann-Hilbert problems are investigated for nonhomogeneous Dirac equations in the upper half space of $\mathbf{R}^m$ ($m \geq 2$) in both the classical and the L^p sense. For each Riemann-Hilbert problem, the uniqueness and the explicit formula of the solution are obtained. The results presented here supplement those in our earlier paper [19], where the same problems for the homogeneous Dirac equation were studied.

15.1 Introduction

Let $D = \sum_{j=1}^{m} e_j \frac{\partial}{\partial x_j}$ denote the Dirac operator over $\mathbf{R}^m$ and $\{e_j\}_{j=1}^{m}$ denote the generators of the real Clifford algebra $\mathbf{R}_{0,m}$. Solutions to the homogeneous Dirac equation $Dw = 0$ in $\mathbf{R}^m$ are called monogenic functions; these functions are, in many respects, higher-dimensional analogues of classical holomorphic functions. The function theory associated with monogenic functions and the applications of this function theory to various branches of mathematics and physics have been investigated by many authors (see, for instance, [1-10], [12-14], [16-20]).

Recently, boundary value problems for Clifford-valued functions have become one of the focal points of research in Clifford analysis. In [17-20], by introducing a decomposition of the Clifford algebra, the Riemann-Hilbert type and Neumann type problems for monogenic functions have been formulated and studied in the unit ball and in the upper half space of $\mathbf{R}^m$. In [4], a half Dirichlet problem has been formulated for monogenic functions. In [7], Gürlebeck and Sprößig describe boundary value problems for the Helmholtz equation $\Delta u + \lambda^2 u = 0$ by decomposing it into $(D +$

0-8493-8481-8/96/$0.00+$.50
© 1996 by CRC Press

$\lambda)(D - \lambda)u = 0$ and making use of the Cauchy kernel for the operator $(D - \lambda)$.

This paper continues the work of [19]. We extend the results of [19] for monogenic functions to those for solutions of nonhomogeneous Dirac equations in a half space of $\mathbf{R}^m$ $(m \geq 2)$. It is well known that solutions of $Dw = 0$ are always C^∞, but this is not true for solutions of the non-homogeneous equation $Dw = F$. For certain F, a C^1 (strong) solution of $Dw = F$ may not exist. In this paper, we also analyze the boundary value problems of Riemann-Hilbert type for weak solutions of $Dw = F$.

15.2 Preliminaries

We consider the real Clifford algebra $\mathbf{R}_{0,m}$ generated by the orthonormal basis $\{e_j\}_{j=1}^m$ of $\mathbf{R}^m$; $\mathbf{R}_{0,m}$ has basis elements $e_A = e_{\alpha_1 \ldots \alpha_h} = e_{\alpha_1} e_{\alpha_2} \ldots e_{\alpha_h}$ for $A = \{\alpha_1, ..., \alpha_h\} \subset M = \{1, ..., m\}$ with $\alpha_1 < \alpha_2 < ... < \alpha_h$ where $e_\phi = e_0 = 1$ is the identity of $\mathbf{R}_{0,m}$ and the multiplication is governed by the rule $e_j e_k + e_k e_j = -2\delta_{jk}, 1 \leq j, k \leq m$ with δ_{jk} the Kroneker delta. The norm of $a = \sum_{A \subset M} a_A e_A \in \mathbf{R}_{0,m}$ is defined by $|a| = (\sum_{A \subset M} a_A^2)^{\frac{1}{2}}$, then

$$|ab| \leq K_{0,m}|a||b| \qquad \text{for} \quad a, b \in \mathbf{R}_{0,m}$$

and the optimal values for the constants $K_{0,m}$ are obtained in [9]. In the sequel a vector $x = (x_1, ..., x_m) \in \mathbf{R}^m$ is identified with the Clifford number $x = \sum_{j=1}^m e_j x_j \in \mathbf{R}_{0,m}$; here $x^2 = -|x|^2$. The Dirac operator in $\mathbf{R}^m$ is defined by $D = \sum_{j=1}^m e_j \frac{\partial}{\partial x_j}$; here $D^2 = -\Delta$, Δ the Laplacian. Let $\mathbf{R}_+^m = \{x | x_m > 0\}$ be the upper half space and $\mathbf{R}_0^m = \{x | x_m = 0\}$ its boundary in $\mathbf{R}^m$; here $\mathbf{R}_0^m$ may be identified with $\mathbf{R}^{m-1}$. The projection on $\mathbf{R}_0^m$ of any $x \in \mathbf{R}_+^m$ is denoted by $Z_x = \sum_{j=1}^{m-1} e_j x_j = (Z_x, 0)$. Thus, $x \in \mathbf{R}^m$ can be written as $x = Z_x + e_m x_m = (Z_x, x_m)$. Let $\mathbf{R}_{0,m-1}$ be the subalgebra of $\mathbf{R}_{0,m}$ generated by $\{e_j\}_{j=1}^{m-1}$. Then we have

$$\mathbf{R}_{0,m} = \mathbf{R}_{0,m-1} \oplus e_m \mathbf{R}_{0,m-1}.$$

Any $a \in \mathbf{R}_{0,m}$ may be written as

$$a = X^{(m)}(a) + e_m Y^{(m)}(a) \qquad \text{with} \qquad X^{(m)}(a), \ Y^{(m)}(a) \in \mathbf{R}_{0,m-1}.$$

It is clear that the above decomposition of $\mathbf{R}_{0,m}$ is a generalization of the classical decomposition of $\mathbf{C} = \mathbf{R} \oplus \sqrt{-1}\mathbf{R}$, and the operators $X^{(m)}$ and

$Y^{(m)}$ acting on $\mathbf{R}_{0,m}$ are generalizations of the operators Re and Im acting on $\mathbf{C}$.

Let $\Omega \subset \mathbf{R}^m$ be a domain. An $\mathbf{R}_{0,m}$-valued function $w \in C^1(\Omega)$ is called (left) monogenic in Ω if $Dw = 0$ in Ω. Recall that monogenic functions are harmonic functions. If $w = X^{(m)}w + e_m Y^{(m)}w$ is an $\mathbf{R}_{0,m}$-valued monogenic function, then $X^{(m)}w$ and $Y^{(m)}w$ are called $\mathbf{R}_{0,m-1}$-valued *harmonic conjugate functions* associated with w.

LEMMA 15.2.1 ([19], Generalized Cauchy-Riemann Equations)
An $\mathbf{R}_{0,m}$-valued function $w \in C^1(\Omega)$ is monogenic in Ω if and only if $X^{(m)}w$ and $Y^{(m)}w$ satisfy the equations

$$D_{m-1}X^{(m)}w - \frac{\partial}{\partial x_m}Y^{(m)}w = 0 \qquad in \ \Omega, \qquad (15.2.1)$$

$$\frac{\partial}{\partial x_m}X^{(m)}w - D_{m-1}Y^{(m)}w = 0 \qquad in \ \Omega, \qquad (15.2.2)$$

where $D_{m-1} = \sum\limits_{j=1}^{m-1} e_j \frac{\partial}{\partial x_j}$ is a Dirac operator in $\mathbf{R}^{m-1}$.

Let

$$P(x,y) = \frac{2x_m}{\omega_m(|Z_x - Z_y|^2 + x_m^2)^{m/2}}, \qquad m \geq 2, \qquad (15.2.3)$$

be the Poisson Kernel for the upper half space $\mathbf{R}_+^m$, where $\omega_m := \frac{2\pi^{m/2}}{\Gamma(m/2)}$ is the surface area of the unit sphere in $\mathbf{R}^m$. It has been shown in [19] that by using Lemma 15.2.1, functions

$$\widetilde{P}(x,y) = \frac{2(Z_x - Z_y)}{\omega_m(|Z_x - Z_y|^2 + x_m^2)^{m/2}} \qquad (15.2.4)$$

and

$$\widehat{P}(x,y) = \frac{2(Z_x - Z_y)}{\omega_m(|Z_x - Z_y|^2 + x_m^2)^{m/2}} + \frac{2Z_y)}{\omega_m(1 + |Z_y|^2)^{m/2}} \qquad (15.2.5)$$

are two harmonic conjugates of the Poisson kernel $P(x,y)$ with respect to the variable x. Moreover, for fixed $x \in \mathbf{R}_+^m$,

$$|\widetilde{P}(x,y)| = O(|Z_y|^{-m+1}) \text{ and } |\widehat{P}(x,y)| = O(|Z_y|^{-m}), \text{ as } |Z_y| \to \infty. \qquad (15.2.6)$$

We now define

$$S(x,y) = P(x,y) + e_m\widetilde{P}(x,y) \qquad (15.2.7)$$

and

$$\widehat{S}(x,y) = P(x,y) + e_m\widehat{P}(x,y). \qquad (15.2.8)$$

Then, for fixed $y \in \mathbf{R}_0^m$, $S(x,y)$ and $\widehat{S}(x,y)$ are monogenic in $\mathbf{R}_+^m$ and their projections $X^{(m)}S(x,y)$ and $X^{(m)}\widehat{S}(x,y)$ are the Poisson kernel for the half space $\mathbf{R}_+^m$. Moreover, for fixed $x \in \mathbf{R}_+^m$,

$$|S(x,y)| = O(|Z_y|^{-m+1}) \text{ and } |\widehat{S}(x,y)| = O(|Z_y|^{-m}), \text{ as } |Z_y| \to \infty. \tag{15.2.9}$$

Notice that $S(x,y)$ may also be expressed in the form

$$S(x,y) = -2E(y-x)e_m|_{y \in \mathbf{R}_0^m} \tag{15.2.10}$$

where

$$E(x) := -\frac{x}{\omega_m |x|^m} \tag{15.2.11}$$

is the fundamental solution of the Dirac operator D on $\mathbf{R}^m$.

For $w, F \in L_{loc}^1(\mathbf{\Omega})$, we say $Dw = F$ (weak) in $\mathbf{\Omega}$ if and only if

$$\int_{\mathbf{\Omega}} [\phi F + (\phi D)w]dx = 0 \text{ for any } \phi \in C_0^\infty(\mathbf{\Omega}), \tag{15.2.12}$$

where $C_0^\infty(\mathbf{\Omega})$ is the space of infinitely differentiable functions with compact support contained in $\mathbf{\Omega}$.

In [8] (see also [6]) it was shown by Hile that if $w \in L_{loc}^1(\mathbf{\Omega})$ satisfies $Dw = 0$ (weak), then w is also a strong solution of $Dw = 0$ in $\mathbf{\Omega}$ and, moreover, $w \in C^\infty(\mathbf{\Omega})$.

We now introduce the following integral operators:

$$(\mathbf{P}f)(x) := \int_{\mathbf{R}_0^m} P(x, Z_y)f(Z_y)dZ_y, \tag{15.2.13}$$

$$(\mathbf{K}f)(x) := \int_{\mathbf{R}_0^m} S(x, Z_y)f(Z_y)dZ_y, \tag{15.2.14}$$

$$(\widehat{\mathbf{K}}f)(x) := \int_{\mathbf{R}_0^m} \widehat{S}(x, Z_y)f(Z_y)dZ_y, \tag{15.2.15}$$

$$(\mathbf{T}f)(x) := \int_{\mathbf{R}_+^m} [E(x-y) - E(x^*-y)]F(y)dy, \tag{15.2.16}$$

where $x^* = (Z_x, -x_m)$ is the symmetric point of x with respect to the hyperplane $\mathbf{R}_0^m$.

REMARK From (15.2.10), we also have

$$(\mathbf{K}f)(x) := \int_{\mathbf{R}_0^m} 2E(y-x)d\sigma_y f(Z_y) \tag{15.2.17}$$

where $d\sigma_y = \sum_{j=1}^m (-1)^{j-1} e_j d\hat{y}_j$ with $d\hat{y}_j = dy_1 \wedge ... \wedge dy_{j-1} \wedge dy_{j+1} \wedge ... \wedge dy_m$.

Thus, the operator $\mathbf{K}$ is just the Cauchy integral operator mutiplied by 2.

The integral on the right-hand side of (15.2.14) is a singular integral for $x \in \mathbf{R}_0^m$. If f is bounded and Hölder continuous on $\mathbf{R}_0^m$, the integral defined by the right-hand side of (15.2.14) is finite in the sense of Cauchy's principal value. Following the technique of V. Iftimie [10], one can prove that the function defined by (15.2.14) can be continuously extended to the hyperplane $\mathbf{R}_0^m$. Moreover,

$$(\mathbf{K}f)^+(\tilde{x}) = \lim_{x \to \tilde{x},\ x \in \mathbf{R}_+^m} (\mathbf{K}f)(x) = f(\tilde{x}) + (\mathbf{K}f)(\tilde{x}), \quad \tilde{x} \in \mathbf{R}_0^m. \quad (15.2.18)$$

The equation (15.2.18) is called the *Plemilj* formula.

Let p be a positive number and B the unit ball in $\mathbf{R}^m$. Define function spaces $h^p(\mathbf{R}_+^m)$ and $L^{p,\nu}(\mathbf{R}^m)$ by

$$h^p(\mathbf{R}_+^m) := \{u|\ \Delta u = 0 \text{ in } \mathbf{R}_+^m \text{ and } \sup_{x_m > 0} \int_{\mathbf{R}_0^m} |u(Z_x, x_m)|^p dZ_x < \infty\}$$

and

$$L^{p,\nu}(\mathbf{R}^m) := \{F(x)|F(x) \in L^p(B), F_\nu(x) = |x|^{-\nu} F(x^{-1}) \in L^p(B)\}.$$

The norm of F in $L^{p,\nu}(\mathbf{R}^m)$ is defined by

$$\|F\|_{p,\nu} := [\int_B |F(x)|^p dx]^{1/p} + [\int_B |F_\nu(x)|^p dx]^{1/p}.$$

For an $\mathbf{R}_{0,m}$-valued function F defined in $\mathbf{R}_+^m$, we write $F \in L^{p,\nu}(\mathbf{R}_+^m)$ if its extension

$$\tilde{F}(x) = \begin{cases} F(x), & x \in \mathbf{R}_+^m \\ 0, x \in \mathbf{R}^m \backslash \mathbf{R}_+^m \end{cases} \quad \in L^{p,\nu}(\mathbf{R}^m)$$

15.3 Riemann-Hilbert Problems for Nonhomogeneous Dirac Equations

We now formulate the proper Riemann-Hilbert problems for nonhomogeneous Dirac equations in the upper half space of $\mathbf{R}_+^m$ ($m \geq 2$) and show that there exists a unique solution in some sense under certain hypotheses. The explicit formula for the solution is obtained as well. To this end we first derive several properties of the integral operator $\mathbf{T}$.

THEOREM 15.3.1
Let F be an $\mathbf{R}_{0,m}$-valued bounded continuous function in $\mathbf{R}_+^m$ with

$$|F(x)| = O(|x|^{-1-\epsilon}) \text{ as } x \to \infty, \text{ for some } \epsilon \in (0,1). \qquad (15.3.1)$$

Then $\mathbf{T}F$ is well defined by (16) for all $x \in \mathbf{R}_+^m \cup \mathbf{R}_0^m$, and it satisfies:

(a) $D[(\mathbf{T}F)(x)] = F(x), x \in \mathbf{R}_+^m$;

(b) $(\mathbf{T}F)(x) = 0, x \in \mathbf{R}_0^m$;

(c) $|(\mathbf{T}F)(x)| = O(|x|^{-\epsilon})$ as $x \to \infty$.

PROOF   Under the assumption (15.3.1), $\mathbf{T}F$ is well defined by (15.2.16) for all $x \in \mathbf{R}_+^m \cup \mathbf{R}_0^m$. Since

$$E(x-y) - E(x^*-y) = 0, \quad \text{for } x \in \mathbf{R}_0^m \text{ and } y \in \mathbf{R}_+^m, \qquad (15.3.2)$$

and since, in the distributional sense,

$$D_x[E(x-y) - E(x^*-y)] = \delta(x-y), \quad \text{for } x,y \in \mathbf{R}_+^m, \qquad (15.3.3)$$

(a) and (b) are easily verified.

Now we prove (c). Let

$$Q(x) := \int_{\mathbf{R}_+^m} E(x-y)F(y)dy.$$

By (15.3.1), there exists a constant $M > 0$ such that

$$|Q(x)| \le M \int_{\mathbf{R}_+^m} \frac{dy}{|x-y|^{m-1}|y|^{1+\epsilon}} \le M \int_{\mathbf{R}^m} \frac{dy}{|x-y|^{m-1}|y|^{1+\epsilon}}.$$

We need to examine the behavior of the right-hand integral when $R := |x| \to \infty$. Without loss of generality, we may estimate the integral at the point $x = (R, 0, ..., 0) \in \mathbf{R}^m$ (by symmetry, an estimate at this point will suffice). Let $y = R\xi = (R\xi_1, ..., R\xi_m)$. Then

$$\int_{\mathbf{R}^m} \frac{dy}{|x-y|^{m-1}|y|^{1+\epsilon}} = \int_{\mathbf{R}^m} \frac{dy}{[(y_1-R)^2+y_2^2+...+y_m^2]^{(m-1)/2}|y|^{1+\epsilon}}$$

$$= R^{-\epsilon} \int_{\mathbf{R}^m} \frac{d\xi}{[(\xi_1-1)^2+\xi_2^2+...+\xi_m^2]^{(m-1)/2}|\xi|^{1+\epsilon}}$$

$$= O(|x|^{-\epsilon}) \text{ as } x \to \infty,$$

since the last integral converges absolutely near its three singularities: ∞, 0 and $(1,0,...,0)$. Hence we have proven that $|Q(x)| = O(|x|^{-\epsilon})$ as $x \to \infty$. Similarly, we also get that $|Q(x^*)| = O(|x|^{-\epsilon})$ as $x \to \infty$. Thus $(\mathbf{T}F)(x) = Q(x) - Q(x^*)$ satisfies the estimate (c). ∎

THEOREM 15.3.2
Let $F \in L^{p,m}(\mathbf{R}_+^m, \mathbf{R}_{0,m})$ with $m < p < \infty$. Then

(a) $D[(\mathbf{T}F)(x)] = F(x)$ *(weak)*, $x \in \mathbf{R}_+^m$:

(b) $(\mathbf{T}F)(x) = 0, x \in \mathbf{R}_0^m$;

(c) $|(\mathbf{T}F)(x)| = O(|x|^{(m/p)-m+1})$, *as* $x \to \infty$;

(d) $(\mathbf{T}F)(\bullet, x_m) \in L^s(\mathbf{R}_0^m)$, *with* $\|(\mathbf{T}F)(\bullet, x_m)\|_{L^s(\mathbf{R}_0^m)} \to 0$ *as* $x \to 0^+$,

where $\frac{m-1}{m-1-(m/p)} < s < \infty$.

PROOF   Following the technique used in [6, Theorem 4.25], one can prove

$$|(\mathbf{T}F)(x)| \leq M(m,p)\|F\|_{p,m}$$

and

$$|(\mathbf{T}F)(x) - (\mathbf{T}F)(y)| \leq M(m,p)\|F\|_{p,m}|x-y|^\alpha, \quad \alpha = \frac{p-m}{p},$$

for any $x, y \in \mathbf{R}^m$ and for some constant $M(m,p)$ depending on m and p. Moreover, for any $R > 0$, there is a constant $M(m,p,R)$ such that

$$|(\mathbf{T}F)(x)| \leq M(m,p,R)\|F\|_{p,m}|x|^{(m/p)-m+1}, \quad \text{for } |x| \geq R.$$

Thus, $(\mathbf{T}F)(x)$ is a Hölder continuous function in $\mathbf{R}^m$ and $(\mathbf{T}F)(x)$ satisfies (c). Using (15.3.2) and (15.3.3), (a) and (b) are easily verified. Finally, (d) follows from (b) and (c).   ∎

THEOREM 15.3.3   (Uniqueness)
If w is a solution of the following Riemann-Hilbert problem

$$
\begin{array}{llll}
Dw = 0 & \text{(weak)} & in & \mathbf{R}_+^m, \\
X^{(m)}w = 0 & & on & \mathbf{R}_0^m, \\
w & \text{is bounded} & in & \mathbf{R}_+^m,
\end{array}
$$

then $w \equiv e_m C$ where C is an $\mathbf{R}_{0,m-1}$-valued constant.

PROOF   By [8] (or [6, Theorem 4.7]), w is also a strong solution of $Dw = 0$ in $\mathbf{R}_+^m$. Hence $X^{(m)}w$ is harmonic in $\mathbf{R}_+^m$. Applying Liouville theorem to the odd extension of $X^{(m)}w$ on the whole space $\mathbf{R}^m$ with respect to the variable x_m, we get $X^{(m)}w = 0$. Then, by applying Lemma 15.2.1, we see that $\frac{\partial}{\partial x_m}Y^{(m)}w = D_{m-1}Y^{(m)}w = 0$ in $\mathbf{R}_+^m$. Thus $Y^{(m)}w$ is an $\mathbf{R}_{0,m-1}$-valued bounded monogenic function of $Z_x = (x_1, ..., x_{m-1}) \in \mathbf{R}^{m-1}$. By Liouville theorem, $Y^{(m)}w$ is an $\mathbf{R}_{0,m-1}$-valued constant.   ∎

THEOREM 15.3.4

Let $f \in C(\mathbf{R}_0^m, \mathbf{R}_{0,m-1})$ be bounded and Hölder continuous with

$$\int_{\mathbf{R}_0^m} \frac{|f(Z_y)|dZ_y}{(1+|Z_y|^2)^{(m-1)/2}} < \infty. \tag{15.3.4}$$

Let $F \in C(\mathbf{R}_+^m, \mathbf{R}_{0,m})$ with $|F(x)| = O(|x|^{-1-\epsilon})$ as $x \to \infty$, for some $\epsilon \in (0,1)$. Then there is a unique classical C^1 solution to the Riemann-Hilbert problem

$$Dw = F \quad in \ \mathbf{R}_+^m, \tag{15.3.5}$$

$$X^{(m)} = 0 \quad on \ \mathbf{R}_0^m, \tag{15.3.6}$$

$$\lim_{x \to \infty} w = 0 \quad , \tag{15.3.7}$$

and the solution can be represented as

$$w(x) = (\mathbf{K}f)(x) + (\mathbf{T}F)(x) \tag{15.3.8}$$

where the operators $\mathbf{K}$ and $\mathbf{T}$ are defined by (15.2.14) and (15.2.16), respectively.

PROOF   Under the hypotheses (15.3.4), it is easy to see that $\mathbf{K}f$ is well defined by (15.2.14) for all $x \in \mathbf{R}_+^m$, $\mathbf{K}f \to 0$ as $x \to \infty$, and $\mathbf{K}f$ is monogenic in $\mathbf{R}_+^m$. Since f is a bounded Hölder continuous function on $\mathbf{R}_0^m$, $(\mathbf{K}f)(x)$ can be continuously extended to the hyperplane $\mathbf{R}_0^m$. If we denote the extension of $\mathbf{K}f$ on $\mathbf{R}_0^m$ by $(\mathbf{K}f)^+$, then the Plemilj formula (15.2.18) and the fact that the second term on the right hand side of (15.2.18) is an $e_m\mathbf{R}_{0,m-1}$-valued function give that

$$X^{(m)}((\mathbf{K}f)^+(Z_x,0)) = f(Z_x) \qquad for \ all \ Z_x \in \mathbf{R}_0^m.$$

By the above argument and Theorem 15.3.1, it is clear that the function defined by (15.3.8) is a solution to the problem (15.3.5)-(15.3.7). The uniqueness of the solution follows from Theorem 15.3.3 and (15.3.7).   ∎

THEOREM 15.3.5

Let $f \in C(\mathbf{R}_0^m, \mathbf{R}_{0,m-1})$ be bounded and $F \in C(\mathbf{R}_+^m, \mathbf{R}_{0,m})$ with $|F(x)| = O(|x|^{-1-\epsilon})$ as $x \to \infty$, for some $\epsilon \in (0,1)$. Then the Riemann-Hilbert problem

$$Dw = F \quad (weak) \ in \quad \mathbf{R}_+^m, \tag{15.3.9}$$

$$X^{(m)}w = f \qquad on \qquad \mathbf{R}_0^m, \tag{15.3.10}$$

$$w \ is \ bounded \ in \ \mathbf{R}_+^m, \tag{15.3.11}$$

has a classical C^1 solution which is unique up to a Clifford constant, and the solution can be represented as

$$w(x) = (\widehat{\mathbf{K}}f)(x) + (\mathbf{T}F)(x) + e_m C$$

where C is an $\mathbf{R}_{0,m-1}$-valued constant and the operators $\widehat{\mathbf{K}}$ and $\mathbf{T}$ are defined by (15.2.15) and (15.2.16), respectively.

REMARK In Theorem 15.3.5, whether or not the solution w may be continuously extended to $\mathbf{R}_0^m$, its projection $X^{(m)}w$ can always be continuously extended to $\mathbf{R}_0^m$ and the extension of $X^{(m)}w$ satisfies the boundary condition (15.3.10).

PROOF [**Proof of Theorem 15.3.5**] By the definition of $\widehat{\mathbf{K}}f$ and (15.2.9), $\widehat{\mathbf{K}}f$ is well defined for all $x \in \mathbf{R}_+^m$ and $\widehat{\mathbf{K}}f$ is monogenic and bounded in $x \in \mathbf{R}_+^m$. Since $X^{(m)}(\widehat{\mathbf{K}}f) = \mathbf{P}f$, it follows from [15, II, Theorem 2.1(b)] that

$$X^{(m)}((\widehat{\mathbf{K}}f)^+)(Z_x, 0) = (\mathbf{P}f)^+(Z_x, 0) = f(Z_x), \quad \text{for all } Z_x \in \mathbf{R}_0^m.$$

Combining the above argument and Theorem 15.3.1, we know that $\widehat{\mathbf{K}}f + \mathbf{T}F$ is a solution to the problem (15.3.9)-(15.3.11). The uniqueness of the solution follows from Theorem 15.3.33 and (15.3.11). ∎

THEOREM 15.3.6
Let $F \in L^{p,m}(\mathbf{R}_+^m, \mathbf{R}_{0,m})$ with $m < p < \infty$ and $f \in L^s(\mathbf{R}_0^m, \mathbf{R}_{0,m-1})$ with $\frac{m-1}{m-1-(m/p)} < s < \infty$. Then the Riemann-Hilbert problem (15.3.5)-(15.3.11) has a unique bounded Hölder continuous (weak) solution w in the following sense:

(a) *$Dw = F$ (weak) in $\mathbf{R}_+^m$;*

(b) *$\lim_{x_m \to 0+} \|X^{(m)}w(\bullet, x_m) - f(\bullet)\|_{L^s(\mathbf{R}_0^m)} = 0$;*

(c) *w vanishes at infinity.*

Moreover, the solution can be represented as

$$w(x) = (\mathbf{K}f)(x) + (\mathbf{T}F)(x).$$

PROOF From [15, IV, Theorem 4.17] and (15.2.12), it follows that $\mathbf{K}f \in h^s(\mathbf{R}_+^m, \mathbf{R}_{0,m})$ is monogenic in $\mathbf{R}_+^m$ and it vanishes at infinity.

It has been shown in [19] (and in [11] for m = 2) that the integral operator $\mathbf{P}$ defined by (15.2.13) is a mapping from $L^s(\mathbf{R}_0^m, \mathbf{R}_{0,m-1})$ to $h^s(\mathbf{R}_+^m, \mathbf{R}_{0,m}), 1 \le s < \infty$. Moreover, for any $f \in L^s(\mathbf{R}_0^m, \mathbf{R}_{0,m-1}), \mathbf{P}f$

vanishes at infinity and

$$\lim_{x_m \to 0^+} ||(\mathbf{P}f)(\bullet, x_m) - f(\bullet)||_{L^s(\mathbf{R}_0^m)} = 0.$$

Now, since $X^{(m)}(\mathbf{K}f) = \mathbf{P}f$, the result follows directly from Theorem 15.3.2 and Theorem 15.3.3. ∎

Acknowledgment

The authors would like to express their gratitude to Professor John Ryan and the University of Arkansas for their invitation and financial support during this conference.

Bibliography

[1] F. Brackx, R. Delanghe and F. Sommen, *Clifford Analysis*, Pitman, London, 1982.

[2] H. Begehr and Z. Xu, *Nonlinear half-Dirichlet problems for first order elliptic equations in the unit ball of* $\mathbf{R}^m (m \geq 3)$, Appl. Anal. **45**, (1992), 3–18.

[3] J. S. R. Chisholm and A. K. Common (Eds.), *Clifford Algebras and their Applications in Mathematical Physics*, D. Reidel, Dordrecht, 1986.

[4] R. Delanghe, F. Sommen and Z. Xu, *Half-Dirichlet problems for powers of the Dirac operator in the unit ball of* $\mathbf{R}^m (m \geq 3)$, Bull. Soc. Math. Belg., 3 Ser. B, **42**, (1990), 409–429.

[5] J. E. Gilbert and M. A. M. Murray, *Clifford Algebras and Dirac Operators in Harmonic Analysis*, Cambridge University Press, Cambridge, 1991.

[6] R. P. Gilbert and J. L. Buchanan, *First Order Elliptic Systems: A Function Theoretic Approach*, Academic Press, New York, 1983.

[7] K. Gürlebeck and W. Spröβig, *Quaternionic Analysis and Elliptic Boundary Value Problems, V. 56*, Akademie-Verlag, Berlin, 1989.

[8] G. N. Hile, *Hypercomplex Function Theory Applied to Partial Differential Equations*, Ph.D. thesis, Indiana University, Bloomington, IN, 1972.

[9] G. N. Hile and P. Lounesto, *Matrix representations of Clifford algebras*, Linear Algebra Appl., **128**, (1990), 51–63.

[10] V. Iftimie, *Functions hypercomplexes*, Bull. Soc. Sci. Math. R. S. Roum., **9 (57)**, (1965), 279–332.

[11] P. Koosis, *Introduction to H_p Spaces*, Cambridge University Press, Cambridge, 1980.

[12] A. Micali, R. Boudet and J. Helmstetter (Eds.), *Clifford algebras and their Applications in Mathematical Physics*, Kluwer Academic, Boston, 1992.

[13] J. Ryan, *Complexified Clifford analysis*, Complex Variables, **1**, (1982), 119–149.

[14] J. Ryan, *Plemelj formulae and transformations associated to plane wave decompositions in complex Clifford analysis*, Proc. London Math. Soc., **(3) 64**, (1992), 70–94.

[15] E. M. Stein and G. Weiss, *Introduction to Fourier Analysis on Euclidean Spaces*, Princeton University Press, Princeton, NJ, 1975.

[16] A. Sudbery, *Quaternionic analysis*, Math. Proc. Camb. Philos. Soc., **85**, (1979), 199–225.

[17] Z. Xu, *On boundary value problems of Neumann type for hypercomplex functions with values in a Clifford algebra*, Rend. Circolo Math. Palermo, Serie II, **No. 22**, (1989), 213–226.

[18] Z. Xu, J. Chen and W. Zhang, *A harmonic conjugate of the Poisson kernel and a boundary value problem for monogenic functions in the unit ball of $\mathbf{R}^n(n \geq 2)$*, Simon Stevin, **64**, (1990), 187–201.

[19] Z. Xu and C. Zhou, *On boundary value problems of Riemann-Hilbert type for monogenic functions in a half space of $\mathbf{R}^m(m \geq 2)$*, Complex Variables, **22**, (1993), 181–194.

[20] C. Zhou, *On boundary value problems of Neumann type for the Dirac operator in a half space of $\mathbf{R}^m(m \geq 3)$*, Complex Variables **23**, (1993), 1–16.

Department of Mathematics,
Ryerson Polytechnic University,
350 Victoria Street,
Toronto, Ontario M5B 2K3,
CANADA
E-mail: zxu@acs.ryerson.ca

Department of Mathematics,
University of Hawaii at Honolulu Community College,
874 Dillingham Boulevard, Honolulu, Hawaii 96817
E-mail: chiping@pulua.hcc.hawaii.edu

16

Regularity and Approximation Results for the Maxwell Problem on C^1 and Lipschitz Domains

Marius Mitrea,* Rodolfo H. Torres,† and Grant V. Welland‡

ABSTRACT Solutions to boundary values problems for Maxwell's equations are described using the layer potential Maxwell operator. The Maxwell, electric and magnetic boundary value problems for time harmonic electromagnetic wave propagation have unique solutions in C^1 domains and satisfy optimal estimates. Sobolev-Besov regularity results hold for domains with Lipschitz boundaries. Also, multipole expansion methods can be used to approximate L^p tangential vector fields on the boundaries of domains which are only Lipschitz.

16.1 Introduction

Our main purpose is to describe extensions of the classical treatment of the boundary value problems for Maxwell equations via layer potential techniques on smooth domains (i.e., C^2 or $C^{1,\alpha}$, $\alpha > 0$) [Mu], [Mu1], [Weyl], [Cal2] to domains with less regular boundaries, i.e., C^1 or Lipschitz. Our work is in the spirit of work that has been published in the last decade. Layer potential techniques have proved to be very effective tools in the study of many boundary value problems for constant coefficients and elliptic and parabolic operators, including also systems of equations, on C^1 ([FJR],[FR], [V1]) and even Lipschitz domains ([B1],[DK],[DKV], [FKV],[V2]). Results are obtained in the case of Lipschitz domains, in much of what follows. The invertibility results require a C^1 boundary.

*Supported in part by ONR Grant N0001-90-J-1343.
†Supported in part by a Rackham Fellowship and NSF grant DMS 9303363.
‡Supported under Airforce Grant number 90-0307.
 1991 *Mathematics Subject Classification*. Primary 35J05, 45P05, 58G20; Secondary 31A25.

The Maxwell equations in the case of time harmonic waves in $\mathbf{R}^3$ with wave number k, $k \in \mathbf{C}$, $\operatorname{Im} k > 0$, are

$$\operatorname{curl} E - ikH = 0,$$

$$\operatorname{curl} H + ikE = 0.$$

See [CK] for the classical treatment of the problems presented here (for smooth domains, some related problems are also discussed in [GS]). We do not give proofs for most of the results presented. They will appear elsewhere in [MTW]. We do give indications of proofs in the last two sections. In one case we give regularity results, where the method uses Clifford analysis. In the other section we give the method of multipole approximation for Lipschitz domains. This is an extension of the result in [Cal2].

The boundary value problems treated are similar to:

$$(M) \begin{cases} \operatorname{curl} E - ikH = 0 \text{ on } D, \\ \operatorname{curl} H + ikE = 0 \text{ on } D, \\ E^*, \ H^* \in L^p(\partial D), \\ N \times E = A \text{ on } \partial D. \end{cases}$$

Here A is a given tangential vector field on ∂D and * stands for the usual nontangential maximal function on D.

We use acoustic layer potentials. The properties of the single- and double-layer acoustic potentials and the boundedness results for the corresponding trace operators are essentially the same as those for the case of Laplace's equation $k = 0$ and they can be obtained from the results in [CMM] on the Cauchy integral on Lipschitz curves. See [BS], [M], [TW].

A natural space for the datum A is $L_T^{p,Div}(\partial D)$, the vector space of all tangential fields in $L^p(\partial D)$ having the surface divergence (in a distributional sense) in $L^p(\partial D)$, $1 < p < \infty$. In this setting, we obtain existence, uniqueness and optimal estimates for the full range $p \in (1, \infty)$, assuming that the boundary of the domain is of class C^1. Solutions are expressed in the form of layer potential operators. More specifically,

$$E(X) := \int_{\partial D} \operatorname{curl}_X \{\Phi(X - Q)B(Q)\} d\sigma(Q), \quad X \in D,$$

and

$$H(X) := -ik \int_{\partial D} \Phi(X - Q)B(Q)d\sigma(Q)$$
$$+ \tfrac{i}{k} \int_{\partial D} \nabla_X \Phi(X - Q) \operatorname{Div} B(Q)d\sigma(Q),$$

$X \in D$, where

$$\Phi(X) := -\frac{e^{ik|X|}}{4\pi|X|}.$$

is the radial fundamental solution for the Helmholtz operator in $\mathbf{R}^3$ and B is a suitably chosen vector field in $L_T^{p,Div}(\partial D)$. In particular, for $p = 2$, as a simple consequence of these integral representation formulas we have that E and H belong to the Sobolev space $W^{1/2}(D)$, and that

$$C^{-1}\|A\|_{L_T^{2,Div}(\partial D)} \leq \|E\|_{W^{1/2}(D)} + \|H\|_{W^{1/2}(D)} \leq C\|A\|_{L_T^{2,Div}(\partial D)}.$$

Particular regularity properties for solutions of Maxwell equations have been considered in [Co]. See section 16.4 for more precise Sobolev space regularity results (see the references in [Co] for the consequences and applications of these results).

The treatment of the various singular layer potential operators on non-smooth boundaries rests on the fundamental results of Calderón [C] and Coifman, McIntosh and Meyer [CMM]. The main idea of the proof is that, even in the C^1 case, the singular integral equations arising on the boundary can still be handled via Fredholm theory. The particular form of the equations and the specific nature of the boundary condition in (M) make this problem different from other elliptic problems already considered in the literature. A major ingredient is the solution of the scalar Helmholtz equation on Lipschitz domains from [M] and [TW].

Similar results are also valid for the complementary domain, $\mathbf{R}^3 \setminus \bar{D}$, under the additional assumption that E and H have an adequate decay at infinity, i.e., they satisfy the Silver-Müller radiation condition.

One can also equivalently reformulate the problem (M) only in terms of the component E, namely as

$$(M')\begin{cases} (\triangle + k^2)E = 0 \ \text{ on } \ D, \\ \text{div } E = 0 \ \text{ on } \ D, \\ E^*, (\text{curl } E)^* \in L^p(\partial D), \\ N \times E = A \in L_T^{p,Div}(\partial D). \end{cases}$$

Consequently, it is interesting to compare it with the version in which the datum A is assumed to be an L^p–integrable tangential vector field with no extra smoothness:

$$(D)\begin{cases} (\triangle + k^2)E = 0 \ \text{ on } \ D, \\ \text{div } E = 0 \ \text{ on } \ D, \\ E^* \in L^p(\partial D), \\ N \times E = A \in L_T^p(\partial D). \end{cases}$$

It is therefore convenient to think of (M') (or (M)) as the "regularity" boundary value problem for the Maxwell equations, in contrast to (D), more resembling of an L^p–type Dirichlet problem for the Maxwell equations.

As a byproduct of our approach, (D) can be solved uniquely (with optimal estimates) on C^1 domains for the full range of p's. The companion field

$H := \frac{1}{ik}\operatorname{curl} E$ of E for (D) has boundary traces in $L^p(\partial D)$ if and only if the datum A actually belongs to the regularity space $L_T^{p,Div}(\partial D)$. Hence, from this point of view, (M') appears to be more physically relevant.

16.2   The vector Helmholtz equation

The analogue of the classical Stratton-Chue integral representation formulae for vector fields satisfying the Helmholtz equation holds on Lipschitz domains (see, e.g., [CK]).

Let D be a Lipschitz domain in $\mathbf{R}^3$ and consider a complex valued vector field $E \in C^2(D)$ for which $\triangle E + k^2 E = 0$ in D, and such that $|||E|||_p < +\infty$, where

$$|||E|||_p := \|E^*\|_{L^p(\partial D)} + \|(\operatorname{div} E)^*\|_{L^p(\partial D)} + \|(\operatorname{curl} E)^*\|_{L^p(\partial D)}.$$

We shall also treat the case when E is defined on $\mathbf{R}^3 \setminus \bar{D}$, in which case we need an adequate decay at infinity, i.e., E satisfies the so-called Silver-Müller radiation condition

$$\operatorname{curl} E(X) \times \tfrac{X}{|X|} + \operatorname{div} E(X) \tfrac{X}{|X|} - ikE(X) = \mathrm{o}(|X|^{-1}), \qquad (R)$$

for $|X| \to \infty$, uniformly in all directions in $\mathbf{R}^3$ (see [CK]).

LEMMA 16.2.1
Consider $1 < p < \infty$ and let $E \in C^2(D)$ (or $E \in C^2(\mathbf{R}^3 \setminus \bar{D})$ with (R)) be a vector field satisfying both the Helmholtz equation and the condition $|||E|||_p < +\infty$.

Then, the nontangential limits of E, $\operatorname{div} E$, and $\operatorname{curl} E$, respectively, exist at almost any point P on the boundary and the limit functions, denoted by the same symbols, are in $L^p(\partial D)$.

Moreover, for all $X \in D$, or $X \in \mathbf{R}^3 \setminus \bar{D}$, respectively,

$$\pm E(X) = -\int_{\partial D} \operatorname{curl}_X \{\Phi(X - Q)N(Q) \times E(Q)\} d\sigma(Q)$$

$$- \int_{\partial D} \nabla_X \Phi(X - Q) \langle N(Q), E(Q)\rangle \, d\sigma(Q)$$

$$- \int_{\partial D} \Phi(X - Q)\{N(Q) \times \operatorname{curl} E(Q) - \operatorname{div} E(Q)N(Q)\} d\sigma(Q).$$

16.3 The singular integral operator M

We give the main ingredients of the integral operator machinery needed to solve boundary value problems in electromagnetism in the form of layer potential operators.

Let A be a tangential vector field in $L^p(\partial D)$. Then, at almost every point P on ∂D, one has

$$\lim_{\substack{X \to P \\ X \in \Gamma_\pm(P)}} N(P) \times \operatorname{curl} \mathcal{S}A(X) = (\mp \tfrac{1}{2}I + M)A(P),$$

where

$$MA(P) := \text{p.v.} \int_{\partial D} N(P) \times \operatorname{curl}_P \{\Phi(P-Q)A(Q)\} d\sigma(Q).$$

Clearly, the vector field MA is always tangential, and by the results in [CMM], M is a well defined, bounded operator from $L_T^p(\partial D)$ into itself, for any $1 < p < \infty$. The formal transpose $M^* : L_T^p(\partial D) \to L_T^p(\partial D)$, has the form

$$M^*B = N \times M(N \times B), \quad B \in L_T^p(\partial D).$$

The following theorems hold.

THEOREM 16.3.1
Let D be a bounded Lipschitz domain in $\mathbf{R}^3$. Then the operators $\pm\tfrac{1}{2}I + M$: $L_T^p(\partial D) \to L_T^p(\partial D)$ are injective for $2 \le p < \infty$ and have dense ranges for $1 < p \le 2$.

THEOREM 16.3.2
For any bounded Lipschitz domain D in $\mathbf{R}^3$ and for any $1 < p < \infty$, M is a bounded operator from $L_T^{p,Div}(\partial D)$ into itself. In particular, M^* : $L_N^{p,Div}(\partial D) \to L_N^{p,Div}(\partial D)$ is also well defined and bounded for any $1 < p < \infty$.

Moreover, if $p_0 < p < q_0$ then a tangential vector field A from $L_T^p(\partial D)$ has the property that $(\pm\tfrac{1}{2}I + M)A \in L_T^{p,Div}(\partial D)$ if and only if A belongs to $L_T^{p,Div}(\partial D)$.

THEOREM 16.3.3
If D is a bounded C^1 domain, then the operator M is compact on any $L_T^p(\partial D)$, $1 < p < \infty$.

Here is an immediate consequence of this result and Theorem 16.3.1.

COROLLARY 16.3.4
Let D be a bounded C^1 domain. Then $\pm\frac{1}{2}I + M$ are invertible on $L_T^p(\partial D)$ for any $1 < p < \infty$.

Also, as a direct consequence of the above corollary and Theorem 16.3.2 we have the following.

COROLLARY 16.3.5
If D is a bounded C^1 domain, then the operators $\pm\frac{1}{2}I + M$ are invertible on $L_T^{p,Div}(\partial D)$, for any $1 < p < \infty$. Furthermore, the operators $\pm\frac{1}{2}I + M^*$ are also invertible on $L_N^{p,Div}(\partial D)$, for all $1 < p < \infty$.

We do not give any further details regarding the structures of the boundary problems of electromagnetism here. Much more detailed information is given in [MTW]. However, before going to the next section we note that the problems of the interior and exterior Maxwell boundary value problems and the interior and exterior electric and magnetic boundary value problems are addressed and solved [MTW] in the context of the above theorems.

16.4   Sobolev-Besov space regularity results

For a bounded Lipschitz domain D in $\mathbf{R}^3$ and for $0 < \alpha < 1 < p, q < \infty$, the Sobolev-Besov space $\Lambda_\alpha^{p,q}(D)$ is defined as the collection of all (classes of) measurable functions u on D for which the quantity

$$\|u\|_{L^p(\partial D)} + \left(\iint_D \left(\iint_D |u(X) - u(Y)|^p dY \right)^{q/p} \frac{dX}{|X-Y|^{3+\alpha q}} \right)^{1/q},$$

denoted in the sequel by $\|u\|_{\Lambda_\alpha^{p,q}(D)}$, is finite. Note that for $p = q = 1/\alpha = 2$ we have that $\Lambda_{1/2}^{2,2}(D) = W^{1/2}(D)$, the Sobolev space of L^2–functions with half of the derivative still L^2–integrable in D ([A]).

We shall also need a couple of results from [M]. Let $\mathcal{A}$ denote the Clifford algebra with three generators, i.e., $\mathcal{A}$ is the 8-dimensional complex vector space with the basis

$$1, e_1, e_2, e_3, e_1 e_2, e_1 e_3, e_2 e_3, e_1 e_2 e_3,$$

where the generators $\{e_j\}_j$ satisfy $e_j^2 = -1$, $e_j e_k = -e_k e_j$ for $j \neq k$.

The Dirac operator $\mathbb{D}$ acting on $\mathcal{A}$–valued functions is defined by $\mathbb{D} := \partial_1 e_1 + \partial_2 e_2 + \partial_3 e_3$.

THEOREM 16.4.1
Let D be a bounded Lipschitz domain in $\mathbf{R}^3$ and $1 < p < \infty$. Then any $\mathcal{A}$ valued function $F \in C^1(D)$ such that $(ikI + D)F = 0$ on D and for which $F^* \in L^p(\partial D)$ belongs to $\Lambda_{1/q}^{p,q}(D)$, with $q := \max\{p, 2\}$. Furthermore, for some positive constant C depending only on the Lipschitz character of D,

$$\|F^*\|_{L^p(\partial D)} \le C\|F\|_{\Lambda_{1/q}^{p,q}(D)}.$$

Actually, for $p = q = 2$, the converse implication is also true. Namely, any function $F \in W^{1/2}(D)$ such that $(ikI + D)F = 0$ has $F^* \in L^2(\partial D)$ and, moreover,

$$\|F\|_{L^2(\partial D)} \approx \|F^*\|_{L^2(\partial D)} \approx \|F\|_{W^{1/2}(D)}.$$

As also noted in [M], an immediate corollary of this theorem is the following.

COROLLARY 16.4.2
Let D be a bounded Lipschitz domain in $\mathbf{R}^3$ and let $p_0 < p < \infty$, $q := \max\{p, 2\}$. Then any function u satisfying the scalar Helmholtz equation in D and $u^* \in L^p(\partial D)$ belongs to the Sobolev-Besov space $\Lambda_{1/p}^{p,q}(D)$. Moreover,

$$\|u\|_{\Lambda_{1/p}^{p,q}(D)} \le C\|u^*\|_{L^p(\partial D)}.$$

Now we can state our results for solutions of the vector Helmholtz equation.

THEOREM 16.4.3
With the notation of the above corollary, any vector field satisfying the Helmholtz equation in D and having the nontangential maximal function in $L^p(\partial D)$ belongs to $\Lambda_{1/p}^{p,q}(D)$.
In particular, the solution

$$E := \int_{\partial D} \nabla \Phi \times (-\tfrac{1}{2}I + M)^{-1} A d\sigma \qquad (16.4.1)$$

of the Dirichlet problem for the Maxwell equations (D) belongs to $\Lambda_{1/p}^{p,q}(D)$ and

$$\|E\|_{\Lambda_{1/p}^{p,q}(D)} \le C\|A\|_{L^p(\partial D)}. \qquad (16.4.2)$$

PROOF The first part is a simple application of the above corollary. As for the field E given by 16.4.1, by the estimate in the Corollary 16.4.2 and by the Coifman, McIntosh and Meyer theorem

$$\|E\|_{\Lambda_{1/p}^{p,q}(D)} \le C\|E^*\|_{L^p(\partial D)} \le C\|A\|_{L^p(\partial D)}.$$

∎

THEOREM 16.4.4

Let D be a bounded Lipschitz domain in $\mathbf{R}^3$ and $1 < p < \infty$, $q :=$ $\max\{p, 2\}$.

(i)   For a vector field E satisfying $\||E\||_p < +\infty$ and the Helmholtz equation in D, one has that E, $\operatorname{curl} E$ and $\operatorname{div} E$ belong to $\Lambda^{p,q}_{1/p}(D)$. In addition, the sum of the $\Lambda^{p,q}_{1/p}$-norms of these quantities is less than a fixed multiple of $\||E\||_p$.

(ii) For $p = 2$ the converse of the above result (including the converse estimate) also holds true.

In particular, the fields E, H satisfying the Maxwell equations in D belong to $W^{1/2}(D)$ if and only if E^*, H^* belong to $L^2(\partial D)$. In this case,

$$\|E\|_{W^{1/2}(D)} + \|H\|_{W^{1/2}(D)} \approx \|E^*\|_{L^2(\partial D)} + \|H^*\|_{L^2(\partial D)}. \tag{16.4.3}$$

PROOF   The key observation is that $(ikI + D)(ikI - D) = -(\triangle + k^2)$, so that if

$$F := (ikI - D)(E_1 e_1 + E_2 e_2 + E_3 e_3) = \operatorname{div} E + ik(E_1 e_1 + E_2 e_2 + E_3 e_3)$$

$$-(\partial_1 E_2 - \partial_2 E_1)e_1 e_2$$

$$-(\partial_1 E_3 - \partial_3 E_1)e_1 e_3$$

$$-(\partial_2 E_3 - \partial_3 E_2)e_2 e_3,$$

then $(ikI + D)F = 0$ if and only if $(\triangle + k^2)E = 0$, whereas $\|F^*\|_{L^p(\partial D)} = \||E\||_p$. With this at hand, the conclusions of the theorem are provided by Theorem 16.4.1.   ∎

Finally, as $|E \times N| \leq |E|$, a simple combination of 16.4.3 and 16.4.2 yields the estimate

$$\|E\|_{W^{1/2}(D)} + \|H\|_{W^{1/2}(D)} \approx \|A\|_{L^{2,Div}_T(\partial D)}$$

stated in the introduction.

16.5   Some approximation results

Let D be the union of finitely many Lipschitz domains in $\mathbf{R}^3$ with disjoint closures, after eventually removing from every domain a finite number of closed, disjoint Lipschitz subdomains.

Recall the radial fundamental solution Φ for the Helmholtz operator in $\mathbf{R}^3$. A *dipole* at $P \in \mathbf{R}^3$ is any vector field on ∂D of the form $\nabla\Phi(P-\cdot) \times e_j$

where $\{e_j\}_j$ is the canonical basis in $\mathbf{R}^3$. Also, *a multipole at $P \in \mathbf{R}^3$ is* defined as any partial derivative ∂^α, $\alpha \in I\!N^3$, applied to a a dipole at P.

We first need the following lemma.

LEMMA 16.5.1
With D as above, if $\nabla \times \mathcal{S}A$ vanishes identically in $\mathbf{R}^3 \setminus \bar{D}$ for some $A \in L^2(\partial D)$, then $A \equiv 0$ on ∂D.

PROOF Note first that for $A \in L_T^{2,Div}(\partial D)$ this lemma is a consequence of uniqueness results for the vector solutions of the Helmholtz equations presented in [MTW].

Let $\{\partial D_j\}_j$ be the components of ∂D, $A_j \in L_T^2(\partial D_j)$ the restriction of A to ∂D_j and $\mathcal{S}_j$ the single layer potential operator associated with ∂D_j. From the hypotheses of the lemma we have that

$$\sum_j \nabla \times \mathcal{S}_j A_j = 0 \quad \text{on } D.$$

Fixing a j_0, taking the vector product with N_{j_0}, the unit normal to ∂D_{j_0}, and then going to ∂D_{j_0} we finally get

$$\pm \tfrac{1}{2} A_{j_0} + N_{j_0} \times \int_{\partial D_{j_0}} \nabla \Phi \times A_{j_0} d\sigma = - \sum_{j \neq j_0} N_{j_0} \times E_j,$$

with E_j smooth vector fields on ∂D_{j_0}. The right-hand side belongs to $L_T^{2,Div}(\partial D_{j_0})$. It follows from Theorem 16.3.1 that A_{j_0} also belongs to this space. Iterating this argument several times we finally obtain that $A \in L_T^{2,Div}(\partial D)$, hence the lemma follows. ∎

THEOREM 16.5.2
Let D be as before and let $1 \leq p \leq 2$. Then any field in $L_T^p(\partial D)$ can be arbitrarily well approximated by the tangential component of a finite sum of dipoles located at points outside $\bar{D}$.

PROOF Let $\mathcal{M}$ be the closure in $L_T^p(\partial D)$ of the linear span of the tangential components of dipoles located in $\mathbf{R}^3 \setminus \bar{D}$. If $\mathcal{M} \neq L_T^p(\partial D)$, first by Hahn-Banach extension theorem and then by the Riesz representation theorem, there exists a nonzero vector field A in $L_T^q(\partial D)$, $1/q + 1/p = 1$, such that

$$\int_{\partial D} \langle \nabla \Phi(P - X) \times V, A(X) \rangle \, d\sigma(X) = 0$$

for all $V \in \mathbf{R}^3$ and all $P \in D$. In turn, this easily implies that $\nabla \times \mathcal{S}A$ identically vanishes in D, so that $A = 0$, by Lemma 16.5.1 since $2 \leq q \leq \infty$. This contradicts our initial assumption, hence the proof is complete. ∎

THEOREM 16.5.3

Let D be as described at the beginning of this section and $1 \leq p \leq 2$. Given a selection of arbitrary, a priori fixed poles $\{P_j\}_j$, one in each component of $\mathbf{R}^3 \setminus \bar{D}$, then any field in $L_T^p(\partial D)$ can be arbitrarily well approximated by the tangential component of a finite sum of multipoles located at $\{P_j\}_j$.

PROOF   Let $\mathcal{M}$ be this time the closure in $L_T^p(\partial D)$ of the linear span of the tangential components of the multipoles located at $\{P_j\}_j$.

As before, having $\mathcal{M}$ only a proper subspace of $L_T^p(\partial D)$ amounts to to the fact that, for some not identically zero field A in $L_T^q(\partial D)$, $1/p + 1/q = 1$, the function $\nabla \times \mathcal{S}A$ together with all its derivatives vanishes at each P_j. By the real analyticity of this function, it must be identically zero in each component of $\mathbf{R}^3 \setminus D$, hence, as in the previous theorem, A has to be zero. This contradiction yields the desired conclusion.

Note that for C^1 domains, the above results are valid for the full range $1 < p < \infty$. Our last remark concerns the series expansion of the elements of $L_T^2(\partial D)$ in fields of multipoles obtained in [Cal2] for smooth domains. More specifically, for D a bounded C^1 domain and $P_0 \in \mathbf{R}^3 \setminus \bar{D}$, starting with

$$\nabla \partial^\alpha \Phi(P_0 - X) \times e_k, \quad k = 1, 2, 3, \ \alpha \in I\!\!N^3$$

we can use the Gram-Schmidt orthonormalization process to obtain a sequence $E_{k,\alpha}(X)$ such that $\{N \times (E_{k,\alpha}|_{\partial D})\}_{k,\alpha}$ is an orthonormal basis for $L_T^2(\partial D)$. Consequently, any $A \in L_T^2(\partial D)$ has the series expansion

$$A = \sum c_{k,\alpha} N \times (E_{k,\alpha}|_{\partial D}).$$

In particular, the series $\sum c_{k,\alpha} E_{k,\alpha}$ converges uniformly on compact subsets of D to the unique solution E of the interior Dirichlet problem D.   ∎

Bibliography

[A]   R. Adams, *Sobolev Spaces*, Academic Press, New York, 1975.

[B1]   R. Brown, *The method of layer potentials for the heat equation in Lipschitz cylinders*, Am. J. Math. **111**, (1989), 339–379.

[BS]   R. Brown and Z. Shen, *The initial–Dirichlet problem for a fourth–order parabolic equation in Lipschitz cylinders*, Ind. Univ. Math. J. **39**, (1990), 1313–1353.

[C]   A. Calderón, *Cauchy integral on Lipschitz curves and related operators*, Proc. Natl. Acad. Sci. U.S.A. **74**, (1977), 1324–1327.

[Cal2] A. Calderón, *The multipole expansion of radiation fields*, J. Rat. Mech. Anal. **3**, (1954), 523–537.

[CMM] R. Coifman, A. McIntosh, and Y. Meyer, *L'intégrale de Cauchy définit un opérateur borné sur L^2 pour les courbes Lipschitiziennes*, Ann. Math. **116**, (1982), 361–387.

[CK] D. Colton and R. Kress, *Integral Equation Methods in Scattering Theory*, John Wiley & Sons, New York, 1983.

[Co] M. Costabel, *A remark on the regularity of Maxwell's equations on Lipschitz domains*, Math. Meth. Appl. Sci. **12**, (1990), 365–368.

[DK] B. Dahlberg and C. Kenig, *Hardy spaces and the L^p–Neumann problem for Laplace's equation in a Lipschitz domain*, Ann. Math. **125**, (1987), 437–465.

[DKV] B. Dahlberg, C. Kenig, and G. Verchota, *Boundary Value Problems for the Systems of Elastostatics in Lipschitz Domains*, Duke Math. J. bf 57, (1988), 795–818.

[FJR] E. Fabes, M. Jodeit, and N. Rivière, *Potential techniques for boundary value problems on C^1 domains*, Acta Math. **141** (1978), 165–186.

[FKV] E. Fabes, C. Kenig, and G. Verchota, *The Dirichlet Problem for the Stokes System on Lipschitz Domains*, Duke Math. J. **5**, (1988), 769–793.

[FR] E. Fabes and N. Rivière, *Dirichlet and Neumann problems for the heat equation in C^1–cylinders*, Proc. Symp. Pure Math. **35**, (1979), 179–198.

[GS] K. Gürlebeck and W. Spröessig, *Quaternionic Analysis and Elliptic Boundary Value Problems*, Birkhauser-Verlag, Basel, 1990.

[M] M. Mitrea, *Boundary value problems and Hardy spaces associated to the Helmholtz equation on Lipschitz domains*, submitted.

[MTW] M. Mitrea, R. Torres, and G. Welland, *Layer potential techniques in electromagnetism*, preprint.

[Mu] C. Müller, *Über die Beugung elektromagnetischer Schwingungen an endlichen homogenen Körpern*, Math. Ann. **123**, (1951) 345–378.

[Mu1] C. Müller, *Zur Methode der Strahlungskapazität von H. Weyl*, Math. Zeit. **56**, (1952) 80-83.

[TW] R. Torres and G. Welland, *The Helmholtz equation and transmission problems with Lipschitz interfaces*, Ind. Univ. Math. J. **42**, (1993), 1457–1485.

[V2]   G. Verchota, *Layer potentials and boundary value problems for Laplace's equation in Lipschitz domains*, J. Funct. Anal. **59**, (1984), 572–611.

[V1]   G. Verchota, *The Dirichlet problem for the biharmonic equation in C^1 domains*, Ind. Univ. Math. J. **36**, (1987), 867–895.

[Weyl] H. Weyl, *Kapazität von Strahlungsfeldern*, Math. Zeit. **55**, (1952), 187–198.

Department of Mathematics,
University of Minnesota,
Minneapolis, Minnesota 55455
E-mail: mitrea@math.umn.edu

Department of Mathematics,
University of Michigan,
Ann Arbor, Michigan 48109
E-mail: rodolfo.torres@math.lsa.umich.edu

Department of Mathematics,
University of Missouri,
8001 Natural Bridge Road,
Saint Louis, Missouri 63121
E-mail: welland@eads.umsl.edu

17

Continuity of Calderón-Zygmund Type Operators on the Predual of a Morrey Space

Josefina Alvarez

ABSTRACT It is known that the Morrey space is the dual of certain atomic space $H^{p,\varphi}$. In this paper, we define a notion of a molecule in terms of p and φ and we show that this molecule belongs to $H^{p,\varphi}$ by obtaining a suitable atomic decomposition. We use this result to prove that a Calderón-Zygmund type operator satisfying certain cancelation condition, maps $H^{p,\varphi}$ continuously into itself. By duality, we obtain a continuity result on the Morrey space.

17.1 Introduction

The space, now called the Morrey space, was introduced by Morrey, (cf.[15]) and it was used by Morrey and other authors in questions related to the calculus of variations and the theory of elliptic PDE's, (cf. [14], [15], [16], [17], [11], [18]). The Morrey space can be placed as part of a family that includes L^p, BMO, and Lipschitz spaces. Several authors studied this family of functional spaces, among them, John and Nirenberg, [12], Meyers, [13], Stampacchia, [22], Campanato, [3, 4, 5, 6], Campanato and Murthy, [7], Spanne, [21], Peetre, [19, 20]. Of particular interest to us are Peetre's results in [19], where he proved the continuity of a class of convolution operators that includes the Hilbert transform. As a consequence, Peetre showed that the Hilbert transform preserves BMO.

It was proved in [1] that the Morrey space contains functions whose distribution function does not decay at infinity. This result contrasts with the estimate proved in BMO by John and Nirenberg.

On the other hand, the Morrey space exhibits some similarities with BMO, concerning duality. Indeed, Zorko showed in [24] that the Morrey

0-8493-8481-8/96/$0.00+$.50
© 1996 by CRC Press

space is the dual of an atomic space.

Recent papers show significant applications of the Morrey space to PDE's (cf. [23]), as well as further properties of this space in connection with the theory of singular integrals, (cf. [23], [8], [10], [2]).

The purpose of this work is to study the atomic space introduced by Zorko in [24]. We define a suitable notion of a molecule and we use this notion to prove the continuity of a Calderón-Zygmund type operator from this atomic space to itself, provided that the operator satisfies certain cancellation condition. By duality, we obtain a continuity result on the Morrey space.

The notation used in this paper is the standard one in the subject. It may be useful to point out that given $1 \leq p \leq \infty$, p' will denote the conjugate exponent, $1/p + 1/p' = 1$. The symbols C_0^∞, S', D', L^p, l^p, etc. will indicate the usual spaces of distributions, sequences, or functions defined on R^n, with complex values. Moreover, $\|f\|_p$ will denote the L^p norm of the function f. With $|A|$ we will denote the Lebesgue measure of a measurable set A, and $B(z, \sigma)$ will be the ball centered at z with radius σ. As usual, the letter C will be an absolute constant, probably different at different occurrences. Other notations will be introduced at the appropriate time.

The rest of this paper is divided into four sections, as follows:

The paper ends with a list of references.

17.2   Definition of the Space $L^{p,\varphi}$

Among the various versions of the Morrey space found in the literature, we will work with the one used by Peetre in [19]. This is also the version considered by Zorko in [24]. More precisely, Peetre defines a family of spaces, $L^{p,\varphi}$, which includes, among others, the Morrey space. We give the definition of the space $L^{p,\varphi}$, now. Later, we impose additional conditions so that we will be working with a Morrey space.

DEFINITION 17.2.1   *Let* $1 < p < \infty, \varphi : (0, \infty) \to (0, \infty)$. *A function* f, *locally in* L^p, *belongs to the space* $L^{p,\varphi}$, *if it satisfies the following condition.* ●

$$\sup_B \inf_{c \in \mathbf{C}} \frac{1}{\varphi(|B|)} \left(\frac{1}{|B|} \int_B |f(x) - c|^p \, dx \right)^{1/p} < \infty$$

The $\sup_B \inf_{c \in \mathbf{C}}$ that appears in the above expression, defines a seminorm in $L^{p,\varphi}$. The space $(L^{p,\varphi}, \| \ \|_{L^{p,\varphi}})$ becomes a Banach space after identifying functions which differ by a constant. The most important example is given by $\varphi(t) = t^{(\lambda-1)/p}$, when $\lambda = 0, L^{p,\varphi}$ becomes L^p. For $0 < \lambda < 1$, the space $L^{p,\varphi}$ is, by definition, the Morrey space, and for $\lambda = 1$, it coincides with one of the equivalent versions of the space BMO. Finally, for $1 < \lambda < 1 + p$, the space $L^{p,\varphi}$ is known to coincide with $Lip_{(\lambda-1)/p}$.

17.3   The Space $\mathbf{L}^{p,\varphi}$ as the Dual of the Atomic Space $\mathbf{H}^{p,\varphi}$

In this section we will summarize the definitions and results we will use from Zorko [24].

DEFINITION 17.3.1   *Given* $1 < p < \infty$ *and given a function* $\varphi : (0, \infty) \to (0, \infty)$, *a* (p, φ) *atom is a function* a *such that*

(i) a *is supported on a ball* B
(ii) $\int a(x)dx = 0$
(iii) $\|a\|_p \leq \dfrac{|B|^{-1/p'}}{\varphi(|B|)}$

DEFINITION 17.3.2   *We denote, by* $H^{p,\varphi}$, *the family of distributions* f *that, in the sense of distributions, can be written as* $f = \sum_{k \geq 1} \lambda_k a_k$, *where* a_k *is a* (p, φ) *atom and* $\sum_{k \geq 1} |\lambda_k| < \infty$.

Clearly, $H^{p,\varphi}$ is a vector space. If we define

$$\|f\|_{p,\varphi} = \inf_{f = \sum \lambda_k a_k} \sum_{k \geq 1} |\lambda_k|$$

then, $\left(H^{p,\varphi}, \| \ \|_{p,\varphi} \right)$ becomes a Banach space, provided that the function $\varphi(t)$ is nonincreasing and the function $t\varphi^{p'}(t)$ is nondecreasing.

Conversely, assuming the above conditions on φ, given a sequence $\{a_k\}$ of (p, φ) atoms and given a sequence $\{\lambda_k\}$ in l^1, the series $\sum_{k \geq 1} \lambda_k a_k$ converges in the sense of distributions to a distribution in $H^{p,\varphi}$.

PROPOSITION 17.3.3
Given $f \in L_{p',\varphi}$ and $g \in H^{p,\varphi}$, the integral $\int f(x)g(x)dx$ is well defined. The correspondence $g \to \int f(x)g(x)dx$ determines a functional L_f in $(H^{p,\varphi})^*$. Moreover, $\|L_f\| = \|f\|_{L^{p,\varphi}}$. Conversely, given $L \in (H^{p,\varphi})^*$, there exists $f \in L_{p',\varphi}$ such that $L = L_f$.

17.4   A Notion of a Molecule in the Space $H^{p,\varphi}$

DEFINITION 17.4.1   Let $1 < p < \infty, \alpha > n(p-1)$, and $\varphi : (0,\infty) \to (0,\infty)$. A $p-$ integrable function $M(x)$ is called a (p,α,φ) molecule if there exists a ball $B = B(z,\sigma)$ such that $M(x)$ satisfies the following conditions:

(i) $\int |M(x)|^p \, dx \leq C |B|^{-p/p'} \varphi^{-p}(|B|)$

(ii) $\int |M(x)|^p |x - z|^\alpha \, dx \leq C |B|^{\alpha/n - p/p'} \varphi^{-p}(|B|)$

(iii) $\int M(x)dx = 0$

It should be observed that conditions i) and ii) imply that the function $M(x)$ is integrable. Indeed,

$$\int |M(x)| \, dx = \int_B |M(x)| \, dx$$

$$+ \int_{\mathbf{R}^n \setminus B} |M(x)| \, dx \leq \left(\int |M(x)|^p \, dx \right)^{1/p} |B|^{1/p'}$$

$$+ \left(\int |M(x)|^p |x - z|^\alpha \, dx \right)^{1/p} x$$

$$x \left(\int_{\mathbf{R}^n \setminus B} |x - z|^{-\alpha p'/p} \, dx \right)^{1/p'}.$$

Using condition i), the first term can be estimated by $C\varphi^{-1}(|B|)$, where $\varphi^{-1}(|B|)$ denotes the reciprocal, $1/\varphi(|B|)$. For the second term we use condition ii), obtaining also $C\varphi^{-1}(|B|)$.

This observation implies that condition iii) makes sense.

PROPOSITION 17.4.2
Let $M(x)$ be a (p,α,φ) molecule with φ nonincreasing and $t\varphi^{p'}$ nondecreasing. Then, $M(x)$ belongs to $H^{p,\varphi}$.

PROOF   Given a molecule M we want to decompose it as an appropriate infinite linear combination of (p,φ) atoms.

Let $B = B(z,\sigma)$ be the ball used in the definition of the molecule. Let $B_j = B(z, 2^j\sigma)$ and let $C_j = B_j \backslash B_{j-1}$, $j = 0, 1, \ldots$ where $B_{-1} = \emptyset$ and $B_0 = B$.

First, we consider $(M - M_j)\chi_{C_j}$, where $M_j = \frac{1}{|C_j|} \int_{C_j} M(x)dx$.

For $m = 1, 2, \ldots$ fixed, let $\sum_{j=0}^{m} (M - M_j)\chi_{C_j} = M\chi_{B_m} - \sum_{j=0}^{m} M_j\chi_{C_j}$.

Let us write the sum on the right-hand side of the above equation as follows.

$$\sum_{j=0}^{m} M_j\chi_{C_j}$$

$$= \sum_{j=0}^{m} \frac{\chi_{C_j}}{|C_j|} \left(\int_{\mathbf{R}^n \backslash B_{j-1}} M(x)dx - \int_{\mathbf{R}^n \backslash B_j} M(x)dx \right)$$

$$= \sum_{j=1}^{m-1} \left(\frac{\chi_{C_{j+1}}}{|C_{j+1}|} - \frac{\chi_{C_j}}{|C_j|} \right) \int_{\mathbf{R}^n \backslash B_j} M(x)dx - \frac{\chi_{C_m}}{|C_m|} \int_{\mathbf{R}^n \backslash B_m} M(x)dx$$

So, we finally have

$$M\chi_{B_m} = \sum_{j=0}^{m} (M - M_j)\chi_{C_j} - \frac{\chi_{C_m}}{|C_m|} \int_{\mathbf{R}^n \backslash B_m} M(x)dx$$

$$+ \sum_{j=1}^{m-1} \left(\frac{\chi_{C_{j+1}}}{|C_{j+1}|} - \frac{\chi_{C_j}}{|C_j|} \right) \int_{\mathbf{R}^n \backslash B_j} M(x)dx$$

Since the function M is integrable, we have that $M\chi_{B_m} \to M$ in L^1 as $m \to \infty$ and $\frac{\chi_{C_m}}{|C_m|} \int_{\mathbf{R}^n \backslash B_m} M(x)dx \to 0$ in L^1 as $m \to \infty$.

Let us now analyze the behavior of each of the other two terms. Let $\alpha_j = (M(x) - M_j)\chi_{C_j}$. We will show that α_j is a multiple of a (p, φ) atom. Indeed, $\operatorname{supp}(\alpha_j) \subseteq B_j$ and

$$\int \alpha_j(x)dx = \int_{C_j} (M(x) - M_j)dx == \int_{C_j} M(x)dx - |C_j| M_j = 0.$$

Moreover,

$$\|\alpha_j\|_p^p \leq C \int_{C_j} |M(x)|^p \, dx = C \int_{C_j} |M(x)|^p |x - z|^\alpha |x - z|^{-\alpha} \, dx$$

$$\leq C |B_{j-1}|^{-\alpha/n} \int |M(x)|^p |x - z|^\alpha \, dx$$

$$\leq C |B_{j-1}|^{-\alpha/n} |B|^{\alpha/n - p/p'} \varphi^{-p}(|B|)$$

$$\leq C 2^{-j(\alpha - np/p')} |B_j|^{-p/p'} \varphi^{-p}(|B_j|).$$

Thus, the function $C^{-1}2^{j(\alpha-n/p')}\alpha_j$ is a (p,φ) atom and

$$\sum_{j\geq 1} C2^{-j(\alpha-n/p')} < \infty.$$

As a consequence, the series $\sum_{j\geq 1}\alpha_j$ converges in D' to a distribution in $H^{p,\varphi}$.

Now, let $\beta_j = \left(\frac{\chi_{C_{j+1}}}{|C_{j+1}|} - \frac{\chi_{C_j}}{|C_j|}\right)\int_{\mathbf{R}^n\setminus B_j} M(x)dx$. We will show that β_j is also a multiple of a (p,φ) atom.

Indeed, $\operatorname{supp}(\beta_j) \subseteq B_{j+1}$ and

$$\int \beta_j(x)dx = \left(\frac{|C_{j+1}|}{|C_{j+1}|} - \frac{|C_j|}{|C_j|}\right)\int_{\mathbf{R}^n\setminus B_j} M(x)dx = 0.$$

Moreover,

$$\|\beta_j\|_p \leq \left(\int_{\mathbf{R}^n\setminus B_j} |M(x)|\,dx\right)\left(\left\|\frac{\chi_{C_{j+1}}}{|C_{j+1}|}\right\|_p + \left\|\frac{\chi_{C_j}}{|C_j|}\right\|_p\right).$$

Let us now estimate each parenthesis. We estimate the first one in the same way we proved that the molecule $M(x)$ is integrable.

$$\int_{\mathbf{R}^n\setminus B_j} |M(x)|\,dx = \int_{\mathbf{R}^n\setminus B_j} |M(x)|\,|x-z|^{\alpha/p}\,|x-z|^{-\alpha/p}\,dx$$

$$\leq \left(\int |M(x)|^p\,|x-z|^\alpha\,dx\right)^{1/p} x$$

$$x\left(\int_{\mathbf{R}^n\setminus B_j} |x-z|^{-\alpha p'/p}\,dx\right)^{1/p'}$$

$$\leq C\,|B|^{\alpha/np-1/p'}\,\varphi^{-1}(|B|)\,|B_j|^{1/p'-\alpha/np}.$$

The second parenthesis can be estimated as

$$\left(\left\|\frac{\chi_{C_{j+1}}}{|C_{j+1}|}\right\|_p + \left\|\frac{\chi_{C_j}}{|C_j|}\right\|_p\right) \leq C\,|C_j|^{1/p-1} \leq C\,|B_j|^{1/p-1}.$$

Thus, we finally have the estimate

$$\|\beta_j\|_p \leq C2^{-j(\alpha/p-n/p')}\,|B_{j+1}|^{-1/p'}\,\varphi^{-1}(|B_{j+1}|).$$

This estimate shows that the function $C^{-1}2^{j(\alpha/p-n/p')}\beta_j$ is a (p,φ) atom. Moreover, $\sum_{j\geq 1} C2^{-j(\alpha/p-n/p')} < \infty$. Thus, the series $\sum_{j\geq 1}\beta_j$ converges in D' to a distribution in $H^{p,\varphi}$.

This completes the proof of the Proposition 17.4.2. ∎

REMARK The proof above also shows that given a (p, α, φ) molecule $M(x)$, we have $\|M\|_{p,\varphi} \leq cC$, where c is a numerical constant not depending on M, and C is the constant that appears in the Definition 17.4.1.

17.5 Continuity on the Spaces $\mathbf{H}^{p,\varphi}$ and $\mathbf{L}^{p,\varphi}$ of Calderón-Zygmund Type Operators

Let us first remark that if $\varphi(t) \geq Ct^{-1/p'}$, then, any linear and continuous operator $T : L^p \to L^q, 1 < p < \infty, 1 \leq q \leq \infty$, will map continuously $H^{p,\varphi}$ into L^q. Indeed, it suffices to observe that given $f = \sum_{j \geq 1} \lambda_j a_j$ in $H^{p,\varphi}$, the series $\sum_{j \geq 1} \lambda_j a_j$ converges in L^p.

We will now study the action of a Calderón-Zygmund type operator on $H^{p,\varphi}$. The precise definition of the operators under consideration is as follows.

DEFINITION 17.5.1 A linear and continuous operator $T : C_0^\infty \to D'$ is called a Calderón-Zygmund type operator, if it satisfies the following two conditions.

(i) The distribution kernel of T concides in the complement of the diagonal in $\mathbf{R}^n \times \mathbf{R}^n$, with a locally integrable function $k(x,y)$ satisfying

$$\sup_{z \in \mathbf{R}^n, \sigma > 0} \sup_{y \in B(z,\sigma)} \frac{1}{|B(z,\sigma)|^{\alpha/n - p/p'}} \tag{17.5.1}$$

$$\int_{\mathbf{R}^n \setminus B(z,2\sigma)} |k(x,y) - k(x,z)|^p |x - z|^\alpha \, dx < \infty$$

for some $\alpha \in \mathbf{R}, 1 \leq p < \infty$, where B denotes any ball in $\mathbf{R}^n$.

(ii) The operator T extends to a continuous operator on L^p.

REMARK (i) It can be readily seen that the pointwise condition

$$|k(x,y) - k(x,z)| \leq C \frac{|y-z|^\delta}{|x-z|^{n+\delta}}$$

if $2|y - z| < |x - z|$, for some $0 < \delta \leq 1$, implies condition (17.5.1), provided that $\alpha < p(n + \delta) - n$.

(ii) Likewise, if $\alpha > n(p - 1)$, condition (17.5.1) implies the so-called Hörmander condition

$$\sup_{z \in \mathbf{R}^n, \sigma > 0} \sup_{y \in B(z,\sigma)} \int_{\mathbf{R}^n \setminus B(z,2\sigma)} |k(x,y) - k(x,z)| \, dx < \infty \tag{17.5.2}$$

(iii)   In the next result we will use the cancellation condition of David
        and Journé (cf. [9]), $T^*(1) = 0$. Given any linear and continuous
        operator $T : C_0^\infty \to D'$ whose distribution kernel $k(x, y)$ satisfies the
        Hörmander condition (17.5.1), and given any function $f \in C^\infty \cap L^\infty$,
        the value $T^*(f)$ can be defined as a linear and continuous functional
        on the subspace of C_0^∞ of those functions having integral zero.

(iv)   Let $T : C_0^\infty \to D'$ be a linear and continuous operator. Suppose
        that T extends to a continuous operator from L^p to itself, for some
        $1 \le p < \infty$. Furthermore, assume that the distribution kernel of T
        satisfies the Hörmander condition (17.5.2). Then, it can be easily
        shown that the operator T maps continuously L^∞ into BMO.

PROPOSITION 17.5.2

Let T be a Calderón-Zygmund type operator in the sense of the Definition
17.5.1, for some $1 < p < \infty, \alpha > n(p-1)$, and assume that $T^*(1) = 0$ in the
sense of BMO. Then, T maps continuously $H^{p,\varphi}$ into itself, provided that
the function $\varphi(t)$ is nonincreasing and the function $t\varphi'(t)$ is nondecreasing.

PROOF   We will first show that the operator T maps (p, φ) atoms into
(p, α, φ) molecules.

Indeed, let $a(x)$ be a (p, φ) atom supported on the ball $B = B(z, \sigma)$.
We will show that $T(a)$ satisfies conditions i), ii), and iii) in the Definition
17.4.1.

First, $\int |T(a)|^p \, dx \le C \int |a|^p \, dx \le C |B|^{-p/p'} \varphi^{-p}(|B|)$. Thus, $T(a)$ sat-
isfies condition i). We will now verify condition ii).

$$\int |T(a)(x)|^p \, |x - z|^\alpha \, dx = \int_{B(z, 2\sigma)} |Ta(x)|^p \, |x - z|^\alpha \, dx$$

$$+ \int_{\mathbf{R}^n \setminus B(z, 2\sigma)} |T(a)(x)|^p \, |x - z|^\alpha \, dx$$

Let us estimate each term in the sum above.

$$\int_{B(z, 2\sigma)} |T(a)(x)|^p \, |x - z|^\alpha \, dx$$

$$\le C |B|^{\alpha/n} \int |T(a)(x)|^p \, dx$$

$$\le C |B|^{\alpha/n - p/p'} \varphi^{-p}(|B|).$$

$$\int_{\mathbf{R}^n \setminus B(z,2\sigma)} |T(a)(x)|^p \, |x-z|^\alpha \, dx$$

$$\leq \int_{\mathbf{R}^n \setminus B(z,2\sigma)} \left(\int_B |k(x,y) - k(x,z)| \, |a(y)| \, dy \right)^p |x-z|^\alpha \, dx$$

$$\leq |B|^{p/p'} \int_B \left(\int_{\mathbf{R}^n \setminus B(z,2\sigma)} |k(x,y) - k(x,z)|^p \, |x-z|^\alpha \, dx \right) |a(y)|^p \, dy$$

$$\leq C |B|^{\alpha/n} \int_B |a(y)|^p \, dy \leq C |B|^{\alpha/n - p/p'} \, \varphi^{-p}(|B|).$$

Thus, condition ii) is also satisfied. Once conditions i) and ii) are satisfied, we can conclude that the function $T(a)$ belongs to L^1.

Finally, we will show that the cancellation condition $T^*(1) = 0$ in the sense of BMO, implies condition iii) in the definition of the molecule.

In fact, according to the definition of the (p,φ) atom, the function $\varphi(|B|)a(x)$ is a $(1,p)$ atom in the Hardy space H^1. Furthermore, according to Remark ii) and iv), the operator T under consideration maps continuously L^∞ into BMO.

Adding up all these observations, we can write

$$0 = (T^*(1), \varphi(|B|)a) = \int T^*(1)(x)\varphi(|B|)a(x)dx = \varphi(|B|) \int T(a)(x)dx,$$

where the parenthesis on the left-hand side denotes the (BMO, H^1) duality, and the last equal sign comes from the (L^∞, L^1) duality.

Thus, we have showed that the operator T maps (p, φ) atoms into (p, α, φ) molecules. Moreover, a careful look at the estimates above will show that the constant C in the definition of a (p, α, φ) molecule is $\leq c \, \|T\|$, where c is a numerical constant not depending on T or a and $\|T\| = \|T\|_{L^p, L^p} + C$, with C being the $\sup_{z \in \mathbf{R}^n, \sigma > 0} \sup_{y \in B(z,\sigma)}$ that appears in (17.5.1), and $\|T\|_{L^p, L^p}$ being the norm of T as a continuous operator from L^p into itself.

Given $f \in H^{p,\varphi}$ and given $\varepsilon > 0$, let us consider an atomic decomposition $f = \sum_{j \geq 1} \lambda_j a_j$, such that $\sum_{j \geq 1} |\lambda_j| < \|f\|_{p,\varphi} + \varepsilon$. We can write $T(f_N) = \sum_{j=1}^N \lambda_j T(a_j)$, where $f_N = \sum_{j=1}^N \lambda_j a_j$. According to the Remark in 17.4.2 and the observation above, we have the estimate $\|T(f_N)\|_{p,\varphi} \leq \sum_{j=1}^N |\lambda_j| \, \|T(a_j)\|_{p,\varphi} \leq c \, \|T\| \, (\|f\|_{p,\varphi} + \varepsilon)$, uniformly on N. This implies that the sequence $\{T(f_N)\}$ is a Cauchy sequence in $H^{p,\varphi}$. Thus, we can take the limit as $N \to \infty$, to obtain $\|T(f)\|_{p,\varphi} \leq c \, \|T\| \, \|f\|_{p,\varphi}$.

This completes the proof of the Proposition 17.5.2. ∎

COROLLARY 17.5.3

Let $T : C_0^\infty \to D'$ *be a linear and continuous operator such that the adjoint* T^* *is a Calderón-Zygmund type operator in the sense of the Definition*

17.5.1, for some $1 < p < \infty, \alpha > n(p-1)$. *Assume also that* $T(1) = 0$ *in the sense of BMO. Then, the operator* T *is continuous from* $L^{p',\varphi}$ *into itself.*

PROOF   Follows from Proposition 17.5.2 by duality. ∎

REMARK   In general, not even continuous functions are dense in the space $L^{p,\varphi}$, (cf. [24]). Thus, it is not obvious how to define the action of an operator like a Calderón-Zygmund operator on that space. One way is by duality, as shown in Corollary 17.5.4. Several recent papers have dealt with *a priori* estimates, (cf. [2], [8], [10]).

Bibliography

[1]   J. Alvarez, *The Distribution Function In The Morrey Space*, Proc. Am. Math. Soc. **83**, (1981), 693-699.

[2]   J. Alvarez and C. Pérez, *Estimates With A_∞ Weights For Various Singular Integral Operators*, Bol. Unione Mat. Ital. **7** 8A, (1994), 123-133.

[3]   S. Campanato, *Proprietà Di Hölderianità Di Alcune Classi Di Funzioni*, Ann. Scuola Norm. Sup. Pisa **17**, (1963), 175-188.

[4]   S. Campanato, *Proprietà Di Una Famiglia Di Spazi Funzionali*, Ann. Scuola Norm. Sup. Pisa **18**, (1964), 137-160.

[5]   S. Campanato, *Teoremi Di Interpolazione Per Transformazioni Che Applicano L^p In $C^{h,\alpha}$*, Ann. Scuola Norm. Sup. Pisa **18**, (1964), 345-360.

[6]   S. Campanato, *Equazioni Ellittiche Del $II°$ Ordine E Spazi $L^{2,\lambda}$*, Ann. Mat. Pura Appl. **69**, (1965), 321-382.

[7]   S. Campanato and M. K. V. Murthy, *Una Generalizzazione Del Teorema Di Riesz-Thorin*, Ann. Scuola Norm. Sup. Pisa **19**, (1965), 87-100.

[8]   F. Chiarenza and M. Frasca, *Morrey Spaces And Hardy-Littlewood Maximal Function*, Rend. Mat. (7) **7**, (1987), 273-279.

[9]   G. David and J. L. Journé, *A Boundedness Criterion For Generalized Calderón-Zygmund Operators*, Ann. Math. **120**, (1984), 371-397.

[10]   G. Di Fazio and M. A. Ragusa, *Commutators And Morrey Spaces*, Bol. Unione Mat. Ital. (7) **5-A**, (1991), 323-332.

[11] D. Greco, *Criteri Di Compattezza Per Insiemi Di Funzioni In n Variabili Independenti*, Ric. Mat. **1**, (1952).

[12] F. John and L. Nirenberg, *On Functions Of Bounded Mean Oscillation*, Commun. Pure Appl. Math. **14**, (1961), 415-426.

[13] G. Meyers, *Mean Oscillation Over Cubes And Hölder Continuity*, Proc. Am. Math. Soc. **15**, (1964), 717-721.

[14] C. Miranda, *Sui Sistemi Di Tipo Ellittico Di Equazioni Lineari A Derivate Parziali Del Primo Ordine In n Variabili Indipendenti*, Atti Acc. Naz. Lincei s. VIII vol. III, (1952).

[15] C. B. Morrey, *Multiple Integral Problems In The Calculus Of Variations And Related Topics*, University of California Pub. 1, (1943).

[16] C. B. Morrey, *Second Order Elliptic Systems Of Differential Equations*, Ann. Math. Studies 3, Princeton University Press, Princeton, NJ, 1954.

[17] C. B. Morrey, *Second Order Elliptic Equations In Several Variables And Hölder Continuity*, Math. Zeit. **72**, (1959).

[18] L. Nirenberg, *Estimates And Existence Of Solutions Of Elliptic Equations*, Commun. Pure Appl. Math. **9**, (1956).

[19] J. Peetre, *On Convolution Operators Leaving $L^{p,\lambda}$ Spaces Invariant*, Ann. Mat. Pura Appl. **72**, (1966), 295-304.

[20] J. Peetre, *On The Theory Of $L^{p,\lambda}$ Spaces*, J. Funct. Anal. **4**, (1969), 71-87.

[21] S. Spanne, *Some Function Spaces Defined Using Mean Oscillation Over Cubes*, Ann. Scuola Norm. Sup. Pisa **19**, (1965), 593-608.

[22] G. Stampacchia, *$L^{p,\lambda}$ Spaces And Interpolation*, Commun. Pure Appl. Math. **17**, (1964), 293-306.

[23] M. E. Taylor, *Analysis On Morrey Spaces And Applications To Navier-Stokes And Other Evolution Equations*, (1993), preprint.

[24] C. Zorko, *The Morrey Space*, Proc. Am. Math. Soc. **98**, (1986), 586-592.

Department of Mathematical Sciences,
New Mexico State University,
Las Cruces, New Mexico 88003-0001
E-mail: jalvarez@nmsu.edu

18

Neumann Type Problems for the Dirac Operator

Chiping Zhou

ABSTRACT We shall formulate Neumann type problems for monogenic functions, and for nonhomogeneous Dirac equations in the upper half space of $\mathbf{R}^m (m \geq 3)$. This is done in both the classical sense and the L^p sense. For each Neumann type problem, a representation of the solution, unique up to a Clifford constant, is obtained.

18.1 Introduction

Let $\mathbf{R}_{0,m}$ denote the real 2^m-dimensional Clifford algebra generated by the orthonormal basis $\{e_j\}_{j=1}^m$ of $\mathbf{R}^m$, where the multiplication is governed by the rule $e_j e_k + e_k e_j = -2\delta_{jk}, 1 \leq j, k \leq m$. An arbitrary element a of $\mathbf{R}_{0,m}$ can be written as $a = \sum_{A \subseteq M} a_A e_A$, where $a_A \in \mathbf{R}$, $e_A = e_{\alpha_1 \dots \alpha_h} = e_{\alpha_1} e_{\alpha_2} \dots e_{\alpha_h}$ for $A = \{\alpha_1, \dots, \alpha_h\} \subset M = \{1, \dots, m\}$. The norm of $a \in \mathbf{R}_{0,m}$ is defined by $|a| = (\sum_A a_A^2)^{\frac{1}{2}}$ and the identity of $\mathbf{R}_{0.m}$ is denoted by $e_\phi = e_0 = 1$.

In [8], it has been shown that

$$|ab| \leq K_{0,m}|a||b| \quad \text{for} \quad a, b \in \mathbf{R}_{0,m}$$

where the best value of the constant $K_{0,m}$ is

$$K_{0,m} = \begin{cases} 2^{m/4}, & \text{if} \quad m = 0, 6 \quad (mod\ 8) \\ 2^{(m-1)/4}, & \text{if} \quad m = 1, 3, 5 \quad (mod\ 8) \\ 2^{(m-2)/4}, & \text{if} \quad m = 2, 4 \quad (mod\ 8) \\ 2^{(m+1)/4}, & \text{if} \quad m = 7 \quad (mod\ 8). \end{cases}$$

0-8493-8481-8/96/$0.00+$.50
© 1996 by CRC Press

Notice that in order to obtain an (algebra)-norm in $\mathbf{R}_{0,m}$, we define

$$|a|_0 = K_{0,m}|a|, \qquad a \in \mathbf{R}_{0,m}$$

Then we have

$$|ab|_0 \leq |a|_0 |b|_0.$$

Let $\mathbf{R}_{0,m-1}$ be a subalgebra of $\mathbf{R}_{0,m}$ generated by $\{e_j\}_{j=1}^{m-1}$. Then we have

$$\mathbf{R}_{0,m} = \mathbf{R}_{0,m-1} \oplus e_m \mathbf{R}_{0,m-1};$$

$a \in \mathbf{R}_{0,m}$ iff $a = X^{(m)}(a) + e_m Y^{(m)}(a)$, where $X^{(m)}(a), Y^{(m)}(a) \in \mathbf{R}_{0,m-1}$.
It is clear that the above decomposition of $\mathbf{R}_{0,m}$ is a generalization of the classical decomposition of $\mathbf{C} = \mathbf{R} \oplus \sqrt{-1}\mathbf{R}$, and the operators $X^{(m)}$ and $Y^{(m)}$ acting on $\mathbf{R}_{0,m}$ are generalizations of the operators Re and Im acting on $\mathbf{C}$.

In what follows each vector $x = (x_1, ..., x_m) \in \mathbf{R}^m$ is identified with the Clifford number $x = \sum_{j=1}^{m} e_j x_j \in \mathbf{R}_{0,m}$; then $x^2 = -|x|^2$. The Dirac operator in $\mathbf{R}^m$ is defined by $D = \sum_{j=1}^{m} e_j \frac{\partial}{\partial x_j}$; then $D^2 = -\Delta$, Δ being the Laplacian. Let $\Omega \subset \mathbf{R}^m$ be an open set. An $\mathbf{R}_{0,m}$-valued function f in the class $C^1(\Omega; \mathbf{R}_{0,m})$ is called (left) monogenic in Ω if $Df = 0$ in Ω. If $w = X^{(m)}w + e_m Y^{(m)}w$ is an $\mathbf{R}_{0,m}$-valued monogenic function, then $X^{(m)}w$ and $Y^{(m)}w$ are called $\mathbf{R}_{0,m-1}$-*valued harmonic conjugate functions* associated to w. For more details about Clifford algebras and monogenic function theory we refer to [1]; for their applications in mathematical physics, harmonic analysis and PDEs we refer to [3], [5], [6], [9].

Recently several boundary value problems for $\mathbf{R}_{0,m}$-valued functions have been studied (see [2], [4], [7], [15-18]). Our main purpose here is to find the solution representation, unique up to a Clifford constant, for the following Neumann type problem in the upper half space of $\mathbf{R}^m (m \geq 3)$:

$$\begin{array}{lll}
Dw = f & \text{in} & \mathbf{R}_+^m \\
\frac{\partial}{\partial n} X^{(m)} w = g & \text{on} & \mathbf{R}_0^m \\
w & \text{is bounded in} & \mathbf{R}_+^m.
\end{array} \qquad \text{(GF)}$$

Here $\mathbf{R}_+^m = \{x | x_m > 0\}$ and $\mathbf{R}_0^m = \{x | x_m = 0\}$ are the upper half space and its boundary in $\mathbf{R}^m$ and $\frac{\partial}{\partial n} = -\frac{\partial}{\partial x_m}$ denotes the outward normal derivative. The following is our approach.

In section 18.2, we first solve the Neumann problem for the harmonic function $u = X^{(m)}w$; and then, by using generalized Cauchy-Riemann equations (Lemma 17.2), we find one of its harmonic conjugate $\tilde{u} = Y^{(m)}w$. Finally, we prove that $w = X^{(m)}w + e_m Y^{(m)}w$ is the solution, unique up to a Clifford constant, to the Neumann problem for monogenic functions (the case of $Dw = 0$) in both the classical sense and the L^p sense. More detailed arguments to Neumann problems for monogenic functions can be found in [18].

In section 18.3, we extend the result for monogenic functions to that for $\mathbf{R}_{0,m}$-valued functions satisfying a nonhomogeneous Dirac equation $Dw = f$. The explicit formula for the solution of the problem (GF) obtained in this section (see Theorem 18.3.1) is new and is much simpler than the one obtained in [18, Theorem 5.2].

18.2 Neumann Problems for Monogenic Functions

Let

$$\Gamma(x) := \tfrac{1}{(2-m)\omega_m} |x|^{2-m}, \qquad m \geq 3,$$

be the usual fundamental solution of the Laplacian on $\mathbf{R}^m$, where $\omega_m := \frac{2\pi^{m/2}}{\Gamma(m/2)}$ is the surface area of the unit sphere in $\mathbf{R}^m$. Then

$$E(x) := -\nabla \Gamma(x) = -\tfrac{x}{\omega_m |x|^m}$$

is the usual fundamental solution of the Dirac operator on $\mathbf{R}^m$, and

$$D_m E(x) = E(x) D_m = \delta(x) \quad and \quad \Delta_m \Gamma(x) = \delta(x). \tag{18.2.1}$$

For $x \in \mathbf{R}_+^m$, its projection on $\mathbf{R}_0^m$ is denoted by Z_x which can be identified with $\sum\limits_{j=1}^{m-1} e_j x_j$ or $(Z_x, 0)$. Thus $x \in \mathbf{R}_+^m$ can be written as $x = Z_x + e_m x_m = (Z_x, x_m)$; and its symmetric point to the hyperplane $\mathbf{R}_0^m$ can be written as $x^* = Z_x - e_m x_m = (Z_x, -x_m)$. The elementary surface measure on $\mathbf{R}_0^m$ will be denoted by dZ_x.

Now define

$$N(x, y) := -[\Gamma(x - y) + \Gamma(x - y^*)]. \tag{18.2.2}$$

By (18.2.1), (18.2.2) and

$$\begin{aligned}
\tfrac{\partial}{\partial n} N(x, y) &= (0, ..., 0, 1) \bullet \nabla_x [\Gamma(x - y) + \Gamma(x - y^*)] \\
&= \tfrac{1}{\omega_m} \big[\tfrac{x_m - y_m}{|x - y|^m} + \tfrac{x_m + y_m}{|x - y^*|^m} \big],
\end{aligned}$$

the following properties can be easily verified:

(i) $-\Delta_x N(x, y) = \delta(x - y)$, for $x, y \in \mathbf{R}_+^m$;

(ii) $N(x, y) = N(y, x)$;

(iii) $\tfrac{\partial}{\partial n_x} N(x, y) = 0$, for $x = Z_x \in \mathbf{R}_0^m$ and $y \in \mathbf{R}_+^m$, where $\tfrac{\partial}{\partial n_x} = -\tfrac{\partial}{\partial x_m}$

denotes the outer normal derivative to the boundary $\mathbf{R}_0^m$ of $\mathbf{R}_+^m$;

(iv) for $y = Z_y \in \mathbf{R}_0^m$, $-\frac{\partial}{\partial x_m} N(x, Z) = \frac{2x_m}{\omega_m(|Z_x - Z_y|^2 + x_m^2)^{m/2}} =: P(x, Z_y)$

which is the Poisson kernel for the half space $\mathbf{R}_+^m$.

Since $N(x, y)$ satisfies (i) - (iii), it is called the Neumann function for the Laplacian over the half space $\mathbf{R}_+^m$ ([10], Definition 9.2).

By using the above Neumann function $N(x, y)$, we obtain the following lemma regarding the representation formula of the solutions of the Neumann problem for the Poisson equation in $\mathbf{R}_+^m (m \geq 3)$.

LEMMA 18.2.1
Let $f \in C(\mathbf{R}_+^m \bigcup \mathbf{R}_0^m)$ and $g \in C(\mathbf{R}^{m-1})$ satisfy

$$|f(y)| = O(|y|^{-2-\epsilon}), \text{ as } y \to \infty, \text{ and } |g(Z_y)| = O(|Z_y|^{-1-\epsilon}), \text{ as } Z_y \to \infty,$$

for some $\epsilon \in (0, 1)$. Then the Neumann problem for the Poisson equation

$$\begin{array}{lll} -\Delta u = f & in & \mathbf{R}_+^m \\ \frac{\partial u}{\partial n} = g & on & \mathbf{R}_0^m \\ u & is\ bounded\ in & \mathbf{R}_+^m \end{array}$$

has a solution which is unique up to a constant, and this solution u can be represented as

$$u(x) = \int_{\mathbf{R}_+^m} N(x, y) f(y) dy$$

$$+ \int_{\mathbf{R}_0^m} N(x, Z_y) g(Z_y) dZ_y + C =: u_1(x) + u_2(x) + C,$$

where C is a constant, and

$$|u_1(x)| = O(|x|^{-\epsilon}) \quad and \quad |u_2(x)| = O(|x|^{-\epsilon}), \quad as\ x \to \infty.$$

The proof of Lemma 18.2.1 is a generalization of the proof in [13, Lecture 13] (where the proof is given for the case of m = 3 and f = 0).

Now we present two lemmas concerning $\mathbf{R}_{0,m-1}$-valued conjugate harmonic functions.

LEMMA 18.2.2 (Generalized Cauchy-Riemann Equations)
Let $\Omega \subset \mathbf{R}^m$ be an open set. Then $w \in C^1(\Omega, \mathbf{R}_{0,m})$ is monogenic in Ω if and only if $X^{(m)}w$ and $Y^{(m)}w$ satisfy the equations

$$D_{m-1} X^{(m)} w - \frac{\partial}{\partial x_m} Y^{(m)} w = 0 \quad in\ \Omega,$$

$$\frac{\partial}{\partial x_m} X^{(m)} w - D_{m-1} Y^{(m)} w = 0 \quad in\ \Omega.$$

Here $D_{m-1} = \sum\limits_{j=1}^{m-1} e_j \frac{\partial}{\partial x_j}$ is a Dirac operator in $\mathbf{R}^{m-1}$.

LEMMA 18.2.3
A harmonic conjugate $\widetilde{N}$ of the Neumann function N for the half space $\mathbf{R}_+^m$ ($m \geq 3$) is given by

$$\widetilde{N}(x,y) = -\frac{1}{\omega_m} \frac{Z_x - Z_y}{|Z_x - Z_y|^{m-1}} [F(\tfrac{m}{2}, \tfrac{x_m + y_m}{|Z_x - Z_y|}) + F(\tfrac{m}{2}, \tfrac{x_m - y_m}{|Z_x - Z_y|})],$$

where $F(\frac{m}{2}, t) = \int_0^t \frac{ds}{(1+s^2)^{m/2}}$ is a bounded function of $t \in \mathbf{R}$.

The proof of Lemma 18.2.2 can be found in [17] and the proof of Lemma 18.2.3 is exactly parallel to that for Theorem 3.1 in [17].
  For $x \in \mathbf{R}_+^m$ and $Z_y \in \mathbf{R}_0^m$, we define

$$T(x, Z_y) : = N(x, Z_y) + e_m \widetilde{N}(x, Z_y) \tag{18.2.3}$$

$$= \frac{2}{(m-2)\omega_m (|Z_x - Z_y|^2 + x_m^2)^{(m-2)/2}} \tag{18.2.4}$$

$$+ e_m \frac{2(Z_y - Z_x)}{\omega_m |Z_x - Z_y|^{m-1}} F(\tfrac{m}{2}, \tfrac{x_m}{|Z_x - Z_y|}).$$

Then, for any fixed $Z_y \in \mathbf{R}_0^m$, $T(x, Z_y)$ is a monogenic function of x in $\mathbf{R}_+^m$; and

$$|T(x, Z_y)| = O(|Z_y|^{-(m-2)}) \quad \text{as } |Z_y| \to \infty, \quad \text{for any fixed } x \in \mathbf{R}_+^m. \tag{18.2.5}$$

REMARK   Different constructions of monogenic functions whose scalar part is the fundamental solution $\Gamma(x)$ of the Laplacian are given in [14] (for $m = 4$) and in [11] and [12] (for general m). For our function T defined by (18.2.3), not only its scalar part but also its $X^{(m)}$ part (i.e., its projection on $\mathbf{R}_{0,m-1}$) is $(-2)\Gamma(x - Z_y)$. The fact that $X^{(m)}T(x, Z_y) = (-2)\Gamma(x - Z_y) = N(x, Z_y)$ is crucial in proving of the following main theorems, and it will be used to form explicit formulae for solutions of Reimann-Hlibert problems.

The main results of this section are stated in the following two theorems.

THEOREM 18.2.4
Let $g \in C(\mathbf{R}_0^m, \mathbf{R}_{0,m-1})$ with

$$|g(Z_y)| = O(|Z_y|^{-1-\epsilon}) \quad \text{as } |Z_y| \to \infty, \text{ for some } \epsilon \in (0, 1). \tag{18.2.6}$$

Then the Neumann boundary value problem

$$\begin{array}{lll} Dw = 0 & in & \mathbf{R}_+^m \\ \frac{\partial}{\partial n} X^{(m)} w = g & on & \mathbf{R}_0^m \\ w & is\ bounded\ in & \mathbf{R}_+^m \end{array} \tag{G}$$

has a classical C^1 solution which is unique up to a Clifford constant, and the solution can be represented as

$$w(x) = \int_{\mathbf{R}_0^m} T(x, Z_y)g(Z_y)dZ_y + C, \qquad (18.2.7)$$

where $T(x, Z_y)$ is defined by (18.2.3) and C is an $\mathbf{R}_{0,m}$-valued constant.

PROOF   By (18.2.3) and (18.2.5), there exists a constant $M > 0$ such that

$$\int_{\mathbf{R}_0^m} |T(x, Z_y)g(Z_y)|dZ_y \le M \int_{\mathbf{R}_0^m} \frac{1}{|Z_x - Z_y|^{m-2}} \frac{dZ_y}{|Z_y|^{1+\epsilon}}. \qquad (18.2.8)$$

We need to examine the behavior of the right-hand side integral when $R := |Z_x| \to \infty$. Without loss of generality, we may estimate the integral at the point $Z_x = (R, 0, ..., 0) \in \mathbf{R}^{m-1}$ (because of symmetry, an estimate at this point will give all that is needed). Let $Z_y = R\xi = (R\xi_1, ..., R\xi_{m-1})$. Then we have

$$\int_{\mathbf{R}_0^m} \frac{dZ_y}{|Z_x - Z_y|^{m-2}|Z_y|^{1+\epsilon}} \qquad (18.2.9)$$

$$= \int_{\mathbf{R}_0^m} \frac{dZ_y}{[(y_1 - R)^2 + y_2^2 + ... + y_{m-1}^2]^{(m-2)/2}|Z_y|^{1+\epsilon}} \qquad (18.2.10)$$

$$= R^{-\epsilon} \int_{\mathbf{R}_0^m} \frac{d\xi}{[(\xi_1 - 1)^2 + \xi_2^2 + ... + \xi_{m-1}^2]^{(m-2)/2}|\xi|^{1+\epsilon}}$$

$$= O(|Z_x|^{-\epsilon}), \text{ as } Z_x \to \infty,$$

since the last integral converges absolutely near its three singularities: ∞, 0 and $(1,0,...,0)$. Hence, by (18.2.7) and (18.2.8), the function w defined by (18.2.6) is uniformly bounded in $\mathbf{R}_+^m$; and it satisfies the equation $Dw = 0$ since $T(x, Z_y)$ is monogenic in x. From (18.2.6), by using the property (iv) of the Neumann function $N(x, y)$, we have

$$\frac{\partial}{\partial n}(X^{(m)}w)(x) = \int_{\mathbf{R}_0^m} \frac{\partial}{\partial(-x_m)} N(x, Z_y)g(Z_y)dZ_y \qquad (18.2.11)$$

$$= \int_{\mathbf{R}_0^m} P(x, Z_y)g(Z_y)dZ_y,$$

where $P(x, Z_y)$ is the Poisson kernel for the half space $\mathbf{R}_+^m$. It is well known that the right-hand side of (18.2.9) approaches $g(Z_y)$ as $x \to Z_x \in \mathbf{R}_0^m$. Thus the function w defined by (18.2.6) solves the problem (G).

Now let w_1 be another solution to the problem (G) and let $v = w - w_1$.

Then v satisfies

$$\begin{array}{lll} Dv = 0 & \text{in} & \mathbf{R}^m_+ \\ \frac{\partial}{\partial n} X^{(m)} v = 0 & \text{on} & \mathbf{R}^m_0 \\ v & \text{is bounded in} & \mathbf{R}^m_+. \end{array}$$

Hence $X^{(m)}v$ is a bounded solution of the classical Neumann problem for Laplace's equation with zero boundary condition. By Lemma 18.2.1, $X^{(m)}v = C_1$ where C_1 is an $\mathbf{R}_{0,m-1}$-valued constant. Applying Lemma 18.2.2 to the function v, we have

$$\frac{\partial}{\partial x_m} Y^{(m)} v = 0 \quad \text{and} \quad D_{m-1} Y^{(m)} v = 0 \quad \text{in } \mathbf{R}^m_0.$$

Hence $Y^{(m)}v$ is an $\mathbf{R}_{0,m-1}$-valued function in the variables $x_1, ..., x_{m-1}$, and it can be regarded as a bounded $\mathbf{R}_{0,m-1}$-valued monogenic function in all of $\mathbf{R}^{m-1}$. Thus, by Liouville's theorem, $Y^{(m)}v \equiv C_2$. Hence $v \equiv C_1 + e_m C_2$ is an $\mathbf{R}_{0,m}$-valued constant. ∎

Let $h^p(\mathbf{R}^m_+)$ be the classical h^p space on the upper half space of $\mathbf{R}^m$; i.e.,

$$h^p(\mathbf{R}^m_+) := \{u| \quad \Delta u = 0 \quad \text{in } \mathbf{R}^m_+ \text{ and } \sup_{x_m > 0} \int_{\mathbf{R}^m_0} |u(Z_x, x_m)|^p dZ_x < \infty\}.$$

THEOREM 18.2.5
Let $g \in L^p(\mathbf{R}^m_0) \bigcap L^\infty(\mathbf{R}^m_0), 1 \le p \le m - 1, m \ge 3$, be an $\mathbf{R}_{0,m-1}$-valued function. Then the Neumann problem (G) has a solution w which is unique up to a Clifford constant in the following sense: w is bounded and monogenic in $\mathbf{R}^m_+$, $\frac{\partial}{\partial(-x_m)} X^{(m)} w(x) \in h^p(\mathbf{R}^m_+)$ and

$$\lim_{x_m \to 0+} ||\frac{\partial}{\partial(-x_m)} X^{(m)} w(\bullet, x_m) - g(\bullet)||_{L^p(\mathbf{R}^m_0)} = 0. \qquad (18.2.12)$$

Moreover, the solution w can be represented as (18.2.6).

PROOF Since $p < m - 1$, we have $q := \frac{p}{p-1} > \frac{m-1}{m-2}$. Hence, from (18.2.3), we have

$$\int_{\mathbf{R}^m_0} |T(x, Z_y) g(Z_y)| dZ_y$$

$$\le M \int_{\mathbf{R}^m_0} \frac{|g(Z_y)|}{|Z_x - Z_y|^{m-2}} dZ_y$$

$$= M \int_{\mathbf{R}^m_0} \frac{|g(Z_x - Z_y)|}{|Z_y|^{m-2}} dZ_y$$

$$\le M\{[\int_{|Z_y| \ge 1} \frac{dZ_y}{|Z_y|^{(m-2)q}}]^{1/q} [\int_{|Z_y| \ge 1} |g(Z_x - Z_y)|^p dZ_y]^{1/p}$$

$$+ \int\limits_{|Z_y| \leq 1} \frac{|g(Z_x - Z_y)|}{|Z_Y|^{m-2}} dZ_y \}$$

$$\leq MK[||g||_{L^p(\mathbf{R}_0^m)} + ||g||_{L^\infty(\mathbf{R}_0^m)}],$$

where the constants M and K are independent of x. So the integral defined by (18.2.6) is well defined and bounded in $\mathbf{R}_+^m$. It is clear that w is monogenic in $\mathbf{R}_+^m$. By the property (iv) of the Neumann function $N(x, y)$,

$$\frac{\partial}{\partial(-x_m)} X^{(m)} w(x) = \int_{\mathbf{R}_0^m} P(x, Z_y) g(Z_y) dZ_y.$$

Applying [17, Theorem 4.4], we get $\frac{\partial}{\partial(-x_m)} X^{(m)} w \in h^p(\mathbf{R}_+^m)$ and (18.2.10). Thus w, as defined by (18.2.6), is a solution to the problem (G). The uniqueness of the solution (up to a Clifford constant) can be proved in the similar manner as in the proof of Theorem 18.2.4. ∎

18.3 Neumann Problems for Nonhomogeneous Dirac Equations

The aim of this section is to give a solution representation formula for the Neumann problem of the nonhomogeneous Dirac equation:

$$\begin{array}{lll} Dw = f & \text{in} & \mathbf{R}_+^m \\ \frac{\partial}{\partial n} X^{(m)} w = g & \text{on} & \mathbf{R}_0^m \\ w & \text{is bounded in} & \mathbf{R}_+^m. \end{array} \qquad \text{(GF)}$$

THEOREM 18.3.1
Let $f \in C(\mathbf{R}_+^m, \mathbf{R}_{0,m})$ be Hölder continuous and $g \in C(\mathbf{R}_0^m, \mathbf{R}_{0,m-1})$, with

$$|f(y)| = O(|y|^{-1-\epsilon}) \; as\, y \to \infty, \; and |g(Z_y)| = O(|Z_y|^{-1-\epsilon}) \; as Z_y \to \infty,$$

for some $\epsilon \in (0, 1)$. Then the Neumann problem (GF) has a solution w which is unique up to a Clifford constant, and the solution can be represented as

$$w(x) = \int_{\mathbf{R}_+^m} [E(x - y) + E(x^* - y)] f(y) dy + \int_{\mathbf{R}_0^m} T(x, Z_y) g(Z_y) dZ_y + C.$$

$$(18.3.1)$$

Here $E(x) = -\frac{x}{\omega_m |x|^m}$ is the fundamental solution of the Dirac operater, T is defined by (3) and C is an $\mathbf{R}_{0,m}$-valued constant.

PROOF   Since

$$D[E(x - y) + E(x^* - y)] = \delta(x - y), \quad \text{for } x, y \in \mathbf{R}_+^m,$$

and

$$\frac{\partial}{\partial X_m}[E(x-y) + E(x^* - y)] = 0, \text{ for } x \in \mathbf{R}_0^m \text{ and } y \in \mathbf{R}_+^m,$$

it can be verified that

$$w_1(x) := \int_{\mathbf{R}_+^m} [E(x-y) + E(x^* - y)]f(y)dy$$

is the unique solution for the Neumann type problem

$$
\begin{aligned}
Dw &= f &&\text{in} &&\mathbf{R}_+^m, \\
\tfrac{\partial}{\partial n}X^{(m)}w &= 0 &&\text{on} &&\mathbf{R}_0^m, \\
\lim_{x \to \infty} w(x) &= 0. &&
\end{aligned}
$$

(In fact, under the assumption for the function f, by using the same proof as for (18.2.8) in Theorem 18.2.4, we can further prove that $|w_1(x)| = O(|x|^{-\epsilon})$ as $x \to \infty$.) Thus, by combining Theorem 18.2.4 with the above result for $w_1(x)$, we have proved that the function w defined by (18.3.1) is the solution, unique up to a Clifford constant, to the Neumann problem (GF). ∎

Acknowledgment

The author is grateful to Professor John Ryan and the University of Arkansas for the invitation and the financial support during this conference.

Bibliography

[1] F. Brackx, R. Delanghe and F. Sommen, *Clifford Analysis*, Pitman, London, 1982.

[2] H. Begehr and Z. Xu, *Nonlinear half-Dirichlet problems for first order elliptic equations in the unit ball of $\mathbf{R}^m (m \geq 3)$*, Appl. Anal. **45**, (1992), 3–18.

[3] J. S. R. Chisholm and A. K. Common (Eds.), *Clifford Algebras and their Applications in Mathematical Physics*, D. Reidel, Dordrecht, 1986.

[4] R. Delanghe, F. Sommen and Z. Xu, *Half-Dirichlet problems for powers of the Dirac operator in the unit ball of $\mathbf{R}^m (m \geq 3)$*, Bull. Soc. Math. Belg., 3 Ser. B, **42**, (1990), 409–429.

[5]   J. E. Gilbert and M. A. M. Murray, *Clifford Algebras and Dirac Operators in Harmonic Analysis*, Cambridge University Press, Cambridge, 1991.

[6]   R. P. Gilbert and J. L. Buchanan, *First Order Elliptic Systems: A Function Theoretic Approach*, Academic Press, New York, 1983.

[7]   K. Gürlebeck and W. Spröβig, *Quaternionic Analysis and Elliptic Boundary Value Problems*, V.56, Akademie-Verlag, Berlin, 1989.

[8]   G. N. Hile and P. Lounesto, *Matrix representations of Clifford algebras*, Linear Algebra Appl., **128**, (1990), 51–63.

[9]   A. Micali, R. Boudet and J. Helmstetter (Eds.), *Clifford algebras and their Applications in Mathematical Physics*, Kluwer Academic, Boston, 1992.

[10] G. F. Roach, *Green's Functions*, 2nd ed., Cambridge University Press, Cambridge, 1982.

[11] J. Ryan, *Complexified Clifford analysis*, Complex Variables **1**, (1982), 119–149.

[12] J. Ryan, *Plemelj formulae and transformations associated to plane wave decompositions in complex Clifford analysis*, Proc. London Math. Soc. **(3) 64**, (1992), 70–94.

[13] S. L. Sobolev, *Partial Differential Equations of Mathematical Physics*, Pergamon Press, London, 1964.

[14] A. Sudbery, *Quaternionic analysis*, Math. Proc. Camb. Philos. Soc. **85**, (1979), 199–225.

[15] Z. Xu, *On boundary value problems of Neumann type for hypercomplex functions with values in a Clifford algebra*, Rend. Circolo Math. Palermo, Serie II **No. 22**, (1989), 213–226.

[16] Z. Xu, J. Chen and W. Zhang *A harmonic conjugate of the Poisson kernel and a boundary value problem for monogenic functions in the unit ball of $\mathbf{R}^n (n \geq 2)$*, Simon Stevin **64**, (1990), 187–201.

[17] Z. Xu and C. Zhou, *On boundary value problems of Riemann-Hilbert type for monogenic functions in a half space of $\mathbf{R}^m (m \geq 2)$*, Complex Variables **22**, (1993), 181–194.

[18] C. Zhou, *On boundary value problems of Neumann type for the Dirac operator in a half space of $\mathbf{R}^m (m \geq 3)$*, Complex Variables **23**, (1993), 1–16.

Department of Mathematics,
University of Hawaii at Honolulu Community College,
874 Dillingham Boulevard, Honolulu, Hawaii 96817
E-mail: chiping@pulua.hcc.hawaii.edu

19

The Hyperholomorphic Bergman Projector and its Properties

E. Ramírez de Arellano, M. Shapiro, and N. Vasilevski*

ABSTRACT The hyperholomorphic Bergman kernel function and the corresponding Bergman projector (integral operator with the Bergman kernel) are studied. This work is a continuation of a series of papers devoted to the hyperholomorphic Bergman function in Clifford and quaternionic analysis.

19.1 Introduction

This article is a continuation of the series of work [SV1], [SV2], [VS] where the Bergman kernel function for quaternionic and Clifford analysis has been studied. This study is based on the intimate connection between the Bergman function and the Green function of the domain.

In the quaternionic case [SV1] it has been found that to study the Bergman function $_\psi\mathcal{B}$ for some fixed class of ψ-hyperholomorphic functions, it is necessary to use another class of φ-hyperholomorphic functions. A new phenomenon (for details see [SV2]) arises in the case of general Clifford analysis: it turns out that a very important method of producing integral representations for $_\psi\mathcal{B}$ does not work for m even, where $m + 1$ denotes the number of independent variables. Hence deeper and more impressive properties can be established only in the case of m odd.

The first section gives the necessary preliminaries, including the description of the main results of previous work.

Section 19.3 deals with the well-known T-operator and with the singular integral operator $^{\varphi,\psi}S$ over a domain in $\Omega \subset \mathbb{R}^{m+1}$. The properties established here allows us, in Section 19.4, to represent the Bergman hyperholomorphic projector $^\varphi B$ (i.e., the integral operator with the kernel $_\varphi\mathcal{B}$) as an

*This work was partially supported by CONACYT Project 1821-E9211, Mexico.

0-8493-8481-8/96/$0.00+$.50

element of the algebra generated by classical Mikhlin-Calderon-Zygmund operator over the domain Ω.

In Section 19.5 we describe the structure of two operator algebras. The first one is generated by $^\psi B$ and multiplication by continuous function operators, the second algebra is generated by the Toeplitz-Bergman operators with right continuous presymbols. Fredholm criteria are given as well.

19.2   Preliminaries

Let $\mathcal{A}_n^-$ denote the universal real Clifford algebra with the quadratic form $-\sum_{k=1}^{n} t_k^2$. In particular, this means that if $i_1, \ldots, i_n$ is the basis of the n-dimensional space on which the algebra $\mathcal{A}_n^-$ is built and i_0 is the unit of $\mathcal{A}_n^-$, then

$$i_k^2 = -i_0, \; i_p i_q + i_q i_p = 0$$

for $k \in \{1, \ldots, n\}, p \neq q$. One can find all necessary properties of this algebra as well as commonly used notations in [BDS], for instance. Let $m \in \{1, \ldots, n\}$. We shall consider functions with values in the algebra $\mathcal{A}_n^-$ and defined in a domain $\Omega \subset \mathbb{R}^{m+1}$.

We denote by $\mathcal{A}_n^{-,k}$ the set of k-vectors from $\mathcal{A}_n^-$ and let

$$\psi := \{\psi^0, \psi^1, \ldots, \psi^m\} \subset \mathcal{A}_n^{-,0} \oplus \mathcal{A}_n^{-,1},$$

and

$$\overline{\psi} := \{\overline{\psi^0}, \overline{\psi^1}, \ldots, \overline{\psi^m}\}.$$

On $\mathcal{C}^1(\Omega, \mathcal{A}_n^-)$ we introduce the operator $^\psi D$ putting

$$^\psi D[f] := \sum_{k=0}^{m} \psi^k \frac{\partial f}{\partial x_k}, \quad D^\psi[f] := \sum_{k=0}^{m} \frac{\partial f}{\partial x_k} \psi^k. \tag{19.2.1}$$

Let the conditions

$$\psi^p \cdot \overline{\psi^q} + \psi^q \cdot \overline{\psi^p} = 2\langle \psi^p, \psi^q \rangle_{\mathbb{R}^{m+1}} = 2\delta_{p,q}$$

be fulfilled for all $p, q \in \{0, 1, \ldots, m\}$, where $\langle \cdot, \cdot \rangle$ is the scalar product in $\mathbb{R}^{m+1}$ and $\delta_{p,q}$ is the Kronecker's delta. Of course this means that ψ is an orthonormalized system in $\mathbb{R}^{n+1} \approx \mathcal{A}_n^{-,0} \oplus \mathcal{A}_n^{-,1}$.

Now the Laplace operator Δ has the following representation:

$$\Delta = {}^\psi D \cdot \overline{{}^\psi D} = \overline{{}^\psi D} \cdot {}^\psi D. \tag{19.2.2}$$

The operator $^\psi D$ will be called the (left) Cauchy-Riemann operator for $\mathcal{A}_n^-$-valued functions, and the elements of the set $^\psi \mathfrak{M}(\Omega, \mathcal{A}_n^-) := \ker {}^\psi D$ will be called left-ψ-hyperholomorphic (in a domain Ω) $\mathcal{A}_n^-$-valued functions of $(m+1)$-dimensional variable. This name may be shortened when misunderstandings cannot arise. It follows from (19.2.2) that if $f \in {}^\psi\mathfrak{M}(\Omega)$, then its coordinate functions are harmonic in Ω.

Let Ω be a bounded domain in $\mathbb{R}^{m+1}$ with smooth boundary $\Gamma = \partial\Omega$. Denote by g the harmonic (or classical) Green's function of Ω. The exact definition and the properties of this function can be found in [HK p. 26], for instance. It is known, in particular, that such a function exists, is unique, and has the form:

$$g(x, \xi) := \theta_{m+1}(\xi - x) + h(x, \xi), \quad (x, \xi) \in \Omega \times \Omega \setminus \mathrm{diag} ,$$

where

$$\theta_{m+1}(x) = -\frac{1}{(m-1) \cdot |\mathbb{S}^m|} |x|^{-m+1} (m \geq 3),$$

is the fundamental solution of the Laplace equation, $|\mathbb{S}^m|$ is the area of the unit sphere in $\mathbb{R}^{m+1}$, h is harmonic in $\Omega \times \Omega$ and is such that

$$h(x, \xi) := -\theta_{m+1}(\xi - x), \quad (x, \xi) \in \Omega \times \partial\Omega,$$

$$h(x, \xi) := h(\xi, x), \quad \text{on } \Omega \times \Omega;$$

$\mathrm{diag} := \{(x, x) | x \in \Omega\}$.

For an arbitrary structural set ψ the Bergman left-ψ-hyperholomorphic kernel function (or Bergman h.h. kernel function, or simply kern-function) $_\psi\mathcal{B}$ has been defined in [SV2], (see also [SV1]), by the formula

$$_\psi\mathcal{B}(x, \xi) := D_\xi^\psi \cdot {}^{\overline{\psi}}D_x[g](x, \xi) = {}^{\overline{\psi}}D_x \cdot {}^\psi D_\xi[g](x, \xi), \quad (x, \xi) \in \overline{\Omega} \times \overline{\Omega} \setminus \mathrm{diag} .$$

The following properties have been established in [SV1] and [SV2]:

1. The function $_\psi\mathcal{B}$ can be continuously extended onto the diagonal of $\Omega \times \Omega$;

2. For any fixed $\xi \in \Omega$ $_\psi\mathcal{B}(\cdot, \xi) \in {}^\psi\mathfrak{M}(\Omega)$,

3. For any fixed $x \in \Omega$ $_\psi\mathcal{B}(x, \cdot) \in \mathfrak{M}^{\overline{\psi}}(\Omega)$,

4. $_\psi\mathcal{B}(x, \xi) = {}_\psi\overline{\mathcal{B}(\xi, x)}$.

5. (Reproducing property of the Bergman h.h. kernel.) Let $\Gamma = \partial\Omega$ be a smooth surface and $f \in {}^\psi\mathfrak{M}(\Omega, A_n^-) \cap \mathcal{C}(\overline{\Omega}, A_n^-)$. Then

$$^\psi B[f](x) := \int_\Omega {}_\psi\mathcal{B}(x, \xi) f(\xi) d\xi$$

$$= f(x), \quad x \in \Omega.$$

The natural name for $^\psi B$ is the Bergman h.h. operator.

6. (Integral representation of the Bergman hyperholomorphic kernel function.) Let m be odd and let ψ and φ be two structural sets with the only condition

$$\sum_{k=0}^{m} \varphi^k \cdot \psi^k = 0, \qquad (19.2.3)$$

then the equality

$$_\psi\mathcal{B}(x,\xi) = \int_\Gamma \mathcal{K}_\psi(\tau - x) \cdot \sigma_{\overline{\varphi},\tau} \cdot {}_{\varphi,\psi}\Lambda(\tau,\xi)$$

$$+ \int_\Omega {}_{\varphi,\psi}\overline{\ell(x,\tau)} \cdot {}_{\varphi,\psi}\ell(\xi,\tau)d\tau, \times(19.2.3')$$

holds for $\forall(x,\xi) \in \Omega \times \Omega$.
Here $\mathcal{K}_\psi$ denotes the Cauchy kernel of the Clifford analysis:

$$\mathcal{K}_\psi(\tau - x) := \frac{1}{|\mathbb{S}^m| \cdot |\tau - x|^{m+1}} \cdot \sum_{k=0}^{m} \overline{\psi^k} \cdot x_k, \qquad (19.2.4)$$

${}_{\varphi,\psi}\Lambda$ and ${}_{\varphi,\psi}\ell$ are of the form

$$_{\varphi,\psi}\Lambda(x,\xi) = {}^\varphi D_\xi \cdot {}^\psi D_x[\theta_{m+1}(\xi - x)] = D_x^\psi \cdot {}^\varphi D_\xi[\theta_{m+1}(\xi - x)]$$

$$= \frac{(m+1) \cdot (\xi - x)_\varphi \cdot (\xi - x)_\psi - |\xi - x|^2 \cdot \sum_{k=0}^{m} \varphi^k \cdot \psi^k}{|\mathbb{S}^m| \cdot |\xi - x|^{m+3}},$$

$$_{\varphi,\psi}\ell(x,\xi) = {}^\varphi D_\xi \cdot {}^\psi D_x[h](x,\xi) = D_x^\psi \cdot {}^\varphi D_\xi[h](x,\xi),$$

$$x_\varphi = \sum_{0}^{m} x_k \varphi_k,$$

$$\sigma_{\overline{\psi},\tau} := \sum_{k=0}^{m} (-1)^k \overline{\psi^k} d\hat{\tau}_k,$$

and $d\hat{\tau}_k$ denotes the differential form $d\tau_0 \wedge d\tau_1 \wedge d\tau_2 \wedge \ldots \wedge d\tau_m$ with the omitted factor $d\tau_k$.

7. Under the same assumptions we have also:

$$_\psi\mathcal{B}(x,\xi) = \int_\Gamma {}_{\varphi,\psi}\overline{\Lambda(\tau,x)} \cdot \sigma_{\varphi,\tau} \cdot \mathcal{K}_{\overline{\psi}}(\tau - \xi) + \int_\Omega {}_{\varphi,\psi}\overline{\ell(x,\tau)} \cdot {}_{\varphi,\psi}\ell(\xi,\tau)d\tau.$$

This follows from the Hermitian symmetry of the function $_\psi\mathcal{B}$.

19.3 The T-operator and the singular integral operator over a domain

The well-known T-operator (or the Vekua operator) is defined by the equality

$$^\psi T[f](x) := -\int_\Omega \mathcal{K}_\psi(\tau - x) \cdot f(\tau) \cdot d\tau$$

with $\mathcal{K}_\psi$ as in (19.2.4).

For an arbitrary pair of structural sets φ and ψ introduce the operator $^{\varphi,\psi}S$:

$$^{\varphi,\psi}S[f](x) := \int_\Omega \overline{_{\varphi,\psi}\Lambda(\tau,x)} \cdot f(\tau) \cdot d\tau =$$

$$= \int_\Omega \frac{(m+1)\cdot(\tau - x)_{\overline{\psi}} \cdot (\tau - x)_{\overline{\varphi}} - |\tau - x|^2 \cdot \sum_{k=0}^m \overline{\psi^k} \cdot \overline{\varphi^k}}{|\mathbb{S}^m| \cdot |\tau - x|^{m+1}}$$

$$\cdot f(\tau) \cdot d\tau.$$

It is clear that both $^\psi T$ and $^{\varphi,\psi}S$ are linear bounded operators on the right $\mathcal{A}_n^-$-modules $L_p(\Omega, \mathcal{A}_n^-)$, $p > 1$, and the Hölder function space $\mathcal{C}^{0,\mu}(\Omega, \mathcal{A}_\mu^-)$, $0 < \mu < 1$.

THEOREM 19.3.1
In the spaces $L_p(\Omega, \mathcal{A}_n^-)$, $p > 1$, and $\mathcal{C}^{0,\mu}(\Omega, \mathcal{A}_n^-)$, $0 < \mu < 1$, the composition $^\psi D \cdot {}^\psi T$ is well defined and in these spaces

$$^\psi D \cdot {}^\psi T = I, \tag{19.3.1}$$

where I is the identity operator.

PROOF Assume that f belongs to $\mathcal{C}^2(\overline\Omega, \mathcal{A}_n^-)$ which is dense in L_p or $\mathcal{C}^{0,\mu}$. For $\tau \neq x$ we have:

$$d_\xi(\theta_{m+1}(\tau - x) \cdot \sigma_{\overline\psi,\tau} \cdot f(\tau)) = (\mathcal{K}_\psi(\tau - x) \cdot f(\tau) + \theta_{m+1}(\tau - x) \cdot {}^\psi D[f](\tau))d\tau, \tag{19.3.2}$$

or in the integral form

$$\int_{\partial\Omega} \theta_{m+1}(\tau - x) \cdot \sigma_{\overline\psi,\tau} \cdot f(\tau) = \int_\Omega \mathcal{K}_\psi(\tau - x) \cdot f(\tau)d\tau \tag{19.3.3}$$

$$+ \int_\Omega \theta_{m+1}(\tau - x) \cdot \overline{{}^\psi D}[f](\tau)d\tau,$$

which implies

$$^\psi T[f](x) = \int_\Omega \theta_{m+1}(\tau - x) \cdot \overline{{}^\psi D}[f](\tau)d\tau - \int_{\partial\Omega} \theta_{m+1}(\tau - x) \cdot \sigma_{\overline\psi,\tau} f(\tau)d\tau. \tag{19.3.4}$$

Applying the operator $^\psi D$ and using the Borel-Pompeiu formula we obtain:

$$^\psi D \cdot {}^\psi T[f](x) = \int_\Omega {}^\psi D_x[\theta_{m+1}(\tau - x)] \cdot {}^\psi D[f](\tau) d\tau$$

$$- \int_{\partial\Omega} {}^\psi D_x[\theta_{m+1}(\tau - x)] \cdot \sigma_{\overline{\psi},\tau} \cdot f(\tau)$$

$$= - \int_\Omega \mathcal{K}_{\overline{\psi}}(\tau - x) \cdot \overline{{}^\psi D}[f](\tau) d\tau$$

$$+ \int_{\partial\Omega} \mathcal{K}_{\overline{\psi}}(\tau - x) \cdot \sigma_{\overline{\psi},\tau} \cdot f(\tau)$$

$$= f(x).$$

Thus for $f \in \mathcal{C}^2(\Omega, \mathcal{A}_n^-)$ we have:

$$^\psi D \cdot {}^\psi T[f] = f. \tag{19.3.5}$$

∎

The equality (19.3.1) follows from (19.3.5), the continuity of the identity operator and the density of $\mathcal{C}^2$ in L_p and $\mathcal{C}^{0,\mu}$.

THEOREM 19.3.2
Let m be odd and let two structural sets φ and ψ satisfy (19.2.3). Then the operator $^\psi D \cdot {}^\psi T$ is well defined in both $L_p(\Omega, \mathcal{A}_n^-), p > 1$ and $\mathcal{C}^{0,\mu}(\Omega, \mathcal{A}_n^-)$ and

$$^\varphi D \cdot \overline{{}^\psi T} = {}^{\varphi,\psi} S.$$

The proof is similar to the previous one. First we check this equality on functions from an appropriate dense subset and then extend it by continuity.

19.4   Bergman hyperholomorphic spaces and Bergman hyperholomorphic projector

Consider any smooth function f in $\overline{\Omega}$ and apply the Bergman h.h. operator $^\psi B$ to it. Using the formula in property 7 from 19.2.3 we have:

$$^\psi B[f](x) = \int_\Omega {}_\psi B(x, \xi) f(\xi) d\xi = \tag{19.4.1}$$

$$= \int_\Omega (\int_\Gamma {}^{\varphi,\psi}\overline{\Lambda(\tau, x)} \cdot \sigma_{\varphi,\tau} \cdot \mathcal{K}_{\overline{\psi}}(\tau - \xi)$$

$$+ \int_\Omega {}_{\varphi,\psi}\overline{\ell(x,\tau)} \cdot {}_{\varphi,\psi}\ell(\xi,\tau)d\tau \,) \, f(\xi)d\xi$$

$$= {}^{\overline{\psi}}D_x \left(\int_\Gamma \mathcal{K}_\varphi(\tau - x) \cdot \sigma_{\varphi,\tau} \cdot {}^{\overline{\psi}}T[f](\tau) \right)$$

$$+ {}_{\varphi,\psi}C[f](x),$$

where ${}_{\varphi,\psi}C$ is the operator with the continuous kernel

$$\int_\Omega {}_{\varphi,\psi}\overline{\ell(x,\tau)} \cdot {}_{\varphi,\psi}\ell(\xi,\tau)d\tau.$$

Now apply to (19.4.1) the Borel-Pompeiu formula and Theorems 19.3.1 and 19.3.2:

$${}^{\psi}B[f](x) = {}^{\overline{\psi}}D \left({}^{\overline{\psi}}T[f](\tau) \right. \tag{19.4.2}$$

$$+ \int_\Omega \mathcal{K}_\varphi(\tau - x) \cdot {}^{\varphi}D_\tau \left[{}^{\overline{\psi}}T[f] \right](\tau)d\tau \right) + {}_{\varphi,\psi} C[f](x)$$

$$= f(x) - ({}^{\overline{\psi}}D_x \cdot {}^{\psi}T) \cdot ({}^{\varphi}D_\tau \cdot {}^{\overline{\varphi}} T)[f](x) + {}_{\varphi,\psi} C[f](x)$$

$$= f(x) - {}^{\overline{\psi} \cdot \overline{\varphi}}S \cdot {}^{\varphi,\psi}S[f](x)$$

$$+ {}_{\varphi,\psi}C[f](x).$$

Thus, finally we have that the following operator equality holds on the smooth functions in $\overline{\Omega}$:

$$^{\psi}B = I - {}^{\overline{\psi},\overline{\varphi}}S \cdot {}^{\varphi,\psi}S + {}_{\varphi,\psi}C. \tag{19.4.3}$$

The right-hand side is well defined both in $L_p(\Omega, \mathcal{A}_n^-)$ and in $\mathcal{C}^{0,\mu}(\overline{\Omega}, \mathcal{A}_n^-)$, being linear and bounded. Thus the following theorem has been proved.

THEOREM 19.4.1
Let m be odd. The Bergman (left) hyperholomorphic operator acts, linearly and boundedly, on $L_p(\Omega, \mathcal{A}_n^-)$, $p > 1$, and on $\mathcal{C}^{0,\mu}(\overline{\Omega}, \mathcal{A}_n^-)$, $0 < \mu < 1$; moreover the following equality holds:

$$^{\psi}B = I - {}^{\overline{\psi},\overline{\varphi}}S \cdot {}^{\varphi,\psi}S + {}_{\varphi,\psi}C, \tag{19.4.4}$$

where φ, ψ are two structural $(m+1)$-tuples with the property (19.2.3), and ${}_{\varphi,\psi}C$ is a compact operator.

DEFINITION 19.4.2 *Given a structural set $\psi, p > 1$ and a domain Ω, we denote by ${}_\psi \mathcal{A}^p(\Omega, \mathcal{A}_n^-)$ the subset of $L_p(\Omega, \mathcal{A}_n^-)$ consisting of all left-ψ-hyperholomorphic functions:*

$$_\psi\mathcal{A}^p(\Omega, \mathcal{A}_n^-) := L_p(\Omega, \mathcal{A}_n^-) \cap {}^\psi\mathfrak{M}(\Omega, \mathcal{A}_n^-).$$

It is clear that $_\psi\mathcal{A}^p$ is a Banach submodule of the right $\mathcal{A}_n^-$-module L_p, and that $_\psi\mathcal{A}^2$ is a Hilbert submodule of the Hilbert $\mathcal{A}_n^-$-module L_2. We shall call $_\psi\mathcal{A}^p(\Omega, \mathcal{A}_n^-)$ the Bergman h.h. space.

THEOREM 19.4.3
Let m be odd. For any Ω, ψ and p under the previous restrictions we have

$$^\psi B(L_p(\Omega, \mathcal{A}_n^-)) = {}_\psi\mathcal{A}^p(\Omega, \mathcal{A}_n^-) \tag{19.4.5}$$

and

$$^\psi B^2 = {}^\psi B. \tag{19.4.6}$$

Hence, in other words the operator $^\psi B$ is a projector of $L_p(\Omega, \mathcal{A}_n^-)$ onto $_\psi\mathcal{A}^p(\Omega, \mathcal{A}_n^-)$.

PROOF We know that $_\psi\mathcal{B}$ has singularities only on the boundary. So if $x \in \Omega$ and $f \in L_p(\Omega, \mathcal{A}_n^-)$ then

$$^\psi D \cdot {}^\psi B[f](x) = 0.$$

Taking also into account Theorem 19.4.1, we obtain that

$$^\psi B[f] \in {}^\psi\mathfrak{M}(\Omega, \mathcal{A}_n^-) \cap L_p(\Omega, \mathcal{A}_n^-),$$

i.e.,

$$^\psi B(L_p) \subset {}_\psi\mathcal{A}^p. \tag{19.4.7}$$

The boundedness of $^\psi B$ together with the density of $^\psi\mathfrak{M}(\Omega, \mathcal{A}_n^-) \cap \mathcal{C}(\overline{\Omega}, \mathcal{A}_n^-)$ in $_\psi\mathcal{A}^p$ and the reproducing property, imply that

$$f = {}^\psi B[f] \tag{19.4.8}$$

for each $f \in {}_\psi\mathcal{A}^p$. Hence $_\psi\mathcal{A}^p \subset {}^\psi B(L_p)$, and (19.4.5) is true. Taking $g \in L_p$ we get a function $f := {}^\psi B[g]$ which belongs to $_\psi\mathcal{A}^p$. Substitution of this functions into (19.4.8) gives (19.4.6). $\blacksquare$

THEOREM 19.4.4
The Bergman h.h. operator is self-adjoint in $L_2(\Omega, \mathcal{A}_n^-)$.

PROOF By a simple calculation, using the fact that kernel $_\psi\mathcal{B}$ is Hermitian:

$$\langle f, {}^\psi B[g] \rangle = \int_\Omega \overline{f(t)} \left(\int_\Omega {}_\psi\mathcal{B}(t, \xi) \cdot g(\xi) d\xi \right) dt$$

$$= \int_\Omega \overline{\left(\int_\Omega \overline{{}_\psi\mathcal{B}(t, \xi)} f(t) dt \right)} \cdot g(\xi) d\xi$$

$$= \int_\Omega \overline{\left(\int_\Omega {}^\psi\mathcal{B}(t,\xi)f(t)dt \right)} \cdot f(\xi)d\xi$$
$$= \langle {}^\psi B[f], g \rangle.$$

∎

COROLLARY 19.4.5
The Bergman h.h. operator is an orthogonal projector in $L_2(\Omega, \mathcal{A}_n^-)$.

REMARK In [GS, Ch.3] for the case of hyperholomorphic functions $f :$ $\Omega \subset \mathbb{R}^3 \cong \mathbb{H}$ the orthogonal decomposition of $L_2(\Omega, \mathbb{H})$ has been constructed. The first summand is the set of hyperholomorphic functions, the second one is described in the terms of some Sobolev space and the Cauchy-Riemann operator. Thus, in fact, the authors give the description of the Bergmann projector in this situation.

19.5 Some algebras generated by the Bergman projector

Let Ω be a bounded domain with smooth boundary $\Gamma = \partial\Omega$. On the (real) spaces $L_p(\Omega, \mathcal{A}_n^-)$, $1 < p < \infty$, for a given function $a \in \mathcal{C}(\overline{\Omega}, \mathcal{A}_n^-)$ introduce the operators $m_r(a)$ and $m_\ell(a)$ as follows:

$$m_r(a) : f \in L_p(\Omega, \mathcal{A}_n^-) \longmapsto f \cdot a \in L_p(\Omega, \mathcal{A}_n^-),$$
$$m_\ell(a) : f \in L_p(\Omega, \mathcal{A}_n^-) \longmapsto a \cdot f \in L_p(\Omega, \mathcal{A}_n^-).$$

These operators are obviously bounded.

The decomposition (19.4.6) and the properties of the Mikhlin-Calderon-Zygmund operators ${}^{\varphi,\psi}S$ and ${}^{\overline{\varphi},\overline{\psi}}S$ imply the following.

LEMMA 19.5.1
For all $a \in \mathcal{C}(\overline{\Omega}, \mathcal{A}_n^-)$ the commutator $[{}^\psi B, m_r(a)] := {}^\psi B \cdot m_r(a) - m_r(a) \cdot {}^\psi B$ is compact.

REMARK At the same time, for an arbitrary $a \in \mathcal{C}(\overline{\Omega}, \mathcal{A}_n^-)$, the commutator $[{}^\psi B, m_\ell(a)]$ is not in general compact.

Denote by $\mathcal{R}$ the algebra generated by all operators of the form

$$A = m_r(a) + m_r(b) {}^\psi B + K, \tag{19.5.1}$$

acting on $L_p(\Omega, \mathcal{A}_n^-)$, where a, $b \in \mathcal{C}(\overline{\Omega}, \mathcal{A}_n^-)$ and K is a compact operator.

Theorem 19.4.3 and Lemma 19.5.1 show that each operator of the algebra $\mathcal{R}$ has the form (4.1) for some a, b and K.

Let us restrict ourselves now onto the case of the space L_2. A standard use of the local principle [D], [S], together with the properties of the Bergman kernel (as it follows from (19.2.3′) the Bergman kernel has singularities only at the points $x = \xi \in \Gamma$), leads to

THEOREM 19.5.2
An operator A of the form (4.1) is Fredholm if and only if

(a)　$a(x)$ *is invertible in* $\mathcal{A}_n^-$ *for all* $x \in \overline{\Omega}$

(b)　$a(x) + b(x)$ *is invertible in* $\mathcal{A}_n^-$ *for all* $x \in \Gamma$.

For a function $a \in \mathcal{C}(\overline{\Omega}, \mathcal{A}_n^-)$ denote by T_a the Toeplitz-Bergman operator with the right presymbol a:

$$T_a = {}^{\psi}Bm_r(a)|_{\psi\mathcal{A}^2(\Omega, \mathcal{A}_n^-)} : f \in {}_{\psi}\mathcal{A}^2 \longmapsto {}^{\psi}B(f \cdot a) \in {}_{\psi}\mathcal{A}^2.$$

Let $T\mathcal{B}(\mathcal{C}(\overline{\Omega}, \mathcal{A}_n^-))$ be the algebra generated by all Toeplitz-Bergman operators T_a with continuous right presymbols acting on the Bergman space $_{\psi}\mathcal{A}^2(\Omega, \mathcal{A}_n^-)$.

As a corollary of Theorem 19.5.2 we have immediately the following.

THEOREM 19.5.3
An arbitrary operator $T \in T\mathcal{B}(\mathcal{C}(\overline{\Omega}, \mathcal{A}_n^-))$ has the form

$$T = T_a + K,$$

where $a \in \mathcal{C}(\overline{\Omega}, \mathcal{A}_n^-)$ and K is a compact operator. The operator T is Fredholm if and only if $a(x)$ is invertible in $\mathcal{A}_n^-$ for all $x \in \Gamma$.

Bibliography

[BDS] F. Brackx, R. Delanghe and F. Sommen, *Clifford Analysis*, Pitman, Boston, 1982.

[D] R.G. Douglas, *Banach Algebra Techniques in Operator Theory,*, Academic Press, New York, 1972.

[GS] K. Gurlebeck and W. Sproessig, *Quaternionic Analysis and Elliptic Boundary Value Problems*, Birkhauser, Basel, 1990.

[HK] W. K. Hayman and P. B. Kennedy, *Subharmonic Functions*, Academic Press, New York, 1976.

[S] I.B. Simonenko, *New general methods of studying of linear equations of the type of singular integral equations*, I, Izvestia AN SSSR, Ser. Matem., v. 29, **3**, (1965), 567–586 (Russian).

[SV1] M. V. Shapiro and N. L. Vasilevski, *On the Bergman kernel function in hypercomplex analysis*, Reporte Interno No. 115, Departamento de Matemáticas, CINVESTAV del I. P. N., Mexico City, 1993, 35p, (to appear in Acta Applicandae Mathematicae).

[SV2] M. V. Shapiro and N. L. Vasilevski, *On the Bergman kernel function in the Clifford Analysis*, Clifford Algebras and their Applications in Mathematical Physics, Deinze 1993, Kluwer Academic, Norwell, MA, 1993, 183–192.

[VS] N. L. Vasilevski and M. V. Shapiro, *On the Bergman kernel function in the quaternionic analysis*, (submitted to Matematicheskie Zametki) (Russian).

Departamento de Matemáticas,
CINVESTAV del IPN,
Mexico, D.F.,
MEXICO
E-mail: eramirez@math.cinvestav.mx
E-mail: nvasilev@math.cinvestav.mx

Departamento de Matemáticas,
ESFM del IPN, Edificio 9, Unidad ALM,
07300, Mexico, D.F.,
MEXICO
E-mail: mshapiro@godel.esfm.ipn.mx

20

Multivector Solutions to the Hyperholomorphic Massive Dirac Equation

William M. Pezzaglia Jr.

ABSTRACT Attention is given to the interface of mathematics and physics, specifically noting that fundamental principles limit the usefulness of otherwise perfectly good mathematical general integral solutions. A new set of multivector solutions to the meta-monogenic (massive) Dirac equation is constructed which form a Hilbert space. A new integral solution is proposed which involves application of a kernel to the right side of the function, instead of to the left, as is usual. This allows for the introduction of a multivector generalization of the Feynman Path Integral formulation, which shows that particular "geometric groupings" of solutions evolve in the manner to which we ascribe the term "quantum particle". Further, it is shown that the role of usual i is subplanted by the unit time basis vector, applied on the right side of the functions.

20.1 Introduction

As a physicist, I am like 'a stranger in a strange land' where familiar names have different meanings. It becomes increasingly clear that the practitioners of hypercomplex analysis, and those of multivector physics do not occupy the same "Clifford" space. This is, I believe, a manifestation of the classical chasm which unfortunately often exists between the disciplines of *applied mathematics* and *theoretical physics*. My attempts to bridge this gap by conversation at this conference often ran aground. Lacking background (or interest) in physics, many mathematicians are reluctant to be drawn into unfamiliar territory (and, of course, vice versa). For brevity I will continue from my own point of view, but it should be clear that many of the statements I make are equally true if you interchange the words

0-8493-8481-8/96/$0.00+$.50
© 1996 by CRC Press

"mathematician" and "physicist". Further, any parochial statements are a function of deliberate hyperbolism.

The interface of mathematics and physics is possibly a *no-man's land*, which is not to be ventured into under threat of being accused, by one's colleagues, of defection to the other side. For example, the theoretical physicist that treads too close to the border runs the risk of having his work dismissed as "mere reformulation", or worse as *just mathematics*, i.e., not 'real' physics. The point is well taken. While learning more math will enable the physicist to communicate better with mathematicians, it will not necessarily lead to new or even better physics. This is because physics is much more than the study of the subset of mathematical equations that is isomorphic to physical phenomena.

Rather than dismiss practitioners of the opposition because they lack sophistication in our own language, it behooves us to acknowledge that mathematics and physics are *different disciplines* with distinct agendas and criteria. An example in point was the 'open problem book' of the conference, in which mathematicians posed concise questions for which answers are yet unknown (or even known to exist, as in Fermat's last theorem) but presumably can be solved through clever logical deduction. It was difficult to come up with an analogous "physics" question. As recently discussed by Romer [17], such questions can usually only be answered by experimental verification (e.g., does the *Top Quark* exist), or an application of a *physical principle*. It was the downfall of Galileo that the 'new science' had new physical laws that could not be derived by logical extension or verified by mathematical proof. Given this, it is somewhat of a mystery that physical principles can be so conveniently expressed in mathematical form [20].

The goal of this paper is twofold. Specifically we consider the more general *meta-monogenic equation*, which is of more interest (to physicists) than the more limited monogenic equation, yet has not received nearly as much attention in Clifford analysis. Secondly, this paper will be used as a vehicle to address the *interface* of mathematics and physics, by providing associations between mathematical concepts and physical principles.

20.2 Algebraic Notation

It is difficult to choose a proper level of presentation for a 'cross-over' paper such that it will be equally readable for both physicists and mathematicians. Conversations with several mathematicians showed which concepts needed to be briefly defined for their benefit and vice versa for the physicists (n.b., the author). Members of either discipline will hence feel these are overly simple remarks.

The goal of this section hence is not to present anything new, rather it is to relate familiar concepts in the "home discipline" to unfamiliar counterparts in the other. This is more than 'calling a rose by another name'. Most of the developments in the rest of the paper have their basis in intrinsically nonmathematical physical principles and cannot be presented as 'stand-alone mathematics'. To start with, the mathematical structure chosen should encode physical principles (e.g., Lorentz covariance). Specifically we seek to use a geometric algebra which will describe the empirically known properties of physical spacetime.

20.2.1 Classical Galilean Space and The Pauli Algebra

Before the theory of relativity, physicists thought "space" was intrinsically three dimensional. Physical quantities [such as the electric field vector $\vec{\mathbf{E}} = \vec{\mathbf{E}}(\vec{\mathbf{X}}, t)$] would be represented as a function of a position *Gibbs vector* $\vec{\mathbf{X}}$ and *scalar* time t,

$$(\vec{\mathbf{X}}, t) = x\sigma_1 + y\sigma_2 + z\sigma_3 + t\sigma_0, \qquad (20.2.1)$$

where $\sigma_0 = 1$ is the unit scalar. The $3 \oplus 1$ notation of eq. (20.2.1) matches our perception, which insists upon seeing time as distinct from 3D space. Equivalently, we say that the domain $(\vec{\mathbf{X}}, t)$ belongs to $\mathbf{R}^3 \oplus \mathbf{R}$; it is an *ordered pair* of a three space *Gibbs vector* and a *scalar*. This structure (n.b. setting $\sigma_0 = 1$) was assumed by many of the contributors at this conference (n.b., plenary presentations).

The Clifford algebra generated by the three mutually anticommuting basis elements $\{\sigma_1, \sigma_2, \sigma_3\}$ is **End $\mathbf{R}^3$**, commonly called the *Pauli algebra*, with matrix representation of 2×2 complex matrices $\mathbf{C}(2)$. While this algebra and the $3 \oplus 1$ notation is sufficient for many physical applications (see for example, [3]), it will lead to conceptual ambiguities which can only be resolved with the full 4D concept (n.b., definition of time reversal).

20.2.2 Minkowski Spacetime and Majorana Algebra

The *principle of Lorentz invariance* demands that $(\vec{\mathbf{X}}, t)$ is an element of a *four-dimensional* domain, rather than $3 \oplus 1$. Algebraically this means that the basis element associated with time [e.g., σ_0 of eq. (20.2.1)] must anticommute with the other three mutually anticommuting basis vectors. In order to avoid confusion, we will introduce a new set of symbols: $\{\mathbf{e}_\mu\}$ as the basis vectors of 4D space, with $\mathbf{e}_4$ (instead of $\mathbf{e}_0$) as the fourth basis element. Hence the coordinate *four-vector* will be expressed in boldface lower case (no overhead arrow),

$$\mathbf{x} = (x, y, z, t) = x^\mu \mathbf{e}_\mu = x\mathbf{e}_1 + y\mathbf{e}_1 + z\mathbf{e}_3 + t\mathbf{e}_4, \qquad (20.2.2)$$

where the Einstein summation notation is assumed on the repeated index: $\mu = 1, 2, 3, 4$.

The *metric* of time is empirically known to be of the opposite sign as that of the 3D positional portion. Mathematically, we would say that the orthogonal space is either $\mathbf{R}^{3,1}$ or $\mathbf{R}^{1,3}$, corresponding to metric signatures $(+++-)$ or $(---+)$, respectively, (physicists refer to these, respectively, as the *east coast* or *west coast* metrics). In the former case, the norm of the position four-vector would be

$$\|\mathbf{x}\| = x_\mu x^\mu = x^2 + y^2 + z^2 - t^2, \tag{20.2.3}$$

where $x_4 = -x^4 = t$, but $x_i = x^i$ for $i = 1, 2, 3$. Physically, this hyperbolic nature is interpreted as a manifestation of the *principle of causality*. For an "event" at the origin to be the source or *cause* of another event at $\mathbf{x}$, the separation four-vector between them must have norm $\|x\|^2 < 0$ in the $\mathbf{R}^{3,1}$ space (or $\|x\|^2 > 0$ in the $\mathbf{R}^{1,3}$ space). Equivalently, we say that the vector connecting an event with its cause is *timelike*, or lies inside the *light cone* (the hypersurface for which $\|x\|^2 = 0$).

The Clifford algebra generated by the four basis vectors in the "east coast" $(+++-)$ metric is known as the *Majorana algebra*, with matrix representation $\mathbf{R}(4) = \text{End } \mathbf{R}^{3,1}$. This algebra is different than that generated by the other metric choice (used, for example, by Hestenes [10]), which has inequivalent matrix representation $\mathbf{H}(2) = \text{End } \mathbf{R}^{1,3}$, i.e., 2×2 quaternionic matrices [16]. Regardless, either algebra has 16 basis elements, but *neither* contains the global commuting i, which is usually required by physicists for standard relativistic quantum mechanics [5]. The standard theories use the "classic" *Dirac algebra*, which has *five* anticommuting generators $\{\gamma^\mu\}$, corresponding to the complex matrix representation $\mathbf{C}(4) = \mathbf{C} \otimes \mathbf{H}(2) = \mathbf{C} \otimes \mathbf{R}(4)$.

The "projection" or *spacetime split* from 4D $\rightarrow$ 3 $\oplus$ 1 can be encoded algebraically in $\mathbf{R}(4)$,

$$-\mathbf{x}\,\mathbf{e}_4 = \vec{\mathbf{X}} + t, \tag{20.2.4}$$

where the 3D Pauli algebra generators σ_j of eq. (20.2.1) are related to the Majorana generators: $\sigma_j = \mathbf{e}_4 \mathbf{e}_j$ for $j = 1, 2, 3$.

20.2.3   Automorphisms and Conservation Laws

Fundamental to physical theories are *conservation laws*, e.g., conservation of energy. These are intimately related via Noether's Theorem [6] to the physical principles being invariant under certain transformation (e.g., rotations). The "allowed" symmetries should correspond mathematically to those generated by the automorphism group of the geometric algebra.

Of particular interest are the orthogonal transformations associated with hyperbolic rotations between space and time. These have the physical interpretation of connecting two reference systems, one at rest and the other moving at velocity $\vec{\mathbf{V}} = \frac{d\vec{\mathbf{X}}}{dt}$. A particle of mass m at rest in the "moving" frame, will be perceived as moving at velocity $\vec{\mathbf{V}}$ in the "rest" frame, with *four-vector momentum* $\mathbf{p}$,

$$\mathbf{p} = p^\mu \mathbf{e}_\mu = \mathcal{L} \, m\mathbf{e}_4 \, \mathcal{L}^{-1}, \qquad (20.2.5a)$$

$$\mathcal{L} = \exp\left(\frac{r\hat{\beta}}{2}\right) = \frac{-\mathbf{p}\mathbf{e}_4 + m}{\sqrt{2m(E+m)}} = \frac{\vec{\mathbf{P}} + E + m}{\sqrt{2m(E+m)}}, \qquad (20.2.5b)$$

$$E = p^4 = -\mathbf{e}_4 \cdot \mathbf{p}, \qquad (20.2.5c)$$

$$\vec{\mathbf{P}} = E\vec{\mathbf{V}} = \mathbf{e}_4 \wedge \mathbf{p}, \qquad (20.2.5d)$$

$$\tanh(r) = \|\vec{\mathbf{V}}\| = E^{-1}\|\vec{\mathbf{P}}\|, \qquad (20.2.5e)$$

$$\hat{\beta} = \frac{\vec{\mathbf{V}}}{\|\vec{\mathbf{V}}\|} = \frac{\vec{\mathbf{P}}}{\|\vec{\mathbf{P}}\|}. \qquad (20.2.5f)$$

The term $\mathcal{L}$ is called the (half-angle) *Lorentz Boost operator*, where r is the *rapidity* associated with the velocity. The timelike unit bivector $\hat{\beta}$ points in the direction of the 3D velocity $\vec{\mathbf{V}}$, (equivalently the 3D momentum $\vec{\mathbf{P}}$). The factor of $\mathbf{e}_4$ in the middle of eq. (20.2.5b) causes the spacetime split [see eq. (2.4)] of the four-momentum, where the fourth component p^4 is physically interpreted as the *energy* E in eq. (20.2.5c).

We define the following algebra involutions,

$$\mathcal{T}(\Gamma) = -\mathbf{e}_1\mathbf{e}_2\mathbf{e}_3 \, \Gamma \, \mathbf{e}_1\mathbf{e}_2\mathbf{e}_3, \qquad (20.2.6a)$$

$$\mathcal{P}(\Gamma) = -\mathbf{e}_4 \, \Gamma \, \mathbf{e}_4, \qquad (20.2.6b)$$

which correspond geometrically to reflections. The first inverts timelike geometry, hence is called *Time Reversal* in physics (although our definition differs from the usual standard found in most textbooks). Basis elements which are invariant under this involution (e.g., $\mathbf{e}_1\mathbf{e}_2$) are called *spacelike*, while those which acquire a minus sign (e.g., $\mathbf{e}_1\mathbf{e}_4$), *timelike*. We will have more use for the second, the contrapositive inversion of 3D space called the *Parity Transformation* [5], under which many (but not all) physical laws are invariant. The composition of the two is the *main involution* of the algebra,

$$\mathcal{P}\mathcal{T}(\Gamma) = \mathcal{T}\mathcal{P}(\Gamma) = -\mathbf{e}_1\mathbf{e}_2\mathbf{e}_3\mathbf{e}_4 \, \Gamma \, \mathbf{e}_1\mathbf{e}_2\mathbf{e}_3\mathbf{e}_4, \qquad (20.2.6c)$$

inverting the odd geometry (i.e., vectors and trivectors).

The anti-involutions associated with eq. (20.2.6b) and (20.2.6c) are, respectively, the "dagger" and "bar" (equivalently the *Hermitian* and *Dirac* conjugates, respectively). They are related,

$$\overline{\Gamma} = -\mathbf{e}_4 \, \Gamma^\dagger \, \mathbf{e}_4, \qquad (20.2.7a)$$

$$\overline{\mathbf{e}}_\mu = -\mathbf{e}_\mu, \qquad (20.2.7b)$$

$$\mathbf{e}_4^\dagger = -\mathbf{e}_4, \quad \mathbf{e}_j^\dagger = +\mathbf{e}_j \ (j = 1, 2, 3), \qquad (20.2.7c)$$

where the "bar" reverses the order of all elements and inverts the basis vectors. The multivector bilinear form $\overline{\Psi}\Psi$ has the advantage of being Lorentz invariant (as the Dirac bar conjugate operation inverts the bivector generators of Lorentz transformations). On the other hand, the scalar part of this form is not in general positive definite, in contrast to the form $\Psi^\dagger\Psi$. Both forms will in general have non-scalar portions, which other authors usually do not fully utilize.

20.3 Functional Solutions of the Massive Dirac Equation

In this section we will present new multivector eigenfunction solutions to the massive meta-monogenic equation. The adjective "massive" should be superfluous, because historically Dirac was describing the electron, a particle with mass. It is therefore a source of confusion for the physicist to encounter in Clifford analysis the use of the term "Dirac equation" applied to the *monogenic* equation $\Box\Psi = 0$, rather than eq. (20.3.2a) below.

20.3.1 Relativistic Quantum Wave Equations

When solution Ψ is restricted to be a bivector, the monogenic equation $\Box\Psi = 0$ would be called the (sourceless) *Maxwell equation*, describing the spin-one *massless photon* (i.e., electromagnetic waves). If Ψ is projected onto a minimal ideal (i.e., a *column spinor*), it would describe a spin-half (again massless) *neutrino*, often called the *Majorana equation*. These are special cases which have many interesting properties which unfortunately vanish once mass is included, or a nonhomogeneous source term is added.

The massive *meta-harmonic* equation over the spacetime domain $\mathbf{R}^{3,1}$ is known as the *Klein-Gordon equation* [6],

$$(\Box^2 - m^2) \, \phi(\mathbf{x}) = 0, \qquad (20.3.1)$$

where $\Box^2 = \partial_x^2 + \partial_y^2 + \partial_z^2 - \partial_t^2$ is the *d'Alembertian* (the 4D generalization of the *Laplacian*, but in non-Euclidean spacetime). Note that some authors use the symbol $\square$ without the square for the d'Alembertian, which makes the use of the same symbol for the factored 'Dirac operator' ambiguous. Physically, eq. (20.3.1) is interpreted to be the relativistic generalization of Schrödinger's quantum wave equation. Historically, ϕ was a complex scalar function describing a charged spinless particle. If however the function ϕ is a vector, then eq. (20.3.1) is called the *Proca equation*, describing a spin-one massive particle field (the "vector boson").

The *meta-monogenic* (hyperholomorphic) massive *Dirac equation* can be obtained by factoring the Klein-Gordon operator of eq. (20.3.1). This minimally requires four anticommuting algebraic entities $\mathbf{e}_\mu$,

$$(\square - m)\ \Psi(\mathbf{x}) = 0, \tag{20.3.2a}$$

$$\Psi(\mathbf{x}) = (\square + m)\ \phi(\mathbf{x}) = (\mathbf{e}^\mu \partial_\mu + m)\ \phi(\mathbf{x}), \tag{20.3.2b}$$

$$\{\mathbf{e}^\mu, \mathbf{e}^\nu\} = 2g^{\mu\nu}, \tag{20.3.2c}$$

where $\square = \mathbf{e}^\mu \partial_\mu$ is the 4D *spacetime gradient* (we hesitate to call it the "Dirac operator", because this term is sometimes used by physicists to refer to the second quantized wavefunction Ψ). Following Greider [8], we note that the *metric tensor* $g^{\mu\nu}$ *must* have the diagonal values $g^{11} = g^{22} = g^{33} = -g^{44} = +1$ of the *east coast* signature $(+ + + -)$ if the use of the commuting i is excluded. It is important to note that most of the main results of this paper, n.b., eq. (20.3.2a), are inaccessible if the other metric $(- - - +)$ with inequivalent algebra $\mathbf{H}(2)$ is used, as assumed by Hestenes [10].

20.3.2 Meta-Monogenic Functions

Real solutions (no i) to the *meta-harmonic* Klein-Gordon eq. (20.3.1) are of the form: $\cos(p_\mu x^\mu)$. Substituting this eigenfunction into eq. (20.3.1) gives the *characteristic equation*,

$$m^2 = -p_\mu p^\mu = E^2 - \|\vec{\mathbf{P}}\|^2, \tag{20.3.4a}$$

known in this case as the *Einstein relation* between mass m, energy E and vector momentum $\vec{\mathbf{P}}$. Note that for a given mass and vector momentum, the energy could be positive or negative, the latter is unphysical. We shall define energy to be positive,

$$\pm p^4 = E = +\sqrt{m^2 + \|\vec{\mathbf{P}}\|^2}. \tag{20.3.4b}$$

Substituting $\phi = \cos(p_\mu x^\mu)$ into eq. (20.3.2b) yields a multivector solution to the Dirac eq. (20.3.2a) which can be expressed in the exponential form [13] [18],

$$\Psi_{\vec{\mathbf{P}}}(x) = \exp(-\hat{\mathbf{p}}\ p_\mu x^\mu)\ \Lambda, \tag{20.3.5a}$$

$$\hat{\mathbf{p}} = m^{-1}\mathbf{p} = m^{-1} p^{\mu}\mathbf{e}_{\mu}, \qquad (20.3.5b)$$

$$\hat{\mathbf{p}}^2 = -1, \qquad (20.3.5c)$$

where Λ is an arbitrary geometric factor (the "spin" degrees of freedom). The unit four-velocity $\hat{\mathbf{p}}$ plays the role of the usual i as the generator of quantum phase (when multiplied sinistrally, i.e., on the left). However, each momentum eigenfunction has its own *unique* generator $\hat{\mathbf{p}}$. Note that eq. (20.3.5a) is invariant under the replacement of $\mathbf{p} \to -\mathbf{p}$, hence we can restrict eq. (20.3.4b) to $p^4 = +E$ without any loss of generality, hence the argument of eq. (20.3.5a) is: $p_{\mu}x^{\mu} = \vec{\mathbf{P}} \cdot \vec{\mathbf{X}} - Et$.

Consider a solution which is a linear combination of eigenfunctions of eq. (20.3.5a), restricted to $\Lambda = 1$. This is a reasonable construct to consider because of the *superposition principle* of quantum theory: *the sum of any two physically interpretable solutions will be a reasonably interpretable solution*. Although each eigenfunction separately will have a unit quadratic form, $\overline{\Psi}\Psi = 1$, the new solution will display non-scalar "interference" terms due to the differing generators $\hat{\mathbf{p}}$. These have been interpreted as a possible useful description as a source for mesonic interactions [15].

20.3.3   Multivectorial Hilbert Space

At this point we introduce a new type of solution which will satisfy special properties regarding physical interpretation. In order to have a probabilistic interpretation of quantum theory, we need a positive definite norm. A restricted choice of factor Λ in eq. (20.3.5a) will make the eigenfunctions unitary [13],

$$\Lambda = \Lambda_{\vec{\mathbf{P}}} = \sqrt{\frac{m}{(2\pi)^3 E}} \; \mathcal{L}, \qquad (20.3.2)$$

where the term $\mathcal{L}$ is the Lorentz Boost operator of eq. (20.2.5b). Using the property of the Lorentz operator, $\mathbf{p}\mathcal{L} = \mathcal{L}m\mathbf{e}_4$ from eq. (20.2.5a), the unimodular meta-monogenic multivector eigenfunction can be expressed in the alternate form,

$$\Psi_{\vec{\mathbf{P}}}(x) = \Lambda_{\vec{\mathbf{P}}} \; \exp\left(-\mathbf{e}_4 \; p_{\mu}x^{\mu}\right), \qquad (20.3.7a)$$

$$= (\Box + m) \; \Phi_{\vec{\mathbf{P}}}(x), \qquad (20.3.7b)$$

$$\Phi_{\vec{\mathbf{P}}}(x) = \frac{\exp\left(-\mathbf{e}_4 \; p_{\mu}x^{\mu}\right)}{(2\pi)^{\frac{3}{2}}\sqrt{2E(E+m)}}. \qquad (20.3.7c)$$

Different than before, each eigenfunction now is seen to have the *same* generator of quantum phase, the (positive parity) unit time vector $\mathbf{e}_4$ when multiplied *dextrally* (right-side applied). Further the meta-monogenic eigenfunction $\Psi_{\vec{\mathbf{P}}}(\mathbf{x})$ can be written in terms of a multivector solution $\Phi_{\vec{\mathbf{P}}}(\mathbf{x})$ to the Klein-Gordon equation, again complex $\mathbf{e}_4$ instead of the usual i. This differs from the work of Hestenes [10] and Benn [4] which both interpret the negative parity geometric pseudoscalar as playing the role of i. Eigenfunctions in that form will not obey the standard parity relation, which ours do,

$$\mathcal{P}\left(\Psi_{\vec{\mathbf{P}}}(\vec{\mathbf{X}}, t)\right) = -\mathbf{e}_4 \ \Psi_{\vec{\mathbf{P}}}(\vec{\mathbf{X}}, t) \ \mathbf{e}_4 = +\Psi_{-\vec{\mathbf{P}}}(-\vec{\mathbf{X}}, t). \tag{20.3.3}$$

The unitary meta-mongenic eigenfunctions are orthonormal and complete,

$$\int d^3\vec{\mathbf{X}} \ \Psi_{\vec{\mathbf{K}}}^\dagger(\vec{\mathbf{X}}, t) \ \Psi_{\vec{\mathbf{P}}}(\vec{\mathbf{X}}, t) = \delta_{\vec{\mathbf{K}}, \vec{\mathbf{P}}}, \tag{20.3.9a}$$

$$\delta(\vec{\mathbf{X}} - \vec{\mathbf{Y}}) = \int d^3\vec{\mathbf{P}} \ \Psi_{\vec{\mathbf{P}}}(\vec{\mathbf{X}}, t) \ \Psi_{\vec{\mathbf{P}}}^\dagger(\vec{\mathbf{Y}}, t). \tag{20.3.9b}$$

Hence these functions are the restricted subset of the solution space that forms a *Multivector Hilbert space*, where again $\mathbf{e}_4$ (*dextrally* applied, i.e. multiplied on the right side) plays the role of the usual i. A general solution in this function subspace (e.g., a "quantum wavepacket") can be expanded,

$$\Psi(\vec{\mathbf{X}}, t) = \int d^3\vec{\mathbf{P}} \ \Psi_{\vec{\mathbf{P}}}(\vec{\mathbf{X}}, t) \ C_{\vec{\mathbf{P}}}, \tag{20.3.10a}$$

$$C_{\vec{\mathbf{P}}} = \int d^3\vec{\mathbf{X}} \ \Psi_{\vec{\mathbf{P}}}^\dagger(\vec{\mathbf{X}}, t) \ \Psi(\vec{\mathbf{X}}, t), \tag{20.3.10b}$$

where the coefficients $C_{\vec{\mathbf{P}}}$ can be thought of as "3D scalars", complex in $\mathbf{e}_4$ and <u>must</u> appear on the right side of eq. (20.3.10a). Note that these coefficients turn out to be independent of time t.

The full 16-degree-of-freedom solution to eq. (20.3.2) is of mixed parity and can be written,

$$\Psi(\vec{\mathbf{X}}, t) = \int d^3\vec{\mathbf{P}} \ \Psi_{\vec{\mathbf{P}}}(\vec{\mathbf{X}}, t) \ C_{\vec{\mathbf{P}}} \ [f_+(\vec{\mathbf{P}}) + \varepsilon \ f_-(\vec{\mathbf{P}})], \tag{20.3.11a}$$

$$\varepsilon = \mathbf{e}_1\mathbf{e}_2\mathbf{e}_3\mathbf{e}_4, \tag{20.3.11b}$$

where $f_\pm$ are multivector functions with four degrees of freedom on the basis set consisting of the scalar and the timelike trivectors: $\{1, \varepsilon\mathbf{e}_1, \varepsilon\mathbf{e}_2, \varepsilon\mathbf{e}_3\}$. These represent the 'spin' degrees of freedom of both parities (indicated by the sign subscript). This full solution can be interpreted as an algebraic representation of the isospin doublet of Dirac bispinors [14][15].

20.4 Integral Meta-Monogenic Solutions

The main development of the paper is the introduction of the PIF=*Path Integral Formulation* solution method to meta-monogenic equations. This will further require us to propose a new right-side applied *dextrad* kernel.

In general, the first-order monogenic equation: $\Box\Psi = 0$ [or the generalized Dirac eq. (20.3.2a)] more directly lends itself to integral solution than the associated second-order harmonic equation [or the generalized Klein-Gordon eq. (20.3.1)]. For example, time independent (i.e. 3D) "static" electromagnetism can be easily reexpressed in Cauchy integral form, even when nonhomogeneous source terms are included. A typical electrostatic example would be to calculate the electric field $\vec{E}$ at any point inside a region knowing the field on the boundry (equivalently knowing the surface charge density $d\sigma = \vec{E} \cdot d\vec{A}$) and the distribution of (scalar) charge $\rho = \nabla\vec{E}$ inside the region [12]. On the other hand, the Bergman kernel would only provide a method of calculating the electric field at a point in the region, if one already knows the complete solution everywhere inside the region. At the moment I cannot come up with any physical example for which this circular statement would be of practical use, although this may be due to my misunderstanding of the concept.

There are additional constraints imposed by fundamental physical laws which further limit the usefulness of integral formulations in describing tangible phenomena. For example, consider the 3D *Helmholtz equation*: $(\nabla^2 + \lambda^2)\phi(\vec{X}) = 0$, of which integral solutions to the factored meta-monogenic form have been recently treated by Gürlebeck and Sprössig [9]. This can be *physically* interpreted as a factored form of the time-independent quantum *Pauli equation* (see recent exposition by Adler and Martin [2]). However, one is then ontologically constrained by the quantum principle which states that only the modulus $\rho = \phi^\dagger\phi$ of the wavefunction $\phi = \sqrt{\rho}\exp(i\theta)$ can be measured; the absolute quantum phase θ cannot be known. Hence it will be impossible to perform a Cauchy boundry integral except for very special situations. This has influenced the way that physicists approach or "formulate" problems.

20.4.1 The Propagating Kernel

One common scenario is to *assume* a physical system's initial configuration, *even though it cannot be directly measured*, [e.g., $\Psi(\vec{X}, t')$ known for all $\vec{X}$ at $t' = 0$], and then to ask how the system must change in time. The *Time Evolution Operator* would have the new multivector form,

$$\Psi(\vec{X}, t) = U(t; t') \; \Psi(\vec{X}, t'), \tag{20.4.1a}$$

$$U(t,t') = U(t - t') = \exp\left[-(t - t')\,(\vec{\nabla} - \mathbf{e}_4 m)\right], \tag{20.4.2b}$$

where $\vec{\nabla} = \mathbf{e}_4 \wedge \square = \mathbf{e}_4 \mathbf{e}^j \partial_j$ is the 3D gradient. It will be assumed that $t > t'$, otherwise one would be talking about 'retrodiction', a different problem. An integral representation of eq. (4.1a) is

$$\Psi(\vec{\mathbf{X}}, t) = \int d^3\vec{\mathbf{X}}'\; F(\vec{\mathbf{X}}, t; \vec{\mathbf{X}}', t')\; \Psi(\vec{\mathbf{X}}', t'), \tag{20.4.3a}$$

$$F(\vec{\mathbf{X}}, t; \vec{\mathbf{X}}', t') = U(t,t')\; \delta(\vec{\mathbf{X}} - \vec{\mathbf{X}}'), \tag{20.4.3b}$$

where the *Dirac Delta Function* $\delta(\mathbf{x})$ is a distribution of measure one at the origin and zero elsewhere. The kernel $F(\mathbf{x}, \mathbf{x}')$ of eq. (20.4.3b) is known in physics as the *transformation function* [1], or sometimes the *Hyugen Propagator* since eq. (20.4.3a) is a description of Hyugen's principle: *each point on a wave front may be regarded as a new source of waves.* The integral is over the 3D hypersurface defined by t' =constant, however it could be generalized to any *spacelike* surface, i.e., a surface made of points that are not *causally connected* to each other, and further that t of eq. (20.4.11a) is greater than the t' of each point on the hypersurface. This amounts to requiring that the surface cannot have any "kinks" in it that would have the surface locally slip inside the "light cone". It is an important interdisciplinary question to ask if the Lipschitz condition for integrability (i.e., see Smith [19], p. 368) is stronger, weaker or equivalent in this case to the physical requirement of causality.

By inspection of eq. (20.4.3b), the kernel is a function only of the *difference* of the coordinates: $F(\vec{\mathbf{X}}, t; \vec{\mathbf{X}}', t') = F(\mathbf{x}, \mathbf{x}') = F(\mathbf{x} - \mathbf{x}')$. Substituting eq. (20.3.9a) for the delta function in equation (20.4.3b) gives a closed integral form,

$$F(\vec{\mathbf{X}}, t); \vec{\mathbf{X}}', t') = \int d^3\vec{\mathbf{P}}\; \Psi_{\vec{\mathbf{P}}}(\vec{\mathbf{X}}, t)\; \Psi^\dagger{}_{\vec{\mathbf{P}}}(\vec{\mathbf{X}}', t'), \tag{20.4.4a}$$

$$= \int \frac{d^3\vec{\mathbf{P}}}{(2\pi)^3 E}\; \exp\left[-\hat{\mathbf{p}}\; p^\mu (x_\mu - x'_\mu)\right]\; \mathbf{p}\, \mathbf{e}_4, \tag{20.4.4b}$$

$$= -(\square + m)\; \Delta(\mathbf{x} - \mathbf{x}')\; \mathbf{e}_4, \tag{20.4.4c}$$

$$\Delta(\mathbf{x}) = \int \frac{d^3\vec{\mathbf{P}}}{(2\pi)^3 E}\; \sin(p^\mu x_\mu). \tag{20.4.4d}$$

From eq. (20.4.4a) it is clear that: $F^\dagger(\mathbf{x}, \mathbf{y}) = F(\mathbf{y}, \mathbf{x})$. The function $\Delta(\mathbf{x})$ is given by [6] to be related to a regular Bessel function inside the light cone. Clearly it is a solution everywhere to the Klein-Gordon eq. (20.3.1), and so it follows from eq. (20.4.4c) that the propagator kernel is *left meta-monogenic*,

$$(\Box - m) \; F(\mathbf{x}, \mathbf{x}') = 0, \qquad\qquad (20.4.5a)$$

$$(\Box + m) \; F(\mathbf{x}', \mathbf{x}) = 0, \qquad\qquad (20.4.5b)$$

where $\Box$ operates on $\mathbf{x}$, but not on $\mathbf{x}'$. What we have derived is a special case of the *Cauchy kernel*. The usual closed boundry integral reduces to eq. (20.4.3a) when one assumes that $\mathbf{\Psi}(\vec{\mathbf{X}}, t) \rightarrow 0$ for: $\|\vec{\mathbf{X}}\| \rightarrow \infty$ or $t \rightarrow \infty$. In quantum theory, one sidesteps the unmeasurable phase by assuming an 'initial' solution at $t' \rightarrow -\infty$ which is a coherent plane wave. This approximates, for example, the collimated beam of electrons shooting down the linear accelerator at SLAC.

20.4.2   Green Function

Another common situation is to know the function over a restricted space-like surface (e.g., a long thin cylinder, such as an antenna), but different than our previous example in that the time dependence of the function is known (e.g., harmonic). This will require the use of a *Green Function* which satisifies: $(\Box - m) \; G(\mathbf{x}) = \delta^4(\mathbf{x})$. Since we have excluded the use of i, the inhomogeneous part will take the new form,

$$G(\mathbf{x}) = \int \frac{d^4\mathbf{k}}{(2\pi)^4 [m^2 + \mathbf{k}^2]} [m \cos(k^\mu x_\mu) + \mathbf{k} \sin(k^\mu x_\mu)], \qquad (20.4.6a)$$

$$= (\Box + m) \; \Theta(t) \; \Delta(\mathbf{x}), \qquad\qquad (20.4.6b)$$

$$= \Theta(t) \; (\Box + m) \; \Delta(\mathbf{x}), \qquad\qquad (20.4.6c)$$

$$= \Theta(t) \; F(\mathbf{x}) \; \mathbf{e}_4, \qquad\qquad (20.4.6d)$$

$$= \Theta(t) \int \frac{d^3\vec{\mathbf{P}}}{(2\pi)^3 E} \; \mathbf{p} \; \exp(-\hat{\mathbf{p}} \; p^\mu x_\mu), \qquad (20.4.6e)$$

where $\Theta(t)$ is a step (Heavyside) function, which takes on the value of $+\frac{1}{2}$ for $t > 0$, and $-\frac{1}{2}$ for $t < 0$. Equation (20.4.6c) follows from eq. (20.4.6b) if we note $\delta(t)\Delta(\mathbf{x}) = 0$. One can go directly from eq. (20.4.6a) to eq. (20.4.6e) by doing the contour integral, however the use of i has been excluded. The generator of the residue can possibly be taken to be the geometry associated with dk^4, again the (minus) time basis vector $-\mathbf{e}_4$ (see for example Hestenes [11]) where there must be attention to its placement. Equivalently we state (without proof) that the unit four-momentum $\hat{\mathbf{p}}$ may be used [inspection of eq. (20.4.6e) shows this to be consistent].

Inspection of eq. (20.4.6e) shows that $\overline{G}(\mathbf{x}) = G(-\mathbf{x})$. Hence this Green function also satisfies the adjoint equation: $G(\mathbf{x})(\Box - m) = \delta^4(\mathbf{x})$, where $\Box$ is understood to operate to the left. Because of this, we may derive the general integral solution,

$$\Psi(\mathbf{x}') = \oint d\Sigma^\mu G(\mathbf{x}', \mathbf{x}) \mathbf{e}_\mu \Psi(\mathbf{x}), \qquad (20.4.1)$$

where $\mathbf{x}$ is an interior point. Acutally it's a bit more complicated than that, one must add a homogeneous term to eq. (20.4.6a) such that the boundry conditions are met.

20.4.3 Path Integral Formulation (PIF)

The Feynman PIF (Path Integral Formulation) of non-relativistic quantum mechanics [7] provides an elegant means of bridging the interpretation gap between classical mechanics and quantum formulations. In the non-relativistic case, it asserts that the propagator kernel of eq. (20.4.3a) can be written,

$$F(\mathbf{x}, \mathbf{x}') = \sum \exp\left(\tfrac{i}{\hbar} \mathcal{S}(\mathbf{x}, \mathbf{x}')\right), \qquad (20.4.8a)$$

$$\mathcal{S}(\mathbf{x}, \mathbf{x}') = \int_{\mathbf{x}'}^{\mathbf{x}} p^\mu dx_\mu. \qquad (20.4.8b)$$

The *classical action* $\mathcal{S}$ is evaluated over the path from $\mathbf{x}$ to $\mathbf{x}'$. The sum in eq. (20.4.8a) is over all possible classical paths: $x^\mu = x^\mu(\tau)$, where $p^\mu = m\frac{dx^\mu}{d\tau}$ and τ is an affine parameter called the *proper time*. Classical paths require $\frac{dx^4}{d\tau} > 0$, i.e., paths in which particles go backward in time are excluded because they violate causality.

Because of the second-order time derivative, the Klein-Gordon eq. (20.3.1) does not easily lend itself to PIF. The first-order Dirac eq. (20.3.2a) is a better candidate. However, the standard derivation of eq. (20.4.8a) from the Schrödinger wave equation [1] required that each eigenfunction have the same generator of quantum phase, a commuting i which is now unavailable to us in the real multivector theory. In fact, we have seen from eq. (20.3.5a) that each eigenfunction (as seen from the left side) has its own unique quantum phase generator $\hat{\mathbf{p}}$, which makes the PIF derivation problematic. In the restricted case of plus parity eigenfunctions, however, we say from eq. (20.3.7a) that they all have the *same* generator $\mathbf{e}_4$ when viewed from the *right* side. We exploit the advantage of multivector wavefunctions (over standard Dirac column spinors) in that we can now consider *right-side applied* operations. Hence we introduce here a new alternative to eq. (4.3a), the *dextrad propagator*,

$$\Psi(\vec{\mathbf{X}}, t) = \int d^3\vec{\mathbf{X}}' \ \Psi(\vec{\mathbf{X}}', t') \ H(\vec{\mathbf{X}}, t; \vec{\mathbf{X}}', t'), \qquad (20.4.9a)$$

$$H(\mathbf{x}, \mathbf{x}') = H(\mathbf{x} - \mathbf{x}') = \int \frac{d^3\vec{\mathbf{P}}}{(2\pi)^3} \exp\left[-\mathbf{e}_4 \, p^\mu(x_\mu - x'_\mu)\right]. \qquad (20.4.9b)$$

It can be easily shown [13] that this propagator may be written in the PIF form of eq. (20.4.8a) with the replacement: $i \to -\mathbf{e}_4$. Further, the other half of the solution space, the negative parity solutions ('antiparticles') will follow the same relation, except requiring the replacement $\mathbf{e}_4 \to -\mathbf{e}_4$ in eq. (20.4.9b) and $i \to +\mathbf{e}_4$ in eq. (20.4.8a). This approach suggest that the particular form of eq. (20.3.7a) is the "special" multigeometric entity which propagates (unchanged) in the manner to which we ascribe the term "quantum particle". Further, the PIF provides a method to introduce interactions into the quantum equation based upon classical mechanics. Because of eq. (20.4.9a), we see that the interactions will couple *dextrally* to the wavefunction (multiplied on the right). In particular, the generator of electromagnetic interactions will be $\mathbf{e}_4$ applied on the right [15].

20.5   Summary

We are not aware of any other work in Clifford analysis that has applied the PIF (Path Integral Formulation) to integral solutions of monogenic equations. Separate from the physical interpretations, the general mathematical relationship between the "right-side applied" *dextrad kernel* introduced in eq. (20.4.9a) and the standard "left-side applied" one of eq. (20.4.3a) [both as integral solutions to eq. (20.3.2a)] should be explored. In particular, it is unknown to the author if a general 'dextrad' form of eq. (20.4.7) exists, nor under what conditions it would reduce to eq. (20.4.9a).

Acknowledgment

The author thanks the organizers of the conference for the arrangement of travel support for himself and two graduate students: John Adams (Physics, San Francisco State University) and Matthew Enjalran (Physics, University of Massachusetts, Amherst). Numerous conversations with J. Ryan, G. Sobczyk and M. Shapiro helped in matching unfamiliar mathematical terminology with physical concepts. Craig Harrison (Philosophy, San Francisco State University) pointed out reference [20], as well as engaged in many discussions. Finally, thanks to thesis advisor G. Erickson (Physics, University of California, Davis), who some 10 years ago guided the author in the development of the original ideas on which this paper is based.

Bibliography

[1] E.S. Abers and B.W. Lee, *Gauge Theories*, Phys. Rep. **C9**, (1973), p.1.

[2] R.J. Adler and R.A. Martin, *The electron g factor and factorization of the Pauli equation*, Am. J. Phys. **60**, (1992), 837-9; W.E. Baylis, 'Comment on *The electron g factor and factorization of the Pauli equation* by R.J. Adler and R.A. Martin', Am. J. Phys. **62**, (1994), 179.

[3] W.E. Baylis and G. Jones, *The Pauli-algebra approach to special relativity*, Phys. Rev. **A45**, (1989), 4293-4302.

[4] I.M. Benn and R.W. Tucker, *The Dirac Equation in Exterior Form*, Commun. Math. Phys. **98**, (1985), 53-63.

[5] J.D. Bjorken and S.D. Drell, *Relativistic Quantum Mechanics*, McGraw-Hill, New York, 1964.

[6] J.D. Bjorken and S.D. Drell, *Relativistic Quantum Fields*, McGraw-Hill, New York, 1965.

[7] R. Feynman and A.R. Hibbs, *Quantum Mechanics and Path Integrals*, McGraw-Hill, New York, 1965.

[8] K. Greider, *A Unifying Clifford Algebra Formalism for Relativistic Fields*, Found. Phys. **14**, (1984), 467-506.

[9] K. Gürlebeck and W. Sprössig, *Quaternionic Analysis and Elliptic Boundary Value problems*, Birkhäuser-Verlag, Basel, Vol. **89** (1990).

[10] D. Hestenes, *Spacetime Structure of Weak and Electromagnetic Interactions*, Found. Phys., **12**, 153 (1982); *Observables, Operators, and Complex Numbers in the Dirac Theory*, J. Math. Phys., **16**, 556-72 (1975).

[11] D. Hestenes, *Multivector Functions*, J. Math. Anal. Appl., **24**, (1968), 467.

[12] J.D. Jackson, *Classical Electrodynamics*, John Wiley & Sons, New York, 2nd ed. 1975.

[13] W.M. Pezzaglia, *A Clifford Algebra Multivector Reformulation of Field Theory*, Ph.D. thesis, UMI-83-26101-mc (University Microfilms microfiche). University of California, Davis, 1993.

[14] W.M. Pezzaglia, *Clifford Algebra Geometric-Multispinor Particles and Multivector-Current Gauge Fields*, Found. Phys. Lett. **5**, (1992), 57-62.

[15] W.M. Pezzaglia, *Dextral and Bilateral Multivector Gauge Field description of Light-Unflavored Mesonic Interactions*, (preprint: CLF-ALG/PEZZ9302, 1993).

[16] I.R. Porteous, *Topological Geometry*, 2nd ed. Cambridge University Press, 1981.

[17] R.H. Romer, *Editorial: Fermat's last theorem*, Am. J. Phys., **61**, (1993), 873.

[18] M. Ross, *Geometric Algebra in Classical and Quantum Physics*, (Ph.D. thesis, University of California, Davis, 1980), (University Microfilms microfiche), p. 61; K. Greider, private communication (July 28, 1977).

[19] K.T. Smith, *Primer to Modern Analysis*, Springer-Verlag, New York, 1983.

[20] E.P. Wigner, *The Unreasonable Effectiveness of Mathematics in Natural Sciences*, in Symmetries and Reflections- Scientific Essays, Ox Bow Press, Woodbridge, CT, 1979, pp. 222-237.

Department of Physics,
California State University,
Hayward, California, 94542-3084
E-mail: bpezzag@mcs.csuhayward.edu

21

Möbius Transformations, Vahlen Matrices, and their Factorization

Pertti Lounesto

ABSTRACT The Lie algebra of the conformal group is spanned by rotations, translations, dilations and transversions. For this reason, some physicists have believed that every element in the identity component of the conformal group could be expressed as a product of four factors including exactly one rotation, one translation, one dilation and one transversion. However, this does not hold, as shown by the counter-example of J. Maks in 1989. It seems that some physicists are still confused about the factorization of conformal transformations. This article points out errors in the recent literature on this topic and gives explicit counter-examples to misplaced claims.

21.1 Similarities and transversions

Recall that a Möbius transformation $\mathbf{x} \to (a\mathbf{x} + b)(c\mathbf{x} + d)^{-1}$ of $\mathbb{R}^{p,q}$, where $a, b, c, d \in C\ell_{p,q}$, can be represented by a Vahlen matrix $\begin{pmatrix} a & b \\ c & d \end{pmatrix}$ in $M_2(C\ell_{p,q})$, see Abłamowicz et al. (1991), Ahlfors (1984, 1985, 1986), Cartan (1908), Elstrodt et al. (1987), Gilbert and Murray (1991), Lounesto and Springer (1989), Maks (1989, 1992), Ryan (1985, 1988) and Vahlen (1902). Here we consider the orthogonal space $\mathbb{R}^{p,q}$ with the scalar product

$$\mathbf{x} \cdot \mathbf{y} = x_1 y_1 + \cdots + x_p y_p - x_{p+1} y_{p+1} - \cdots - x_{p+q} y_{p+q}, \qquad p + q = n$$

and its Clifford algebra $C\ell_{p,q}$ satisfying $\mathbf{x}\mathbf{y} + \mathbf{y}\mathbf{x} = 2\mathbf{x} \cdot \mathbf{y}$. More precisely, the entries a, b, c, d of $\begin{pmatrix} a & b \\ c & d \end{pmatrix}$ are products of vectors, and if invertible, belong to the Lipschitz group (formerly called the Clifford group) $\Gamma_{p,q}$. We

0-8493-8481-8/96/$0.00+$.50
© 1996 by CRC Press

shall also need the spin group

$$\mathbf{Spin}_+(p,q) = \{s \in \mathbf{\Gamma}_{p,q} \cap C\ell_{p,q}^+ \mid s\tilde{s} = 1\}$$

sitting in the even subalgebra $C\ell_{p,q}^+$. The identity component of the Möbius group is generated by the rotations, translations, dilations and the transversions which are represented, respectively, as follows

$a\mathbf{x}a^{-1}$	$a \in \mathbf{Spin}_+(p,q)$	$\begin{pmatrix} a & 0 \\ 0 & a \end{pmatrix}$
$\mathbf{x} + \mathbf{b}$	$\mathbf{b} \in \mathbb{R}^{p,q}$	$\begin{pmatrix} 1 & \mathbf{b} \\ 0 & 1 \end{pmatrix}$
$\mathbf{x}\delta$	$\delta > 0$	$\begin{pmatrix} \sqrt{\delta} & 0 \\ 0 & 1/\sqrt{\delta} \end{pmatrix}$
$\dfrac{\mathbf{x} + \mathbf{x}^2\mathbf{c}}{1 + 2\mathbf{x}\cdot\mathbf{c} + \mathbf{x}^2\mathbf{c}^2}$	$\mathbf{c} \in \mathbb{R}^{p,q}$	$\begin{pmatrix} 1 & 0 \\ \mathbf{c} & 1 \end{pmatrix}.$

THEOREM 21.1.1   (J. Maks 1989)
Consider four Vahlen matrices which represent one rotation, one translation, one dilation and one transversion. A product of these four matrices, in any order, always has an invertible entry in its diagonal (there are $4! = 24$ such products).

PROOF   For instance, in the product

$$\begin{pmatrix} a & 0 \\ 0 & a \end{pmatrix}\begin{pmatrix} 1 & \mathbf{b} \\ 0 & 1 \end{pmatrix}\begin{pmatrix} \sqrt{\delta} & 0 \\ 0 & 1/\sqrt{\delta} \end{pmatrix}\begin{pmatrix} 1 & 0 \\ \mathbf{c} & 1 \end{pmatrix}$$

$$= \begin{pmatrix} a\sqrt{\delta} + a\mathbf{bc}/\sqrt{\delta} & a\mathbf{b}/\sqrt{\delta} \\ a\mathbf{c}/\sqrt{\delta} & a/\sqrt{\delta} \end{pmatrix}$$

the lower right-hand diagonal element $a/\sqrt{\delta}$ is invertible. To complete the proof of the fact that a product of a rotation, a translation, a dilation and a transversion, in any order, is such that the Vahlen matrix representing it always has an invertible entry in its diagonal, one can/must check the claim in all the remaining 23 cases. ∎

21.2   The counter-example of J. Maks

Consider the Minkowski space-time $\mathbb{R}^{3,1}$ and its Clifford algebra $C\ell_{3,1}$ generated by $\mathbf{e}_1, \mathbf{e}_2, \mathbf{e}_3, \mathbf{e}_4$ satisfying $\mathbf{e}_1^2 = \mathbf{e}_2^2 = \mathbf{e}_3^2 = 1$, $\mathbf{e}_4^2 = -1$. The Vahlen matrix

$$W = \tfrac{1}{2} \begin{pmatrix} 1 - e_{14} & -e_1 + e_4 \\ e_1 + e_4 & 1 + e_{14} \end{pmatrix}$$

with entries in $C\ell_{3,1} \simeq M_4(\mathbb{R})$ is such that all its entries are non-invertible. Thereby, the Vahlen matrix W is not a product of just one rotation, one translation, one dilation and one transversion (in any order). However, the Vahlen matrix W is connected to the identity by the following path (here β grows from 0 to $\pi/4$)

$$W = W_{\pi/4}, \qquad W_\beta = \exp\left\{ \beta \begin{pmatrix} 0 & -e_1 + e_4 \\ e_1 + e_4 & 0 \end{pmatrix} \right\}$$

see Maks (1989) p. 41, Abłamowicz and Lounesto and Maks (1991) p. 745 and Maks (1992) p. 62.

REMARK This error occured for instance in Hestenes and Sobczyk 1987, page 218, rows 11-13, formula (5.50).

21.3 Recent mistakes in factorization

In an article Hestenes (1991, pages 90-91) mentioned the results of Maks, but mistakenly believed that the counter-example would invalidate the classical theorem of Liouville (1850) saying that a Möbius transformation can be written as a product of translations, dilations and inversions. Further in the text (page 91, rows 10-12) Hestenes tried to exclude the exception of Maks in a mistaken belief that one can get a closed group without including the exceptional cases of Maks.

Recall that the above Vahlen matrix W is in the identity component of the fourfold covering group of the conformal group of the Minkowski spacetime (since it has an even diagonal and its pseudo-determinant equals 1), but even then W is not a product of just one rotation, one translation, one dilation and one transversion (in any order). However, this Vahlen matrix W is a product of a transversion, a translation and a transversion as follows

$$W = \begin{pmatrix} 1 & 0 \\ \frac{1}{2}(e_1 + e_4) & 1 \end{pmatrix} \begin{pmatrix} 1 & \frac{1}{2}(-e_1 + e_4) \\ 0 & 1 \end{pmatrix} \begin{pmatrix} 1 & 0 \\ \frac{1}{2}(e_1 + e_4) & 1 \end{pmatrix}.$$

Thereby, the Vahlen matrix W satisfies those assumptions which precede the formula (5.89) in Hestenes (1991) and we have a counter-example to the statement presented in rows 10-12 on page 91.

Hestenes (1991) factored W into a product of two *diversions* as follows

$$W = \tfrac{1}{\sqrt{2}} \begin{pmatrix} 1 & -e_1 \\ e_1 & 1 \end{pmatrix} \tfrac{1}{\sqrt{2}} \begin{pmatrix} 1 & e_4 \\ e_4 & 1 \end{pmatrix}$$

(see page 90, row 4, formula (5.88) which has an obvious misprint in the last term). Hestenes seemed to believe that his factorization of W into a product of two 'diversions' gets closer to the heart of the matter than Maks'. In this way, Hestenes (1991) made another error on page 91 rows 1-2 when he mistakenly claimed that a *diversion is a more fundamental counter-example*. Hestenes meant here that a 'diversion' cannot be expressed as a product of just one transversion, one dilation, one translation and one rotation. However, this is false since the first factor above is a product of just one transversion, one dilation and one translation as follows

$$\tfrac{1}{\sqrt{2}} \begin{pmatrix} 1 & -e_1 \\ e_1 & 1 \end{pmatrix} = \begin{pmatrix} 1 & 0 \\ e_1 & 1 \end{pmatrix} \begin{pmatrix} 1/\sqrt{2} & 0 \\ 0 & \sqrt{2} \end{pmatrix} \begin{pmatrix} 1 & -e_1 \\ 0 & 1 \end{pmatrix}$$

and one can insert the identity rotation as the last factor. Thereby, contrary to what Hestenes believed, a 'diversion' is a product of just one transversion, one dilation, one translation and one rotation, and as such cannot serve as a *more fundamental counter-example*.

21.4   How to factor Vahlen matrices?

A general theory on the factorization of conformal transformations can be found on pages 40-41 of Maks (1989), and good literature on this topic in general is Ahlfors (1985). In addition to the literature on Möbius transformations as represented by Vahlen matrices mentioned at the beginning of this article, there is good literature on Möbius transformations represented by higher-dimensional Clifford algebras, see Anglès (1985, 1988), Crumeyrolle (1990), Deheuvels (1985) and Fillmore and Springer (1990).

Acknowledgment

This article was written on the initiative of David Hestenes (Arizona State University).

Bibliography

We give a lengthy bibliography of good references in this topic, because there have been widespread misconceptions.

[1] R. Ablamowicz, P. Lounesto and J. Maks, *Conference Report, Second Workshop on "Clifford Algebras and Their Applications in Mathematical Physics," Montpellier, France, 1989*, Found. Phys. **21** (1991), 735-748.

[2] L.V. Ahlfors, *Old and new in Möbius groups*, Ann. Acad. Sci. Fenn. Ser. A I Math. **9** (1984), 93-105.

[3] L.V. Ahlfors, *Möbius transformations and Clifford numbers*, (I. Chavel and H. M. Farkas, Eds.), *Differential Geometry and Complex Analysis*, Springer, Berlin, 1985.

[4] L.V. Ahlfors, *Möbius transformations in $\mathbb{R}^n$ expressed through 2 × 2-matrices of Clifford numbers*, Complex Variables Theory Appl. **5** (1986), 215-224.

[5] P. Anglès, *Construction de rêvetements du groupe conforme d'un espace vectoriel muni d'une "métrique" de type (p,q)*, Ann. Inst. H. Poincaré Sect. A **33** (1985), 33-51.

[6] P. Anglès, *Real conformal spin structures on manifolds*, Studia Sci. Mat. Hung. **23** (1988), 115-139. (This article contains in disguise Hestenes' conformal split on page 116 row −1 formula (A_1).)

[7] E. Cartan (exposé d'après l'article allemand de E. Study), *Nombres complexes*, In J. Molk (red.), Encyclopédie des sciences mathématiques, Tome **I**, vol. 1, Fasc. 4, art. **I5** (1908), pp. 329-468. (Clifford algebras pp. 463-467, conformal transformations pp. 465-466.) Reprinted in E. Cartan, Œuvres Complètes, Partie II, Cauthier-Villars, Paris, 1953, pp. 107-246.

[8] A. Crumeyrolle, *Orthogonal and Symplectic Clifford Algebras, Spinor Structures*, Kluwer, Dordrecht, 1990. (Conformal group on pages 149-164.)

[9] R. Deheuvels, *Groupes conformes et algèbres de Clifford*, Rend. Sem. Mat. Univ. Politec. Torino **43** (1985), 205-226.

[10] J. Elstrodt, F. Grunewald, and J. Mennicke, *Vahlen's groups of Clifford matrices and spin-groups*, Math. Z. **196** (1987), 369-390.

[11] J.P. Fillmore and A. Springer, *Möbius groups over general fields using Clifford algebras associated with spheres*, Int. J. Theor. Phys. **29** (1990), 225-246.

[12] J. Gilbert and M. Murray, *Clifford Algebras and Dirac Operators in Harmonic Analysis*, Cambridge Studies in Advanced Mathematics,

Cambridge University Press, Cambridge, 1991. (Conformal transformations pp. 37-38, 278-296.)

[13] D. Hestenes and G. Sobczyk, *Clifford Algebra to Geometric Calculus*, D. Reidel, Dordrecht, 1984, 1987.

[14] D. Hestenes, *Design of linear algebra and geometry*, Acta Applic. Math. **23** (1991), 65-93.

[15] J. Liouville, *Extension au cas des trois dimensions de la question du tracé géographique. Applications de l'analyse à géometrie*, G. Monge, Paris, (1850), 609-616.

[16] P. Lounesto and A. Springer, *Möbius transformations and Clifford algebras of Euclidean and anti-Euclidean spaces*, In *Deformations of Mathematical Structures*, (J. Lawrynowicz. Ed.), Kluwer Academic, Dordrecht, 1989, pp. 79-90.

[17] J. Maks, *Modulo (1,1) periodicity of Clifford algebras and the generalized (anti-)Möbius transformations*, Thesis, Technische Universiteit Delft, 1989.

[18] J. Maks, *Clifford algebras and Möbius transformations*, In A. Micali et al. (Eds.), *Clifford Algebras and their Applications in Mathematical Physics*, Kluwer Academic, Dordrecht, 1992, pp. 57-63.

[19] J. Ryan, *Conformal Clifford manifolds arising in Clifford analysis*, Proc. R. Irish Acad. Sect. A **85** (1985), 1-23.

[20] J. Ryan, *Clifford matrices, Cauchy-Kowalewski extension and analytic functionals*, Proc. Centre Math. Anal. Aust. Natl. Univ. **16** (1988), 284-299.

[21] K. Th. Vahlen, *Über Bewegungen und complexe Zahlen*, Math. Ann. **55** (1902), 585-593.

Institute of Mathematics,
Helsinki University of Technology,
SF-02150 Espoo,
FINLAND
E-mail: lounesto@dopey.hut.fi